GENETICS OF FLOWERING PLANTS

Gregor Johann Mendel
(1822-1884)

Karl Erich Correns
(1864-1933)

Erwin Baur
(1875-1933)

Otto Renner
(1883-1960)

GENETICS OF
FLOWERING PLANTS

Verne Grant

Columbia University Press
New York and London

OTHER BOOKS BY THE SAME AUTHOR

Natural History of the Phlox Family (1959)

The Origin of Adaptations (1963)

The Architecture of the Germplasm (1964)

Flower Pollination in the Phlox Family.
With Karen A. Grant (1965)

Hummingbirds and Their Flowers.
With Karen A. Grant (1968)

Plant Speciation (1971)

Library of Congress Cataloging in Publication Data

Grant, Verne.
    Genetics of flowering plants.

    Bibliography: p.
    1. Plant genetics.   2. Angiosperms.   I. Title.
[DNLM: 1. Genetics.   2. Plants.   QH433 G763g]
QH433.G7   582'.13'0415   74–13555
ISBN 0–231–03694–9

For Karen

PREFACE

The primary objective of this book is to present a summary of our knowledge and understanding of the genetics of higher plants. There has been no book-length treatment of basic plant genetics in recent decades. The need for such a treatment has been apparent to many observers of the scene and has led directly to the present work.

The date of publication of Sansome and Philp's *Recent Advances in Plant Genetics* was 1939. This is perhaps the last previous book with the same center of focus as the present work. Yet *Genetics of Flowering Plants* bears little resemblance to *Recent Advances in Plant Genetics,* so greatly has the subject changed during the period since 1939. Basic plant genetics received good coverage in genetics textbooks a generation or two ago, from Baur's classic text (1911–1930) to that of Sirks (1956), but the growth of newer fields of genetics has tended to crowd plant genetics out of the more recent textbooks, and this source of information has also become inadequate.

Let me attempt to place *Genetics of Flowering Plants* in perspective in the current book literature in the same general field. The main emphasis in the present work is on the nature and action of genes, gene systems, linkage systems, and genetic systems in higher plants.

This is not a book on plant breeding (for which see references for Kuckuck and Mudra, 1950; Kappert and Rudorf, 1958–1962; Allard, 1960; and Brewbaker, 1964). Nor is it a book on quantitative genetics (see Falconer, 1960; Hiorth, 1963), or horticultural genetics (see Crane and Lawrence, 1952), although of course it overlaps with all these fields. Similarly this book does not emphasize plant cytogenetics (for which see John and Lewis, 1965; Brown, 1972) or chromosome evolution in plants (see Darlington, 1963; Stebbins, 1971), though these topics necessarily come into our picture. Certain important topics of plant genetics related to plant evolution were covered recently in a companion volume (Grant, 1971) and require no duplicate treatment here. The valuable monograph of Clausen and Hiesey (1958) contains a wealth of information on many of the topics treated in the present work.

Genetics of Flowering Plants grew out of an earlier book of mine, *The Architecture of the Germplasm* (1964), and bears the marks of its parentage in many places throughout the text. Nevertheless, the center of focus is different in the two books. *The Architecture of the Germplasm* deals with the organization of the diploid genotype, in the light of evidence from several kingdoms of organisms, whereas the present book focuses on the organization and workings of plant genotypes. Many of the examples and stories used in the earlier book are, however, equally relevant to the theme of *Genetics of Flowering Plants,* and have accordingly found a place in it.

The literature of plant genetics is very scattered. It is spread across scores of current journals and through decades of back journal files. The dispersed condition of this literature renders it almost inaccessible to nonspecialists. One of my tasks in writing this book has been to bring together and synthesize a fair sample of the scattered information in plant genetics, with bibliographical references to the original sources. It was not my intention, however, to produce a complete compendium of examples and references on every topic taken up in the book.

Many of the older genetic experiments are as sound and as significant today as they ever were, but are now half forgotten and largely ignored. Another task in covering the subject of the book was to dig out some of these buried studies of earlier periods in genetics. This policy will hopefully give some feeling for the historical depth of our subject.

A minor theme of the book is the presentation of a series of classic experiments in plant genetics, beginning with that of Mendel in Chapter 1, and continuing with others of comparable elegance which are less well known outside specialist circles. It might have been sufficient for the main purpose of the book to have stated simply the conclusions established by these experiments. But I have described also the methods and logical rationale of the experiments which lead to the stated conclusions. In this way I hope to convey some of the atmosphere of experimental plant genetics, which is quite different from that of other more generally known fields such as microbial genetics and molecular genetics.

Finally, I have tried wherever possible to relate the findings of plant genetics, both old and new, to modern concepts in other branches of genetics.

University of Texas
July 1973

VERNE GRANT

ACKNOWLEDGMENTS

Several persons had the kindness to read and criticize particular chapters in manuscript. Frederick Seaman, one of my students, read Chapter 2; Theodosius Dobzhansky read Chapter 4; Robert K. Selander read Chapter 24; and Walter V. Brown read Chapter 25. These chapters benefited from their comments.

Other friends and colleagues provided first-hand information on particular topics: Lyle L. Phillips on gene segregation in synthetic amphiploids in Gossypium; Daniel U. Gerstel on variegation in Nicotiana; James L. Walters on distribution of translocation types in Paeonia; John B. Hair on karyotypes of some New Zealand plants; and David P. Bloch on the problem of chromosome organization.

Illustrations were kindly furnished by various colleagues, namely: the late T. H. Goodspeed (Fig. 15); W. M. Hiesey (Fig. 22); Daniel U. Gerstel (Figs. 32, 33); John B. Hair (Figs. 34, 36); Alva Day Whittingham (Fig. 42); Malcolm A. Nobs (Fig. 48); and James L. Walters (Figs. 52, 54). John Wiley and Sons granted permission to reuse a number of illustrations from *The Architecture of the Germplasm.*

Karen A. Grant read and discussed the first draft of the entire manuscript, section by section; her comments were always very helpful.

Mr. Robert J. Tilley of Columbia University Press took an interest in the book from the start (in 1971), and facilitated its way through the press.

To all of these individuals I wish to express my sincere appreciation.

V.G.

CONTENTS

PART III: LINKAGE SYSTEMS

PART IV: GENETIC SYSTEMS

PART I : GENES

THE MENDELIAN MECHANISM

INTRODUCTION   The resemblance between parents and offspring is due to the transmission of a material substance which causes the development of similar characteristics in successive generations. The doctrine of a material basis of heredity was stated by various nineteenth-century biologists: by Mendel (1866), Darwin (1868), Weismann (1883, 1893), de Vries (1889), and others. The nature and mode of organization of this heredity material, however, were to be revealed only gradually and in successive stages, almost entirely in the twentieth century.

The successive revelations have come out of breeding experiments, cytogenetic studies, and biochemical investigations with organisms as diverse as higher animals, flowering plants, fungi, single-celled eukaryotes or protistans, and prokaryotes or monerans. Plants occupied a preeminent position as experimental organisms in the critical early period of modern genetics. The historical trend in genetic work, however, as Mettler and Gregg (1969) point out, has been toward the utilization of ever smaller and shorter-cycled organisms, thus passing from plants to Drosophila to Neurospora to bacteria and viruses.

3

The experimental work on higher plants yielded some of the most important concepts of genetics, from the gene through gene systems to the genotype. These and other concepts originating in plant genetics have entered the mainstream of general genetics. Of course plant genetics has benefited reciprocally as a recipient of basic discoveries such as DNA, made first on members of other kingdoms.

The process of generalizing across two or more kingdoms of organisms, although it has proven fruitful in many cases, is also strewn with pitfalls, and various schools of workers have managed to fall into some of them from time to time. There are universals in genetics; there are generalizations valid within a circumscribed class of organisms; and there are generalizations of intermediate scope which hold for one or more kingdoms but not for all forms of life. One can transfer a basic concept, like that of the gene, from one kingdom to another almost with the stroke of a pen; but the concept, once transplanted, may apply to a different set of phenomena and may come to have an altered meaning in its new kingdom. Herein lies the beginning of confusion.

The present book will attempt to review some of the more important concepts in plant genetics, as regards genes, gene systems, linkage systems, and the unified genotype, in conjunction with the experimental evidence on which these concepts are based. It is desirable to view the gene systems and integrated genotypes in a somewhat broader context. Therefore we will follow the ramifications of our central subject into certain peripheral areas, like gene action and development, but we can only treat these tangential subjects in a very general way here.

THE DIPLOID-HAPLOID CYCLE It is convenient at the outset to recapitulate the normal life cycle of a diploid plant. The individual begins as a single-celled zygote derived from the fusion of two gametes, contributed by its female parent and its male parent. Each gamete carries one set of chromosomes and one set of genes, and is accordingly haploid. The zygote thus receives one set of chromosomes and genes from its mother through the female gamete and another set from its father through the male gamete, and is therefore diploid in its chromosomal and genetic constitution.

The zygote develops into a mature body by a series of mitotic cell divisions which perpetuate its original chromosomal and genetic constitution. There are some exceptions to this rule which do not concern us here. The individual plant then retains throughout its lifetime the particular combination of maternal and paternal chromosomes and genes which it inherited at zygote formation. When it reaches sexual maturity it produces gametes of its own by the process of meiosis. At meiosis, the homologous maternal and paternal genes and chromosome segments become separated and assorted to different haploid gametes.

Each kind of gene may exist in two or more alternative forms or alleles. If a gene A is present in two allelic forms, A and a, different individuals can exist with the genetic constitutions AA, aa, and Aa. An individual or genotype possessing two identical alleles of the same gene or genes, like the AA and aa individuals above, is said to be homozygous, whereas an individual or genotype carrying dissimilar alleles (Aa) is heterozygous.

An AA individual produces gametes all of the constitution A. Therefore if AA individuals interbreed with other AA individuals, or reproduce by self-fertilization, their progeny consist uniformly of AA genotypes. Similarly, aa genotypes produce all a gametes and all aa progeny. Homozygotes are true-breeding.

But the heterozygote Aa, derived from the union of two types of gametes, A and a, produces the same two types of gametes at meiosis in equal numbers. The random union of the A and a male gametes with A and a female gametes at fertilization then yields daughter genotypes AA, Aa, and aa in regular ratios. If the gene A controls some visible characteristic present in alternative states, like round versus wrinkled seeds, the progeny of the heterozygote will differ among themselves, or segregate for this character. A heterozygote or hybrid gives variable or segregating offspring.

SEGREGATION IN *PISUM SATIVUM* Mendel's experiment with the garden pea, *Pisum sativum,* dealt with variation in seven characters in hybrid progenies. The characters in question and their alternative states were as follows:

Form of seeds: round or wrinkled

Color of cotyledons: yellow or green

Color of seed coat: white vs. grayish-brown; correlated with white vs. violet flowers

Shape of seed pods: inflated or constricted

Color of seed pods: green or yellow

Position of flowers: distributed along length of stem or bunched at top of stem

Length of stem: long or short

Mendel crossed true-breeding pea plants exhibiting the contrasting states for each character, and studied the statistical frequency of the contrasting character conditions in the hybrid progeny in the F_1, F_2, F_3, and subsequent generations. He carried different lines to the F_5, F_6, or F_7 generation (Mendel, 1866, 1965).

From the results of these experiments, Mendel was able to infer the essential facts set forth in the preceding section of this chapter as regards the genetical constitution of the parental plants, the hybrids, and their gametes. The cytological facts were added later, by others.

Let us review Mendel's experimental method and results in regard to the inheritance of seed color. The character in question is actually cotyledon color, but it is seen as seed color, and we will refer to it as the latter. Mendel selected 10 true-breeding plants with either green or yellow seeds as parents. He made the cross of green × yellow in both reciprocal directions. Fifty-eight flowers were cross-pollinated to produce F_1 seeds.

The F_1 seeds, borne in pods on the mother plant, were all yellow, irrespective of the direction of the cross. These F_1 seeds were now sown to produce F_1 plants. The flowers on the F_1 hybrids yielded an F_2 generation of seeds by self-fertilization.

The F_2 generation of seeds showed a segregation into the two parental classes of yellow and green seeds. These alternative types were present in the numbers: 6022 yellow seeds and 2001 green seeds. These numbers correspond to a ratio of 3.01 : 1.

The F_3 generation was derived by self-fertilization of various plants grown from either green or yellow F_2 seeds. The F_2 plants from green seeds bred true on selfing, giving uniformly green seeds in F_3. Some of the F_2 plants grown from yellow seeds also bred true for the yellow-seeded condition in F_3 and later generations. But other F_2 plants grown from yellow seeds, on self-fertilization, yielded progenies of F_3 seeds which segregated for cotyledon color. Yellow and green seeds reappeared again in the F_3 generation derived from these F_2 parents in a 3 : 1 ratio.

The true-breeding and the segregating yellow-seeded F_2s were found to occur in definite proportions. Out of 519 plants raised from yellow seeds, 353 or about ⅔ produced segregating progenies of F_3 seeds, whereas 166 or about ⅓ of the yellow-seeded F_2 plants were constant for seed color in F_3. This division between segregating and nonsegregating families occurred in subsequent generations from F_4 to F_7.

Mendel obtained similar results for the six other character differences which he studied in the pea. His F_2 data are summarized in Table 1. The average ratio for all segregating characters combined in F_2 is actually 2.98 : 1 or essentially 3 : 1 (Mendel, 1866, 1965).

The 3 : 1 ratio for yellow and green seeds, and the further 1 : 2 ratio for constant or segregating yellow seeds noted above, combine to give a basic 1 : 2 : 1 ratio in the F_2 generation and in the hybrid fractions of subsequent generations. Symbolizing yellow and green seeds by B and b, respectively, Mendel wrote the basic segregation ratio as $B + 2Bb + b$ (we write it $BB + 2Bb + bb$ today).

This formula expresses the hypothesis that the hybrid Bb, though phenotypically yellow-seeded itself, produces two types of pure gametes, B and b, in equal numbers. Mendel tested this deduction by backcrossing the yellow-seeded hybrid Bb reciprocally to the parental type bb, and finding, as expected, a 1 : 1 ratio of yellow and green in the backcross generation. The data supporting this point are included with other similar data in the section of Mendel's paper on "the reproductive cells of the hybrids."

Years later Correns (1900), without prior knowledge of Mendel's work, found the same pattern of segregation for seed color in the pea. His F_2 population of seeds consisted of 1394 yellow and 453 green seeds, which is practically a 3 : 1 proportion. Correns's

TABLE 1

F_2 SEGREGATION RATIOS OBSERVED BY MENDEL FOR SEVEN PAIRS OF CHAR-
ACTER DIFFERENCES IN *PISUM SATIVUM* (MENDEL, 1866, 1965)

Character	No. of each type in F_2	Ratio
Seed form	5474 round 1850 wrinkled	2.96 : 1
Cotyledon color	6022 yellow 2001 green	3.01 : 1
Seed coat color	705 grayish-brown 224 white	3.15 : 1
Pod shape	882 inflated 299 constricted	2.95 : 1
Pod color	428 green 152 yellow	2.82 : 1
Flower position	651 axial 207 terminal	3.14 : 1
Stem length	787 long 277 short	2.84 : 1

F_3 and F_4 generations divided into segregating and constant families
in the same manner as had been described by Mendel. Correns
thus independently discovered the phenomenon of regular segregation on the basis of parallel experiments with the same plant
materials.

Mendel's results with respect to seed color in peas were confirmed at the same time by Tschermak (1900). Shortly thereafter,
in the years immediately following the rediscovery of Mendel's
laws, these results were confirmed again by Bateson and others.

INDEPENDENT ASSORTMENT IN *PISUM SATIVUM* Mendel next
undertook to determine whether the seven pairs of character differences are correlated or independent in inheritance. Is the segregation
in one character linked with, or is it separate from, the segregation in
other characters?

The analysis of this problem begins with the character combination of seed form and cotyledon color (Mendel, 1866, 1965).
The cross was: female parent plants with round yellow seeds × pol-

len parents with wrinkled green seeds. The experimental results can be summarized as follows:

P   round yellow × wrinkled green

F_1 round yellow

F_2 315 round yellow
        101 wrinkled yellow
        108 round green
        32 wrinkled green

F_3 from wrinkled green F_2, nonsegregating
        from round green F_2, segregating for seed form, constant for
        seed color
        from wrinkled yellow F_2, segregating for seed color, constant
        for seed form
        from round yellow F_2, segregating for both characters

Mendel thus observed four classes of phenotypes in the F_2 and F_3 progeny of a dihybrid cross. The phenotypes appeared in a 9 : 3 : 3 : 1 or dihybrid ratio in F_2. The four phenotypic classes could be subdivided further into constant versus hybrid types. Mendel therefore concluded that "the offspring of the hybrids appeared . . . under nine different forms," or in other words, that nine genotypes were present.

The numerical relations between these nine classes of genotypes, moreover, could be resolved into a combination of two separate 1 : 2 : 1 ratios, $AA + 2Aa + aa$ and $BB + 2Bb + bb$ indicating the independence of the two segregating characters. Their independence and separableness were confirmed by the appropriate backcrosses of the hybrid $AaBb$ to the parental types (Mendel, 1866, 1965).

Mendel went on to consider the results of a trihybrid cross involving the characters of seed form, cotyledon color, and seed coat color. The parental character combinations are round, yellow, gray-brown seeds in the female parents, and wrinkled, green, white seeds in the pollen parents. The F_2 generation derived from this cross contained all possible combinations of the three pairs of character differences, both the parental and the various new combinations, or recombinations. Mendel stated in a summary fashion, without presenting data, that the other four characters studied also showed independence, or independent assortment as we call it now.

TABLE 2
GENE SYMBOLS AND LINKAGE GROUPS FOR THE CHARACTERS STUDIED BY
MENDEL IN *PISUM SATIVUM* (DATA FROM LAMPRECHT, 1961A, 1961B)

Character	Mendel's symbolism	Modern gene symbols	Linkage groups
Seed form	A	R	VII
Seed color (cotyledon color)	B	I	I
Seed coat color (and flower color)	C	A	I
Pod shape	D *	V	IV
Pod color		Gp	V
Flower position		Fa	IV
Stem length	G *	Le	IV

* Used in letter to Nägeli, 1867.

Modern studies confirm this general conclusion with one exception. The relevant part of Lamprecht's summary of the factorial genetics of *Pisum sativum* is shown in Table 2. Mendel's seven monofactorial characters are determined by genes located in four linkage groups. When two genes occur in the same linkage group they are usually loosely linked, with fairly free recombination. Pod shape and stem length in linkage group IV, however, show definite linkage, with a restricted rate (13%) of recombination between them (Lamprecht, 1961a, 1961b).

PARTICULATE INHERITANCE Mendel's theory of heredity is founded on the phenomena of segregation and independent assortment in statistically precise ratios. From these phenomena he deduced the existence of different kinds of gametes containing the determinants of different pairs of character differences. These determinants are present in the gametes in single form for any given character (*A* or *a*) and in all possible combinations for two or more characters (*AB, Ab, aB, ab*). The different classes of gametes, produced in equal numbers, unite at random in fertilization to give the various classes of phenotypes which appear in regular proportions in hybrid progenies. Heredity is a mosaic of separable determinants.

In his paper Mendel speaks frequently of "pairs of differentiating characters," and "different kinds of egg and pollen cells." He alludes only rarely and then briefly to what might be in the reproductive cells. The dissimilar reproductive cells must contain dissimilar "elements." There are hypothetical "differing cell elements"; they have a "material composition and arrangement," and they combine at fertilization to determine "the development of the new individual" (Mendel, 1866, 1965, conclusions). These few passages herald the discovery of the gene.

It was important to find out whether the phenomena of regular segregation and independent assortment occurred in other organisms besides Pisum. Mendel himself investigated several other plants as well as honeybees and mice with varying degrees of thoroughness and success. The preliminary results with Phaseolus, Mirabilis, Matthiola, and Zea, but not with Hieracium, conformed to the theoretical expectations.

Upon the rediscovery of Mendel's laws in 1900, a number of workers—Correns, de Vries, Tschermak, Bateson, Baur, Cuenot—sought and found confirmation in a variety of organisms. Among the plants furnishing confirmation in that early period were Pisum, Oenothera, Lychnis, Solanum, Antirrhinum, Datura, and Zea; and among animals, mice and chickens. Years later the Mendelian principles were found to apply also to invertebrate animals like Drosophila and to fungi like Neurospora. The Mendelian mechanism giving rise to regular and balanced segregations for each gene pair is general among the various eukaryotic kingdoms of organisms but not in the prokaryotic monerans, in which other arrangements exist for combining and recombining the genetic material.

HISTORICAL COMMENTS The story has often been told of the long lag between Mendel's discovery in 1866 and the rediscovery in 1900. The reasons for the lack of interest in and comprehension of Mendel's work in his own time have also been discussed. Why was the scientific world so unreceptive in the 1860s and 1870s?

In the first place, we must remember that Mendel was an

obscure amateur working at an obscure institution and publishing in an obscure journal. Academic scientists as a group tend to look for signs of status in the author of a paper, and often ignore the work of persons without prestigious connections. There are individual exceptions to this tendency, of course, which are very important for the progress of science, but Mendel's paper did not happen to find its way into the hands of any such exceptional scientist during Mendel's lifetime.

Furthermore, this paper was highly original in both its approach and its conclusions and, as such, was out of step with the prevailing beliefs about heredity and variation. A paper about peas, filled with statistical and combinatorial notations, must have appeared very strange to late-nineteenth-century biologists accustomed to descriptions and verbal concepts. The originality of Mendel's paper, not to mention its revolutionary implications for biological thought, was almost enough in itself to guarantee its early neglect and tardy acceptance, in view of the well-known phenomenon of resistance to new ideas in science.

Sturtevant (1965) suggests that among all Mendel's contemporaries Galton would have been able to understand and appreciate Mendel's paper on its intrinsic merits. Perhaps Darwin should be included in this small company too, for Darwin grasped the value of Galton's work. In any event, the first strong advocates of Mendel's principles of heredity were to be Correns and Bateson in 1900.

It may be suggested finally that Mendel could have helped his own cause by following up the 1866 paper with a short book. The book medium would have enabled him to present the new theory of heredity in a form which nineteenth-century biologists could have understood. Such a book might have begun by reviewing the works and concepts of the early plant hybridizers from Koelreuter to Gärtner, thus establishing a familiar point of contact with many readers, and then gone on to state Mendel's revolutionarily different theory of particulate hereditary units. At this point and against this background Mendel could have introduced his beautiful and decisive experimental results.

It is worth recalling that the Darwin and Wallace papers of 1858 likewise attracted virtually no attention, whereas Darwin's book of 1859 stimulated an enormous amount of interest and controversy, leading directly to a needed scientific revolution.

For further information on Mendel and early Mendelism, the reader is referred to the books by Roberts (1929), Iltis (1932), Dunn (1965), Stubbe (1965), Sturtevant (1965), and Olby (1966). Collections of historical papers are given by the Editors of Genetics (1950), Krizenecky and Nemec (1965), and Stern and Sherwood (1966).

THE CHROMOSOME THEORY OF HEREDITY   A great deal of cytological and embryological work was done in the late nineteenth century, clearing up many details in the chromosomal and cellular aspects of the life cycle. This work paved the way for the ready acceptance of Mendelian heredity in 1900 and for the formulation of the chromosome theory of heredity shortly thereafter.

By the end of the nineteenth century several students of heredity who did not then know of Mendel's work, particularly Roux, Weismann, Hertwig, and Wilson, had come to the conclusion that the hypothetical determinants of heredity are borne on the chromosomes. Wilson stated in the first edition of *The Cell in Development and Inheritance* (1896, pp. 326–27):

> In its physiological aspect . . . inheritance is the recurrence, in
> successive generations, of like forms of metabolism; and this is
> effected through the transmission from generation to generation
> of a specific substance or idioplasm [hereditary material] which
> we have seen reason to identify with chromatin [chromosomes].

The course of events in meiosis preceding gamete formation became particularly interesting at the turn of the century when considered in the light of Mendel's laws. During meiosis the members of each chromosome pair become segregated into separate gametes, and furthermore the homologous members of one chromosome pair are segregated to separate daughter gametes independently of the distribution of other chromosome pairs. Consequently the gametes are haploid, segregate in a 1 : 1 ratio for any given homologous chromosome segment, and carry different combinations of nonhomologous chromosomes.

A series of parallelisms could be noted between the behavior of the chromosomes and the behavior of the Mendelian genes. Both the chromosomes and the genes possess the ability to duplicate themselves in cell division. Both the chromosomes and the genes are found in pairs in the body cells and singly in the gametes. The occurrence of a linear division of the chromosomes during mitosis could explain equally the constancy of the chromosomal and the genetic constitution in the cells of the body. Homologous members of a chromosome pair separate from one another at gamete formation, and so do the homologous alleles of a gene pair. These parallelisms could be explained on the hypothesis that the chromosomes are the bearers of the genes.

The chromosome theory of heredity, foreshadowed by Wilson and other earlier authors, was put together as a whole by the cytologists Sutton (1902, 1903) and Boveri (1904).

RECOMBINATION OF LINKED GENES According to Sturtevant (1965) the first known cases of linkage were those reported by Correns in Matthiola in 1900 and by Bateson and Punnett in Lathyrus in 1905. The best analyzed cases came later (1910–1915) in the work on *Drosophila melanogaster* by Morgan and his school. The phenomenon of linkage required a modification of Mendel's law of independent assortment of different genes, but it is the modification needed wherever two or more genes are borne on the same chromosome.

Prophase of meiosis in higher plants finds the homologous chromosomes paired in bivalents. The chromosomes are paired lengthwise, segment by homologous segment. At this stage, each chromosome consists of two chromatids or sister strands, and the bivalent is consequently four-stranded. Crossing-over takes place between homologous chromosomes, while the bivalent is in the four-stranded condition, and involves an exchange of chromatin between nonsister strands belonging to opposite homologs (see Westergaard, 1964; Darlington, 1965; John and Lewis, 1965).

Crossing-over occurs between two of the four strands at any point along the length of the bivalent. The result of single crossing-over as seen in the final products of meiosis is a set of four nuclei,

TYPE OF CROSSING OVER

MEIOTIC PRODUCTS

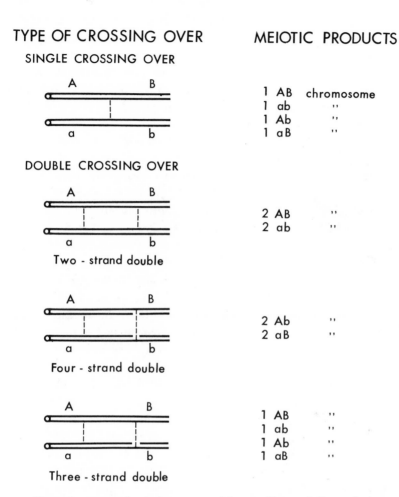

SINGLE CROSSING OVER

1 AB chromosome
1 ab   ,,
1 Ab   ,,
1 aB   ,,

DOUBLE CROSSING OVER

2 AB   ,,
2 ab   ,,

Two - strand double

2 Ab   ,,
2 aB   ,,

Four - strand double

1 AB   ,,
1 ab   ,,
1 Ab   ,,
1 aB   ,,

Three - strand double

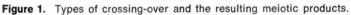

Figure 1. Types of crossing-over and the resulting meiotic products.

two of which contain crossover chromosomes, and the other two of which contain noncrossover chromosomes. One noncrossover chromosome is like one of the parental chromosomes in its genic contents; the other noncrossover chromosome represents the other parental type; one crossover chromosome carries a new combination of the parental segments and genes; the second crossover chromosome has the complementary combination of parental genes (see Fig. 1, top diagram).

Crossing-over may take place in one or more other segments of the same bivalent. The second crossover may or may not involve the same two strands that crossed over at the first point. If the same two strands participate in two exchanges, we have two-strand double crossing-over. The other possibilities are four-strand double crossing-over, with exchanges between different chromatids at each crossover point, and three-strand double crossing-over, where one strand is common to both crossovers. These relations can be seen more clearly with the aid of diagrams (Fig. 1).

We cannot go into the fine details of crossing-over here. Many aspects of the process are controversial. Some, including the actual mechanism of crossing-over, are still poorly understood. The frequency and distribution of crossovers are known to be affected by various environmental and genetic factors. Considerable evidence suggests that homologous pairing may begin in somatic cells prior to meiosis (Brown and Stack, 1968; Stack and Brown, 1969). General reviews are given by Rees (1961), Westergaard (1964), Darlington (1965), and John and Lewis (1965). Some aspects of crossing-over are better known in fungi than in higher plants, and there has consequently been a tendency to generalize from the former to the latter. But meiosis also differs in the two kingdoms. Westergaard (1964) therefore warns against generalizing too readily from the lower to the higher forms.

The spatial distribution of crossovers along the length of the bivalent is determined by chance as well as by various genetic and environmental factors. Any two genes, *P* and *Q*, located close together on the chromosome will in general cross over less frequently than two genes, *P* and *R*, standing farther apart on the same chromosome. The crossover frequencies under consideration now are the average frequencies in a population of meiosing cells.

For example, the double heterozygote *PQ/pq* may form 90% parental-type gametes (*PQ* and *pq*) and 10% recombination gametes (*Pq* and *pQ*), as a result of the relatively uncommon occurrence of crossing-over between *P* and *Q*. By comparison the double heterozygote *PR/pr* produces the crossover types (*Pr* and *pR*) in the frequency of 25%. These average values of crossing-over can be translated into units of measurement known as crossover units on a linkage map. The genes P and Q are 10 crossover units apart, whereas the map distance between P and R is 25 units.

Recombination frequencies range down to 1% or less for genes which are prevented from crossing-over by very close proximity or other factors, and up to a maximum of 50% for genes lying far apart on the chromosome. The frequency of recombination between loosely linked genes thus approaches the frequency of recombination due to independent assortment.

The upper limit of recombination between linked genes is 50% rather than, say, 100%, because of the fact that crossing-over takes place between two strands in a four-strand bivalent. Let us return to the diagrams in Figure 1. A single crossover between A and B gives 50% recombination-type gametes (Fig. 1, top). If A and B are far apart on the chromosome, so that a single crossover always occurs between them, they will exhibit 50% recombination. With increasing distances between A and B the chances of double or multiple crossing-over in the intervening regions also increase. Assume in the case of double or multiple crossing-over that any two nonsister strands are involved at each crossover point. Then the net result in a population of cells with different types of crossing-over is 50% recombination between A and B. The three cells with three types of double crossing-over shown in Figure 1 yield 12 daughter chromosomes, 6 of which are recombinants for A and B.

Empirical data support the conclusion that 50% is the upper limit of recombination. This conclusion is illustrated by data on recombination between the gene P and four other genes located at successively greater distances from P on chromosome 1 in maize (Rhoades, 1950, 1955). The map distances and corresponding recombination values can be compared in Table 3. Although P and bm_2 are 128 map units apart, they show approximately 50% recombination (Rhoades, 1955).

LINKAGE MAPS In the preceding section we considered a hypothetical case of three linked genes, P, Q, and R, with two known map distances, P and Q being known to lie 10 crossover units apart, and P and R 25 units apart. This information does not tell

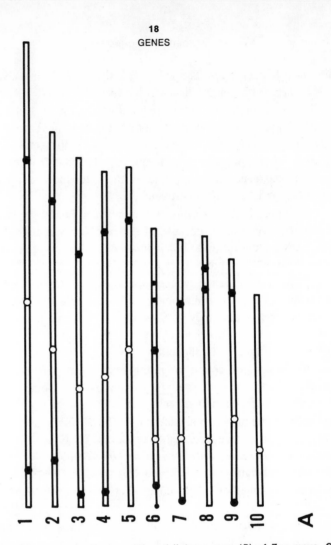

Figure 2. Chromosome diagram (*A*) and linkage map (*B*) of *Zea mays*. Centromeres are shown as hollow circles and knobs as solid circles. (From Rhoades, 1950.)

where *Q* stands in relation to *R*. But if further tests reveal that the map distance between *Q* and *R* is 15 crossover units, the linear order of the three genes on the chromosome must be *P–Q–R*.

In this way a genetic map can be drawn up which shows the linear sequence and relative distances apart of all the known genes in any given linkage group. By the use of further cytogenetic techniques each linkage group can be associated with a particular

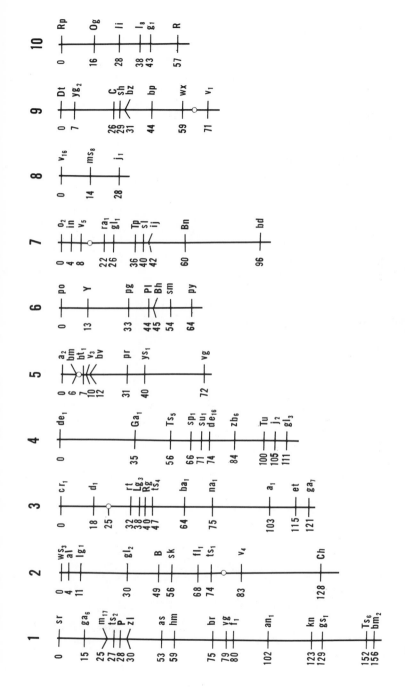

TABLE 3

RECOMBINATION VALUES BETWEEN GENES LOCATED VARYING DISTANCES
APART ON CHROMOSOME 1 IN *ZEA MAYS* (RHOADES, 1955)

Gene pair	Map distance, in units	Recombination value, %
P–as	25	25
P–f_1	52	41
P–an	75	45
P–bm_2	128	49

chromosome of the complement, and the various gene loci can be located in different regions of the chromosome.

The methods of genetic mapping were first worked out, and then carried out in the greatest detail, with *Drosophila melanogaster* (see Morgan, Sturtevant, Muller, and Bridges, 1915; Morgan, 1926). The genetic map of this fruit fly is shown in many textbooks.

TABLE 4

PHENOTYPIC CHARACTERISTICS DETERMINED BY SEVERAL GENES IN *PISUM SATIVUM* (LAMPRECHT, 1961A, 1961B)

Linkage group	Gene	Character	Effect of mutant or abnormal allele (A dominant, a recessive)
I	A	anthocyanin formation	A with anthocyanin, a no anthocyanin
	Am	flower color	am inhibitor of flower color
	I	cotyledon color	i green or cream cotyledons (in interaction with O)
	Lf	flowering time	lf early flowering
	O	herbage color	o yellowish green plant
III	Ca	seed-coat color	Ca grayish seed-coat (in interaction with A)
IV	Fa	branching of upper stems	fa fasciated stem
	Fw	resistance to Fusarium	Fw resistant
	Le	internode length	le short internodes
	V	pod shape	v small patches of sclerenchyma in pods
V	Gp	pod color	gp yellow pods
VII	R	seed form	r wrinkled seeds
	Tl	tendril formation	tl tendrils transformed into leaflets

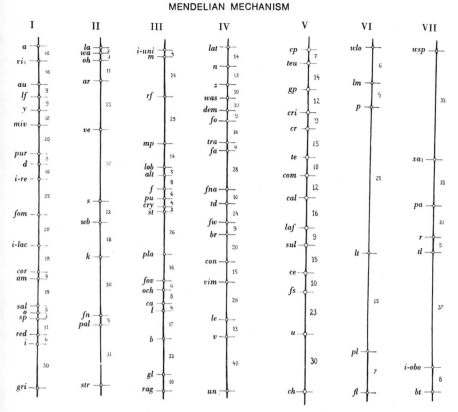

Figure 3. Linkage map of *Pisum sativum*. (From Lamprecht, 1961a, 1961b.)

The best mapped plant is *Zea mays* (Emerson, Beadle and Fraser, 1935; Rhoades and McClintock, 1935; Rhoades, 1950; Neuffer, Jones and Zuber, 1968). The 10 chromosomes of the haploid complement of maize are shown diagrammatically in the upper part of Figure 2. We see at first glance the relative lengths of the 10 chromosomes; the centromeres, indicated by hollow circles, show the relative lengths of the two arms in each chromosome. The position of the knobs, densely staining bodies which furnish important cytological markers on these chromosomes, is indicated by the solid circles. The linkage map of maize is given in the other part of Figure 2. The zero ends of the maps lie in the distal region of the short arm of each chromosome. The locations of numerous genes are shown directly on the linkage map, which in turn can be correlated approximately with the chromosome diagrams. The genetic map of maize shown here is taken from the review of Rhoades (1950); for more recent additions to the gene map, see Neuffer, Jones and Zuber (1968).

In concluding this chapter it is fitting to return once more to the garden pea. The linkage map of *Pisum sativum* (= *P. arvense*) is shown in Figure 3. The chart gives the map location of 92 genes on the seven chromosomes of Pisum (Lamprecht, 1961a, 1961b). The phenotypic characters determined by several of these genes are listed in Table 4.

CHAPTER TWO

GENE ACTION

INTRODUCTION
THE CHEMICAL NATURE OF THE GENETIC MATERIAL
MOLECULAR STRUCTURE OF DNA
GENE SPECIFICITY AND REPLICATION
GENIC CONTROL OF METABOLIC PROCESSES IN NEUROSPORA
GENES AND ENZYMES
TRANSCRIPTION AND TRANSLATION
THE GENETIC CODE
TWO ANALOGIES

INTRODUCTION Our understanding of gene action has been derived largely from studies of bacteria, fungi, and other microorganisms. Some key discoveries have also been made in various animals, including man. Plant genetics has been a benefactor more than a primary contributor in this field.

The biochemical genetic work on microorganisms has yielded some basic principles of general validity, principles which can be taken over by plant genetics; other phenomena and terminologies in microbial genetics have a more restricted scope and a more doubtful place in plant and animal genetics. These circumstances set the natural limits for this chapter. We will confine our attention here to the general principles of gene action. For reviews of biochemical and microbial genetics as such the reader is referred to Hayes (1968) and Watson (1970).

THE CHEMICAL NATURE OF THE GENETIC MATERIAL Miescher's research on the chemical composition of salmon sperm in the late nineteenth century revealed that the cell nucleus consists of both protein and nucleic acid. The two components were later found by Feulgen and others to occur together in the chromosomes within the nucleus. Before 1944 it was generally assumed that the genetic activity of the chromosomes resided in the protein component rather than in the nucleic acid. A representative statement of this widely held view is given, among others, by Goldschmidt (1938).

Proteins are built up from 20 different types of amino acids arranged in particular sequences in polypeptide chains, and consequently they are enormously complex and varied in their molecular structure. The synthesis of a protein macromolecule with its complex and specific structure requires a chemical organizer of correspondingly great specificity. Until recent times the only known substances with the necessary properties were the proteinase enzymes, which are themselves proteins. The nucleic acids, by contrast, were believed to be monotonously simple and similar in their chemical structure.

It was logical to conclude that the genetic material in the chromosomes is proteinase. The different genes would be different kinds of proteinases with the dual ability to replicate themselves and to synthesize various other particular protein molecules. A typical view as presented by Goldschmidt in 1938 held that a chromosome is a long chain of proteinase, stabilized by nucleic acid and consisting of a series of different proteinase sites controlling the synthesis of different protein products (Goldschmidt, 1938, p. 315; also 1955).

The traditional view was first challenged in 1944 by the discovery that hereditary transformations in bacteria are brought about by the nucleic acid fraction rather than by the proteins of the donor strain (Avery, Macleod and McCarty, 1944).

The crucial experiments of Avery and his co-workers were carried out with the pneumonia bacterium, Pneumococcus, and made use of two true-breeding strains differing in cell coat and virulence. Type III pneumococci have a polysaccharide cell coat, are virulent, and form large glistening colonies on artificial culture medium. Type II pneumococci lack the polysaccharide cell coat, are not infectious, and produce small rough colonies in artificial cultures. The two strains breed true to type in normal laboratory cultures. When, however, Type II pneumococci are grown in a medium containing a sterile soluble extract made from killed cells of Type III, the Type II bacteria acquire the morphological and virulence characters of Type III. The transformation is hereditary (Avery, Macleod and McCarty, 1944).

The Pneumococcus cell contains proteins, nucleic acid, lipids, and polysaccharides. Which one of these components is active in hereditary transformation? The protein component was at first suspected of being the transforming substance, in line with the prevailing views of that time. But when highly refined samples of the transforming substance were analyzed, they were found, contrary to expectation, to be deoxyribonucleic acid (DNA). No protein could be detected in the biologically active extract of Pneumococcus (Avery, Macleod and McCarty, 1944).

These indications were verified in experiments with bacteriophage carrying radioactive tracers (Hershey and Chase, 1952; Hershey, 1953, 1956). In one strain of the bacteriophage the protein component was labeled by means of radioactive sulfur, while in another strain the DNA component was labeled with radioactive phosphorus. The viruses were allowed to reproduce in bacterial cells, and their progeny were then tested for radioactivity. The virus strain with radioactive phosphorus yielded highly radioactive daughter viruses. By contrast, the daughter viruses derived from the parental strain with radioactive sulfur showed only slight traces of radioactivity, indicating that only a trace of the protein content is carried over from generation to generation. The genetic continuity of the bacteriophage is shown to be lodged primarily in its DNA component (Hershey and Chase, 1952; Hershey, 1953, 1956).

An early study of artificially induced gene mutations in *Zea mays* indicated that the genetically active part of nucleoprotein in higher plants is the DNA rather than the protein constituent (Stadler and Uber, 1942; Beadle, 1957b). Gene mutations were induced by irradiating the maize pollen with ultraviolet rays of different wavelengths. The mutant phenotypes were then recovered in corn grains developing from female flowers pollinated with the irradiated pollen (Stadler and Uber, 1942).

Some wavelengths in the ultraviolet region of the spectrum are more mutagenic than others. Stadler and Uber (1942) found that the shorter ultraviolet waves, particularly those with lengths between 2500 and 2800 Å, were the most mutagenic in maize, whereas the longer ultraviolet waves, those 3100 Å or longer, were nearly ineffective in inducing mutations. It is also known that DNA and protein absorb ultraviolet radiation in different ranges of the spectrum.

Knowing the characteristic absorption spectra of DNA and protein, and knowing the relative mutagenic effectiveness of the different wavelengths, these workers were therefore able to determine whether the genetically effective waves were those absorbed by the protein or those absorbed by the nucleic acid in the corn nucleus. The strong mutagenic wavelengths of ultraviolet radiation were identified spectroscopically as those absorbed by DNA. The genetically ineffective wavelengths, on the other hand, were in the range absorbed by protein (Stadler and Uber, 1942; Beadle, 1957b).

MOLECULAR STRUCTURE OF DNA   The molecular structure of DNA was worked out by Watson and Crick (1953) from chemical evidence of Chargaff (1950) and x-ray diffraction studies of Wilkins and co-workers (Wilkins, Stokes and Wilson, 1953). The well-known Watson-Crick model of DNA structure is illustrated in Figure 4, following Crick (1954), and is described briefly below with reference to this illustration.

DNA is a long, spirally coiled, polymeric chain of nucleotides. Each nucleotide unit consists of a deoxyribose pentose sugar, a phosphate, and a ringlike purine or pyrimidine base (Fig. 4A). The sugar molecules are linked through the phosphates into a sugar-phosphate chain, and the bases are attached to the sugars as side chains (Fig. 4A, C). The sugar and phosphate groups are identical throughout the length of the chain. The internal diversity arises from the bases. There are four common bases in DNA: adenine, guanine, thymine, and cytosine (Fig. 4B). These four bases can occur in any linear order on a particular sugar-phosphate chain.

The DNA macromolecule consists of two parallel coiled chains, forming the so-called double helix (Fig. 4D). The two sugar-phos-

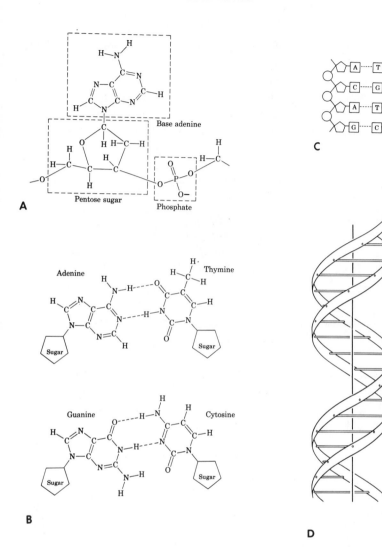

Figure 4. Molecular structure of DNA. (*A*) Nucleotide unit consisting of a pentose sugar, phosphate group, and base. (*B*) The four types of bases. They are shown as forming two base pairs linked by hydrogen bonds. (*C*) A short segment of two parallel sugar-phosphate chains with their complementary pairs of bases. (*D*) The double helix. The ribbons represent the sugar-phosphate chains, the horizontal rods are the base pairs, and the vertical line indicates the axis of the macromolecule. (Rearranged from Crick, 1954, reproduced by permission of *Scientific American*.)

phate chains are linked by weak hydrogen bonds between their respective bases (Fig. 4B). These linkages give rise to a linear series of base pairs (Fig. 4B, C).

There are two possible base pairs: adenine-thymine (A-T) and guanine-cytosine (G-C). This limitation is apparently due to spatial relations in the double helix. The purines (A and G) are large, the pyrimidines (T and C) are small (Fig. 4B). These two classes of molecules fit together along the length of the double helix, and hydrogen bonding between them is possible, in the combination A-T and G-C. Consequently, although the linear sequence of bases on one of the two chains is unrestricted theoretically, the sequence on the opposite chain is predetermined and complementary. If, for example, the sequence of bases on one chain is A-C-A-G, the sequence at corresponding points on the parallel chain is necessarily T-G-T-C, to form a stable linear order of base pairs (Fig. 4C).

The bases lie 3.4 Å apart on the sugar-phosphate chain. The chains comprising the helix make one complete turn in a distance of 34 Å measured along their length. Therefore 10 base pairs are included in every complete turn of the helix (Watson and Crick, 1953; Crick, 1954).

The genetic material of viruses and of such bacteria as *Escherichia coli* is present in a continuous double helix. The total length of this DNA thread, or "chromosome" as molecular biologists inappropriately refer to it, can be estimated from recombination frequencies and genetic map distances combined with the physical measurements given above. The length can then be expressed in either angstrom units or numbers of base pairs. The double helix of bacteriophage T4 contains about 150,000 base pairs, and the DNA strand of *E. coli* contains about 3 million base pairs (Hayes, 1968).

GENE SPECIFICITY AND REPLICATION The Watson-Crick model of DNA structure provides a molecular basis for the two primary properties of the gene, namely, self-duplication and the control of specific biochemical reactions, and for their third characteristic property, mutation.

The specific linear order of base pairs on the DNA double helix can be perpetuated by a copying process. Due to the precise and complementary nature of base pairing, each chain in the double helix can serve as a template for the building up of a new complementary polymeric chain, and the two new double helices will normally be identical to one another and to the original strand in their base sequence. Consider the sequence of base pairs in Figure 4C, which is

A-T
C-G
A-T
G-C

The left strand (ACAG), when functioning as a template in the synthesis of a new DNA molecule, will build up the sequence

A-T
C-G
A-T
G-C

and the right strand (TGTC) will likewise resynthesize

A-T
C-G
A-T
G-C

This scheme requires a supply of DNA precursors in the cell during replication. It requires that the weak hydrogen bonds in the original double helix be broken before replication so that each strand can function singly in synthesis. And it requires a physical separation of the two paired strands in the original double helix, presumably by unwinding, which does present mechanical difficulties for any long segment of DNA (see Hayes, 1968, pp. 239 ff. for further discussion).

Replication of DNA has been followed in vitro by Kornberg and his co-workers. The starting solution contains a primer of natural DNA together with a supply of the four bases, other raw materials, and the enzyme DNA polymerase. In this medium the quantity of DNA increases to an amount 20 times that of the original primer. The newly synthesized DNA is like the primer DNA in the proportions of the four bases (Kornberg, 1961; Hayes, 1968).

In the metabolic phase of gene activity, the specific sequence of bases in a segment of the DNA macromolecule is capable of conferring a correspondingly specific effect upon some biochemical reaction. Any one sequence of bases in a gene may direct in a specific way the course of a biochemical reaction within the cell and the formation of some chemical product. Molecular biologists liken the four bases to the letters of a four-letter alphabet. The different combinations of the four letters which are possible in any segment of DNA of genic or subgenic proportions will then spell out an enormous diversity of words and phrases, such as ATC-CAG or TTC-GTC, and these different words code for different biochemical reactions and products.

By the same token, a mutation in the gene would be a change in the base order, with an altered effect. Let ATC-CAG change to AAC-CTG. The change is permanent, being preserved in subsequent cycles of DNA replication. And it determines the formation of a new type of biochemical product. Some gene or point mutations may produce proteins with an altered amino acid sequence, but with no further detectable phenotypic effects, whereas other point mutations may lead to far-reaching changes in phenotype.

The mutational change may have its origin in an occasional error of base pairing in the parental DNA molecule. A normal base pair is A-T. But the adenine base may occasionally pair improperly with cytosine to form A-C. Then at the next replication the cytosine member forms the normal complementary base pair G-C, and the latter is replicated true to type thereafter. Following the original mismating, there is a permanent base pair substitution, a substitution of G-C for A-T, at a specific point in the DNA chain in half of the daughter molecules (Jukes, 1966; Hayes, 1968).

The three genic properties of replication, specificity, and mutation are all well illustrated by the normal and mutant types of hemoglobin in man.

The hemoglobin molecule, like other proteins, is built up from amino acids linked by peptide bonds into long polypeptide chains. The sequence of amino acids is specific in these chains. Human hemoglobin consists of four polypeptide chains—two alpha chains and two beta chains—together with associated iron groups. The alpha chains contain 141 amino acids each, and the beta chains 146 amino acids each. The sequence of amino acids in each chain is known. The alpha chains are identical in molecular structure, as are the beta chains; the alpha and beta chains are similar and homologous but have different amino acids at many corresponding positions in the chain (Ingram, 1956, 1963).

Normal hemoglobin, designated hemoglobin A, transports oxygen in the bloodstream in the usual fashion. In addition there are many mutant types of hemoglobin in man, some of which are associated with various forms of anemia. We will consider two common mutant types, hemoglobins S and C. The differences between the three types of hemoglobin are due to differences in a single gene, Hb, present in three allelic forms, hb^A, hb^S, and hb^C (Neel, 1949; Ingram, 1956, 1963).

The mutant hemoglobins S and C differ from normal hemoglobin A by one amino acid substitution in the beta chains. In hemoglobin A the amino acid in the number 6 position of the beta chain is glutamic acid. The corresponding position is occupied by valine in hemoglobin S, and by lysine in hemoglobin C (Ingram, 1963).

The alleles hb^A, hb^S, and hb^C direct the synthesis of very similar hemoglobin macromolecules and must themselves be quite similar in their base order. The only differences that need to be postulated between these alleles are those specifying the type of amino acid in position 6 in the beta chain. Base substitutions involving one or a few bases at a particular point in the DNA chain of the gene Hb will account for the differences between normal hemoglobin and the mutant types S and C (Ingram, 1963).

Many other rare mutant forms of hemoglobin are known. These rare mutants are also inherited as single-gene characters, and differ from normal hemoglobin, so far as is known, by single amino acid substitutions at some particular point in either the beta or the alpha chain (Ingram, 1963; Dayhoff and Eck, 1968).

GENIC CONTROL OF METABOLIC PROCESSES IN NEUROSPORA

Combined genetic and physiological studies of the red bread mold, *Neurospora crassa,* by Beadle, Tatum, Srb, Horowitz and others, have shown that individual genes control particular steps in a metabolic pathway. The experimental approach is as follows.

Normal or wild-type strains of Neurospora possess the ability to synthesize most of the complex substances they require for growth. The mold can be grown on an agar medium containing sugar, biotin (vitamin B_1), and certain inorganic salts. From these simple materials it synthesizes the 20 amino acids and builds up a far greater variety of proteins. Similarly, the mold in its wild-type form synthesizes several kinds of vitamins, polysaccharides, and fats. Mutant strains of Neurospora differ from normal strains in being unable to carry out the same syntheses (Beadle and Tatum, 1941).

The mutant strains fail to survive on the minimal medium of sugar, vitamin B_1, and salts, which supports the normal strains. However, the mutant molds can often be made to grow on a complete medium containing the higher sugars, amino acids, and vitamins. This shows that the gene mutations block biosynthesis at some point or points along a metabolic pathway and that the gene-controlled metabolic deficiency can be bypassed artificially by feeding the more complex metabolites to the mutant organisms (Beadle and Tatum, 1941).

It is possible to identify the step in a metabolic pathway which is blocked in any given mutant type. Suppose that Neurospora requires product D for growth. Suppose further that the normal strain of Neurospora can synthesize D from the raw materials present in a minimal medium, but that an assortment of mutant strains of Neurospora grow only if D is furnished in a complete medium. Suppose finally that the synthesis of D is known from biochemical studies to take place in a series of successive stages, $A \rightarrow B \rightarrow C \rightarrow D$, where A is some precursor present in the minimal medium. The problem now is to find out which step in this pathway is blocked in each of the various mutant types unable to synthesize D.

The blocked step can be determined by culturing the mutant strains on a series of progressively more complete media. If B is added to the minimal medium in a form in which it can be absorbed by the mold, and if a mutant strain can survive and hence by inference synthesize D on the B-enriched medium, one can conclude that the nutritional mutant in question is deficient for the ability to carry out the metabolic step $A \rightarrow B$, but can carry out the subsequent steps $B \rightarrow C$ and $C \rightarrow D$. This mutant strain can be designated arbitrarily as type I (Beadle and Tatum, 1941).

Other mutant strains of the mold may be unable to grow even when provided with $A + B$ in their medium, but survive on a medium to which C is added. These mutant types are deficient for the ability to carry out the metabolic step after B, for otherwise they would have been screened out with the type I mutants. They can, however, perform the step after C. The inhibition in this case therefore exists at the stage after B and before C, or in other words at $B \rightarrow C$. These mutants are evidently different from the type I mutants, and can be distinguished from the latter as type II.

Still other mutants may fail to grow on an A, a B, or a C medium, but will grow on the fully complete medium containing D. These type III mutants apparently lack the ability to carry out the final metabolic step, $C \rightarrow D$.

In practical laboratory work it is often convenient to work backward from the complete medium to the minimal medium, rather than forward from the minimal medium. Thus D can be furnished in the first tests to separate out the whole group of D-deficient mutants. In the second series of screening tests these mutants are grown on a medium containing C but not D. Some mutant molds will grow on the C medium while others will not; the former are type III and the latter are a collection of other types. The unknown mutant types are next assayed by culturing on a medium containing B but not C or D; those which grow on the B medium are type II and those which do not are type I.

By crossing a normal individual of Neurospora with an individual of mutant type I, and growing the progeny of this cross, it is possible to show that the difference between the parental strains in ability to carry out the reaction $A \rightarrow B$ is due to differences in the alleles of a single gene. By the same standard genetical technique we can show that the metabolic differences between type II and normal molds and between type III and normal molds are likewise inherited as single-gene differences. Finally, by intercrossing types I, II, and III in every possible combination, and testing for and excluding the possibility of allelism among the various mutants, it can be established that each type carries a mutation at a different gene locus.

As of 1950, in the early period of Neurospora genetics, 484 nutritional mutants had been obtained by irradiation of the spores and by other means, had been isolated in pure cultures derived from single spores, and had been put through a series of screening tests designed to identify their metabolic block. Of these mutant types, 405, or 84%, were found to require single specific chemical substances for normal growth. Most of the genes tested by this method are thus seen to control particular steps in a metabolic pathway (Horowitz, 1950).

The mutant types deficient for the ability to synthesize the amino acid arginine furnish a good example. Arginine is built up in a series of steps passing through several precursors: glutamic acid → proline (actually a substance close to proline) → ornithine → citrulline → arginine. Ornithine is converted to citrulline in two successive steps, and citrulline to arginine in one step (Srb and Horowitz, 1944; Wagner and Mitchell, 1964; Hayes, 1968).

Normal strains of Neurospora can synthesize arginine from its early precursors. Fifteen mutant strains, however, could not synthesize arginine, but required it in their growth medium. The mutant strains were intercrossed and tested for allelism. These tests revealed the existence of arginine-deficiency mutations in seven different gene loci, corresponding to different steps in the metabolic pathway (Srb and Horowitz, 1944; Wagner and Mitchell, 1964; Hayes, 1968).

One mutant type is unable to carry out the conversion of citrulline into arginine. Hence one gene of Neurospora controls the single synthetic step, citrulline → arginine. Two mutant types correspond to the successive steps between ornithine and citrulline. Four other genes whose mutant alleles were analyzed for nutritional ability were found to control various steps in the early part of the metabolic pathway from glutamic acid to ornithine (Srb and Horowitz, 1944; Wagner and Mitchell, 1964; Hayes, 1968).

Another series of nutritional mutants in Neurospora is correlated with different steps in tryptophan synthesis (see Hayes, 1968, for a review).

GENES AND ENZYMES   It is well known that metabolic processes are controlled by enzymes which are highly specific as to substrate and action. Different steps in a metabolic reaction chain are also controlled by different genes, as we have just seen in Neurospora. Therefore, as Beadle and Tatum (1941) pointed out, "gene and enzyme specificities are of the same order," and a good correlation appears to exist between gene action and enzyme action. These considerations have led to the fruitful one gene–one enzyme hypothesis (Beadle and Tatum, 1941; Horowitz, 1950).

According to this hypothesis, the primary function of the gene in many or most cases is to control the production or action of a single enzyme. The difference between the normal and mutant alleles of a given gene may be such as to determine the action or failure of a specific enzyme. A series of functionally related genes, such as those concerned with arginine synthesis, then control different enzymes involved in a metabolic reaction chain.

The enzymes, once formed by gene action, enable the cell to carry out the numerous biochemical reactions and synthesize the many substances necessary for life. An outstanding characteristic of enzymes is their high degree of specificity. An enzyme in green mold breaks down the *dextro* form of tartaric acid but not the *levo* form of the same compound. Enzyme specificity is attributed to the three-dimensional configuration of the enzyme molecule, which enables it to fit together with certain substrates only. Apparently the enzyme of green mold has a molecular structure which fits *dextro*-tartaric acid but not the mirror-image molecules of *levo*-tartaric acid. The catalytic action of enzymes depends on this ability to fit together with particular kinds of substrate molecules. The enzyme brings specific substrate molecules together, probably by forming a temporary chemical combination with them, so that the substrate molecules are in the proper relative position for a biochemical reaction to go forward rapidly.

Direct evidence that genes act through enzymes has been obtained in a number of cases. A classic example is the hereditary disease in man known as alcaptonuria. The urine of alcaptonuric persons turns black on exposure to air. The blackness is due to the abnormal urinary accumulation of an organic acid, which normally is broken down into simpler components by an enzyme present in the blood. The difference in the urine of normal persons and alcaptonuric persons can be reduced to the presence or absence of a certain enzyme. Since alcaptonuria is inherited as a single recessive character, the presence or absence of the critical enzyme can be traced ultimately to the normal or mutant alleles, respectively, of a single gene (Garrod, 1923).

Correlations have been established between single genes and particular enzymes in many other cases. Genes are known which affect the enzymes controlling sugar fermentation in yeasts, cyanide production in white clover (*Trifolium repens*), and the hydrolysis or breakdown of the alkaloid atropine in rabbits. Nutritional mutants deficient in the ability to carry out a particular metabolic step, which is probably enzymatically controlled, have been found not only in Neurospora but also in other fungi such as Aspergillus, Penicillium, and yeast, and in bacteria such as *Escherichia coli* (Wagner and Mitchell, 1955, 1964).

The technique of gel electrophoresis makes it possible to detect differences between enzymes by their relative mobility in an electric field. Different types of enzymes, when introduced into a gel from a tissue extract and exposed to an electric current, separate in the electric field and form characteristic bands as they migrate at different rates through the field. This technique can be combined with genetic methods to assay genic and allelic differences between individuals or species. The combined electrophoretic and genetic approaches have been applied widely in recent years in many groups of plants, insects, and other organisms (see Gottlieb, 1971). One of the results obtained has been the detection of many single-enzyme loci in higher plants and in other groups. For example, six loci in *Avena barbata* control different esterase, phosphatase, and peroxidase enzymes (Marshall and Allard, 1969; Clegg and Allard, 1972).

The one gene–one enzyme hypothesis, as Hayes points out, has provided a common focal point for genetics and biochemistry. It has enabled us to study and understand the gene in terms of biochemical function as well as in terms of inheritance (Hayes, 1968, p. 92).

The one gene–one enzyme theory should be construed in a broad rather than a strictly literal sense, according to Hayes (1968). Thus construed it allows for various deviations from the behavior of the Neurospora nutritional mutants in any more inclusive sampling of genes. Actually the experimental technique in the Neurospora studies was such as to select for mutant types deficient in single precursor substances and hence for genes controlling single metabolic steps. Some nutritional mutants turned up in Neurospora which did not fit into this pattern. In other organisms certain genes are known to direct the synthesis of proteins like hemoglobin without the participation of an enzyme. On the other hand, there exist enzymes which require the operation of more than one gene. Some students have broadened the original concept by restating the relationship as one gene–one polypeptide chain.

TRANSCRIPTION AND TRANSLATION Enzyme-controlled biochemical reactions take place mainly in the cytoplasm, at some distance from the genes in the nucleus, but enzyme activity is generally slight in the nucleus itself. Some intermediary is required to complete the connection between genes in the nucleus and enzymes in the cytoplasm. This intermediary is ribose nucleic acid (RNA), which occurs in both the nucleus and the cytoplasm. RNA moves from nucleus to cytoplasm and becomes concentrated in the sites of active protein synthesis in the ribosomes.

RNA, like DNA, consists of long sugar-phosphate chains with side chains of bases. In RNA, however, the sugars are ribose rather than deoxyribose, and each contains one more oxygen atom. Furthermore, the RNA macromolecule is single stranded. The bases in RNA are four: adenine (A), cytosine (C), guanine (G), and uracil (U). The first three bases are the same as their counterparts in DNA; the fourth, uracil, corresponds to thymine in DNA. In the pairing of an RNA strand with a DNA strand, therefore, the complementary base pairs are C-G and U-A.

RNA conveys the genetic message from the genes to the sites

of enzyme activity and protein synthesis. As molecular geneticists put it, DNA makes RNA, and RNA makes protein (including enzymes). The first step is known as transcription. The base order in a segment of DNA is transcribed, by means of base pairing between this DNA segment and a segment of RNA, into a complementary base order in the RNA. The new RNA chain is built up by polymerization. The second step is translation. The RNA molecule carrying the transcribed genetic message translates this message into protein structure.

Three types of RNA take part in the processes of transcription and translation: messenger RNA (mRNA), transfer RNA (tRNA), and ribosomal RNA (rRNA).

Messenger RNA receives the genetic message of DNA in transcribed and complementary form, as a result of complementary base pairing of the RNA with one strand of the DNA, and then carries this message to the ribosomes. Transfer RNA consists of an array of small molecules of different types corresponding to the different kinds of amino acids. The various tRNA molecules hook onto the different amino acid units. The ribosomes with their ribosomal RNA function as a sort of workbench for the assembly of the amino acids into polypeptide chains.

The short tRNA molecules line up successively in a linear sequence on the long mRNA chain on the ribosome. The specific linear sequence of the tRNA molecules is probably determined by complementary base pairing between these molecules and corresponding sections of the mRNA chain. The amino acids carried by the different tRNA molecules are then also aligned in the proper sequence to produce a given type of polypeptide chain (see Hayes, 1968; Watson, 1970).

The direct role of RNA in protein synthesis is well illustrated by the findings of Fraenkel-Conrat and his co-workers on tobacco mosaic virus (TMV) (Fraenkel-Conrat 1956, 1962; Fraenkel-Conrat, Singer and Williams, 1957). The rodlike particles of TMV, which live parasitically in plant tissues, consist of a long narrow core of RNA surrounded by a protein shell. Different strains of TMV are distinguished by the type of protein in the shell. The so-called HR strain differs from the common form of TMV in the proportions of various amino acids in the protein sheath.

The protein and RNA components of TMV can be separated by gentle chemical treatment, so that, when mixed together again in solution, they can combine to form a more normal virus particle capable of growing and multiplying in plant tissue. By a simple extension of this scheme, a new "hybrid" virus can be assembled from the protein and RNA components of different parental strains. Fraenkel-Conrat and co-workers separated the protein and RNA constituents of normal TMV and of TMV-HR, and recombined these constituents in new mixtures to produce new infectious virus particles.

The daughter virus particles descended from reconstituted "hybrid" viruses with new combinations of protein and RNA were found to possess the type of protein which is characteristic for a particular strain of RNA. The type of protein contributed to the reconstituted hybrid virus was not perpetuated in its descendants. Thus, reconstituted virus particles consisting of RNA from the HR strain and protein from the common strain of TMV gave rise to progeny conforming to the HR type in both their RNA and protein components. The specific type of protein is built up anew in each generation of TMV particles under the direction of the specific type of RNA carried by those particles (Fraenkel-Conrat, Singer and Williams, 1957; Fraenkel-Conrat, 1962).

THE GENETIC CODE   The four types of bases in DNA and RNA have to code for 20 types of amino acids in protein. A one-to-one relation between single bases and amino acids would provide the specificity for only 4 amino acids. Two-base sequences like AA, AT, and AC could code for 16 types of amino acids, and would still be insufficient. But sequences of three bases, or triplets, like AAA or AAT, could specify as many as 64 amino acids if each triplet were to code for a different amino acid; and this variety in DNA and RNA would be more than enough to control the existing diversity of amino acids.

A linear sequence of triplets on a segment of a DNA chain (ATT-CAG-TGC-etc.) could then specify, through an RNA intermediary, the sequence of amino acids in an entire polypeptide chain.

Ingenious biochemical methods have been devised and used by Nirenberg, Ochoa, Khorana, and others to identify the triplets which specify different amino acids. The methods make use of simple synthetic forms of RNA built up from known types and proportions of bases. The various kinds of synthetic RNA are introduced into cell-free protein-synthesizing systems in vitro, and the proteins which are built up are then assayed for their amino acid composition. In this way the RNA triplet UUU was correlated with the amino acid phenylalanine. This first identification was later extended to the other triplets of RNA. And the early methods were later improved in various ways and made more precise (see Hayes, 1968).

The genetic code expressed in terms of RNA is summarized in Table 5. This code appears to be universal, applying in plant and animal cells as well as in bacteria and viruses. The triplets of DNA in the genes which determine the various RNA triplets are those with the complementary base order; it will be recalled that the possible base pairs between DNA and RNA are:

DNA		RNA
A	–	U
T	–	A
C	–	G
G	–	C

Table 5 shows that all 64 possible triplets of RNA have known functions. Some amino acids are specified by one or two triplets. In other cases four or even six different triplets specify the same amino acid. In such cases, however, the synonymous triplets are closely related in their base composition, being alike in the first base or second base or both. One result of the existence of two or more synonymous triplets, all of which code for the same amino acid, is that many mutational changes can take place in the triplets without causing an amino acid substitution in the polypeptide chain.

TWO ANALOGIES In describing the relation between gene structure and gene function, molecular geneticists have employed an analogy with alphabetical codes and information theory, and this very useful analogy has been adopted by biologists generally. The genetic information contained in the genes is said to be encoded in

TABLE 5
THE GENETIC CODE: TRIPLETS OF BASES IN MESSENGER RNA WHICH SPECIFY
THE DIFFERENT AMINO ACIDS (FROM HAYES, 1968)

Amino acid	RNA triplets	Amino acid	RNA triplets	Amino acid	RNA triplets
Alanine	GCU	Glycine	GGC	Proline	CCA
	GCC		GGA		CCG
	GCA		GGG		
	GCG			Serine	UCU
		Histidine	CAU		UCC
Arginine	CGU		CAC		UCA
	CGC				UCG
	CGA	Isoleucine	AUU		AGU
	CGG		AUC		AGC
	AGA		AUA		
	AGG			Threonine	ACU
		Leucine	UUA		ACC
Aspartic acid	GAU		UUG		ACA
	GAC		CUU		ACG
			CUC		
Asparagine	AAU		CUA	Tryptophan	UGG
	AAC		CUG		
				Tyrosine	UAU
Cysteine	UGU	Lysine	AAA		UAC
	UGC		AAG		
				Valine	GUU
Glutamic acid	GAA	Methionine	AUG		GUC
	GAG				GUA
		Phenylalanine	UUU		GUG
Glutamine	CAA		UUC		
	CAG			Termination of	UAA
		Proline	CCU	chain	UAG
Glycine	GGU		CCC		UGA

a four-letter alphabet consisting of the letters A, C, G, and T. These
letters are combined into 64 three-letter words. A long series of
these words the length of a gene then gives the proper instruc-
tions to build up a particular type of polypeptide chain or protein
molecule. Or, in the parlance of the field, the "genetic message"
enciphered in the "codons" of DNA is "transcribed" and then
"translated" into the chemical structure of protein. And the genetic
information enciphered in the whole genome can be compared
with the verbal information in a large encyclopedia.

The analogy can be extended so as to encompass the entire

field of biology. The DNA code spells out "the language of life," provides a "grammar of biology." It follows that the proper approach to the study of life is to attempt to decipher the code and read out the genetic message.

Analogies are useful as ways of visualizing an invisible process or relationship, but they become hazardous when taken too literally, usually as a result of long-continued and uncritical repetition. The code analogy in molecular biology does appear to be taken rather literally in many quarters. Explicit statements in the literature to the effect that the genetic code is a figure of speech and nothing more are conspicuous by their scarcity.

Therefore it may be pertinent to remind ourselves that there is no genetic code in the real sense of the term. DNA does not spell anything and does not write sentences, let alone encyclopedias. The concept of a biochemical language of life is a second-order metaphor built on previous metaphors.

In an earlier work (Grant, 1964) I employed a different analogy for illustrative purposes, which of course should not be pushed too far either. The relation between gene structure and gene function may be likened to a segment of a phonograph record when played on a phonograph. Both the gene and the record groove have a specific physical structure which determines a specific expression. In both entities the structure and its expression can be copied exactly in reproduction. This analogy has the merit that it focuses attention on the role of surface configuration in determining the specificity of primary gene action and of enzyme action.

The analogy can be extended, too. The individual gene can be compared with a musical phrase on the record. The gene is built up from smaller units—nucleotides, triplets, and subgenes—and similarly the phrase is built up from notes, chords, and measures. The individual gene is organized into units of a higher order—gene systems and linkage systems—just as musical phrases are organized into parts and movements. At each level of organization in either case a specific physical structure produces an equally specific result.

DEVELOPMENT

INTRODUCTION Embryonic development in multicellular organisms comprises the exceedingly complex series of stages from the single-celled zygote through the embryo to the mature body. The zygote, a single cell, gives rise by orderly and integrated processes to an adult organism consisting of numerous differentiated organs and tissues. The process of development is well known for different groups of higher organisms, in the descriptive sense. But it is still rather poorly understood in a causal sense.

Development is most complex in the higher vertebrates, in which the end product of development, the terrestrial or aerial vertebrate body, also reaches the pinnacle of complexity. Development takes a simpler form in the vascular plants, with their open system of growth and relatively less differentiated plant body, but the process is still tremendously complex here. Other major groups such as the insects present still other developmental features. There are, in other words, important differences between kingdoms and classes, as would be expected, but there are also many underlying common features.

The prominent differences between major groups restrict our ability to formulate valid broad generalizations. Much important experimental embryological work on frogs and salamanders, for instance, has revealed modes of developmental control which have a wide applicability in the vertebrates but not necessarily in higher plants, and conversely, plants with their transportable chemical growth regulators present some unique features of their own. The more basic laws of development are those applying to the features of embryonic development which are common to different kingdoms. This is the realm in which the student of plant embryogeny and morphogenesis can make use of the concepts of animal embryology, and vice versa.

The main features of development which are common to plants and animals can perhaps be summarized as follows. In all cases a single-celled zygote gives rise by series of mitotic cell divisions to the multicellular adult individual. The daughter cells derived by mitosis from the zygote change into the forms characteristic of different specialized tissues. And different parts of the developing body become molded into different organs. There is growth in size, cell differentiation, and morphogenesis, respectively, and these three aspects of the overall process occur simultaneously (see Waddington, 1957, 1962; Simpson and Beck, 1965).

Furthermore, the course of development from the embryo to the various mature organs follows canalized pathways which resist environmental pressures to become diverted from the normal track. There is a certain norm of development which Waddington (1957) refers to as developmental canalization. Finally, the body develops as a functionally integrated whole. There is coordination between the various regions and organs.

The purpose of this and the next chapter is to discuss some basic aspects of development from the genetical viewpoint. The emphasis will be on principles and problems rather than details. Furthermore, the discussion has special reference to problems of development in higher plants. Our objective, in other words, is limited. A complete review of plant development goes well beyond the scope of this book. For further readings see Waddington (1957, 1962), Stebbins (1968), Galston and Davies (1970), and Markert and Ursprung (1971).

GENOTYPE AND PHENOTYPE It is appropriate to begin a discussion of genetic aspects of development by introducing the twin concepts of genotype and phenotype. These two basic concepts grew out of Johannsen's selection experiment with the kidney bean, *Phaseolus vulgaris*. This plant is predominantly self-fertilizing and homozygous.

The foundation stock for Johannsen's (1903) experiment was a mixture of beans of different sizes and weights. The original sample of bean seeds was assembled from many individual plants. This sample exhibited continuous variation in seed weight over a wide range, from 15 to 90 cg. Johannsen selected for high weight of seeds in one series of lines and for low weight in another series. At first there was a response to the selection, and lines were quickly established which produced either heavy seeds or light seeds. Considerable variation in seed size remained in the selected lines, however, and the differences between them were differences in averages and ranges for seed weight.

Johannsen now subdivided the bean population into 19 separate inbred lines or pure lines derived from 19 different parent plants in the original mixture. Each pure line continued to show variation in seed size. Johannsen practiced selection for seed size within the different pure lines. The procedure was to plant a large seed and a small seed from the seed output of the same parental plant, raise these daughter plants to maturity in a fairly uniform environment, and compare their seed output as to size. Thus in a large-seeded line two beans at opposite ends of the range, one weighing 84 cg and the other 46 cg, were chosen to propagate daughter plants; the beans produced by these daughter plants had average weights of 73.0 and 74.4 cg, respectively (see table 6). Similar results were obtained in other replicate progeny tests in both large-seeded and small-seeded lines. The lack of differentiation between the paired samples of bean seeds derived from the same mother plant con-

TABLE 6

SEED SIZE IN DIFFERENT FAMILIES OF SEEDS IN ONE PURE LINE OF *PHA-SEOLUS VULGARIS* THROUGH SIX GENERATIONS OF SELECTION (JOHANNSEN, 1909, 1926)

Year	Weight of parent seeds, cg	Average weight of progeny seeds, cg
1902	70	64.9
	60	63.2
1903	80	70.9
	55	75.2
1904	87	56.9
	50	54.6
1905	73	63.6
	43	63.6
1906	84	73.0
	46	74.4
1907	81	67.7
	56	69.1
Means	79.2	66.2
	51.7	66.7

tinued through several generations of selection (Table 6). Selection among variants within any given pure line was shown to be ineffective (Johannsen, 1903, 1909, 1926).

Johannsen concluded that there was an initial store of genetic variation in the original seed mixture, which was sorted out into different pure lines by selection, and an additional component of environmentally induced variation not subject to control by selection. The observed variations in bean size, in other words, reflect differences at two separate levels of determination. The first cause of variation stems from genotypic differences between separate lines of bean plants. But any given genotype will give rise to a certain range of phenotypes, depending on environmental variables.

The genotype is the sum total of genetic determinants, and the phenotype is its particular character expression. Each genotype can be thought of as a "norm of reaction" which can engender a range of phenotypes in different environments (Johannsen, 1909, 1911, 1926). The problem of development as viewed from the genetical standpoint is then to trace the connection between genotype and phenotype. How does a certain phenotype arise from a particular genotype?

FROM GENE ACTION TO PHENOTYPIC CHARACTER   The phenotype is the sum total of characters of an organism. Lewis and John (1963) draw a useful distinction between the exo-phenotype and the endo-phenotype. The exo-phenotype is the set of externally expressed characters, such as flower color and leaf shape, whereas the endo-phenotype comprises the internal features of the organism, like chromosome form and biochemical products.

In Chapter 2 we saw that the course of gene action can be traced from the molecular structure of DNA segments to the formation of enzymes or other proteins. This is the sequence which is understood in a causal sense. The known products of gene action which can be connected with gene structure belong to the endo-phenotype.

Gene structure and primary gene action are the beginning but only the beginning of the long series of developmental steps which culminate in the exo-phenotypic characters. The ultimate phenotypic expression of a gene is removed from the primary biochemical action of the gene by many steps, and the later steps in this series are mostly not well understood in genetical terms. Only rarely is it possible to trace out the whole chain of gene-determined causes and effects involved in development.

We noted in the preceding chapter that different alleles of the gene Hb in man determine the formation of different kinds of hemoglobin molecules. The action of the Hb gene does not end with hemoglobin synthesis, however, but continues in the formation of red blood cells with different properties depending on the type of hemoglobin present. The Hb^A allele produces normal, round red blood cells, the Hb^S allele sickle-shaped cells, and the Hb^A/HB^S heterozygote a mixture of both types of blood cells. Individuals of the homozygous constitution Hb^S/Hb^S exhibit sickle-cell anemia, whereas Hb^A/Hb^A individuals do not have this disease. There are still other phenotypic effects further removed from the primary biochemical action of the Hb gene. The heterozygote Hb^A/Hb^S has superior resistance to malaria (Allison, 1956).

A few analogous cases are known in higher plants, in which single-gene mutants determine simple metabolic defects. Among these are the mutants for cysteine deficiency in Arabidopsis, vitamin deficiency in Arabidopsis, thiamine deficiency in tomato, disturbance of magnesium metabolism in celery, disturbance of iron metabolism in maize, and gibberellin deficiency in maize (see Simmonds, 1965b, for brief discussion and literature references). In each case an altered exo-phenotype can be related to a simple gene-controlled metabolic block.

Many other cases of mutant exo-phenotypes in plants cannot be reduced to simple biochemical causes. Simmonds studied 12 mutant types in diploid potatoes (*Solanum tuberosum stenotomum, S. t. phureja*, and other cultivars) in an attempt to trace the altered phenotypic expression back to simple metabolic or hormonal deficiencies. The attempt was unsuccessful. None of the 12 mutant genes tested in this study has early effects on development which can be identified as simple nutritional or hormonal defects (Simmonds, 1965b).

Four of the 12 mutant types were analyzed morphogenetically or physiologically. In two of these the earliest detectable aberration is an anatomical defect. Thus the mutant Shorty (*sh*) has a reduced cell elongation which leads ultimately to dwarf plants with short internodes. The mutant *dx* at maturity is stunted, with small leaves and brittle stems, and the mutation is sometimes lethal. This syndrome of mutant phenotypic characters can be traced back to weak and malformed vessels and tracheids in the xylem (Simmonds, 1965b).

In common barley (*Hordeum vulgare*), the mutant type Hooded differs from normal-awned barley by a malformation of the lemmas in the spikelets. In normal barley the lemma tapers into a long awn. Hooded barley has in place of the awn a short tip bearing various lateral appendages. The difference is a single-gene one, normal barley being *kk* whereas the Hooded mutant carries the incompletely dominant allele *K*. Ontogenetic studies reveal that the mutant allele *K* acts at an early stage in the development of the lemma. It alters the orientation and the rate of mitotic divisions in meristematic tissues. In normal barley the mitotic divisions are oriented along a linear axis for growth of the awn in length, whereas in hooded barley the cells proliferate in all directions to form a cushionlike area in the future lemma tip (Stebbins and Yagil, 1966; Stebbins, 1968).

PLEIOTROPY In biological as well as physical systems, single causes often produce multiple effects. A single initial action determines multiple primary effects; each primary effect becomes the starting point for several or many secondary effects; and the compounding of effects continues through subsequent steps in the process.

The role of gene action in development also displays these features. Gene action follows a series of branching pathways on the way to phenotypic character expression. A single gene generally produces several different phenotypic effects. The term pleiotropy denotes the occurrence of diverse phenotypic effects originating from the action of a single gene.

The gene S in *Nicotiana tabacum* affects the form of several different plant parts—leaves, calyx, corolla, anthers, and capsule. The dominant allele S produces leaves with long petioles and acuminate tips, calyces with long slender teeth, corolla lobes with slight points, long anthers, and capsules with a long narrow form (Fig. 5). Plants homozygous for the recessive allele s have the contrasting leaf and floral characters shown in Figure 5: unpetioled and oval leaves, short-toothed calyces and unpointed corollas, and roundish anthers and capsules. These pleiotropic effects could be due to a single mode of action of the gene S during the development of different organs. The allele S causes all of the lateral organs to grow long and narrow, but the allele s determines a short broad growth of these same organs (Stebbins, 1959).

The Compacta mutant of the European columbine (*Aquilegia vulgaris*), which is inherited as a single-gene recessive, differs from the normal form in several vegetative and floral characters. In habit, the Compacta mutant is shorter and bushier, with more branches. Its stems are more brittle than those of the normal type of columbine. The flower buds of the Compacta type are erect, in contrast to the drooping flower buds of normal *A. vulgaris*. Finally, the petal blades and sepals are smaller in the Compacta mutants. Anatomical studies reveal that secondary thickening of cell walls

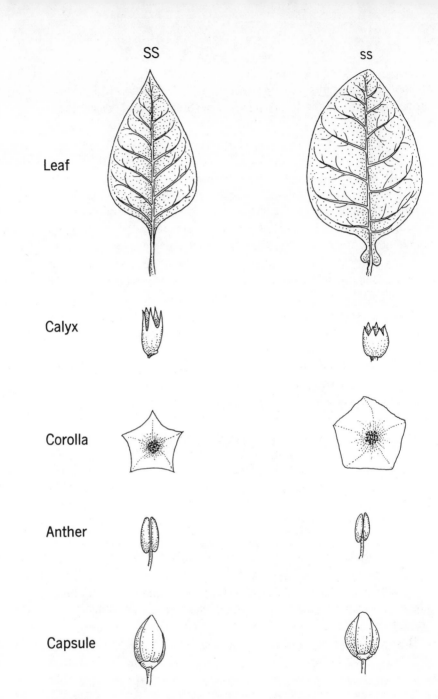

SS   ss

Leaf

Calyx

Corolla

Anther

Capsule

Figure 5. Pleiotropic effects of two alleles of the gene *S* on different organs in *Nicotiana tabacum*. (From Stebbins, 1959, reproduced by permission of Pergamon Press.)

takes place precociously in the Compacta type. The manifold phenotypic effects of the Compacta gene appear to be different direct or indirect results of this early thickening of the cell walls. The early cell-wall thickening inhibits cell elongation and thus brings about a dwarf, compact habit of growth. The brittle stems and the erect rather than drooping position of the flower buds are still other effects of the same basic alteration in early development (Anderson and Abbe, 1933).

Pleiotropy may connect some characters which are outwardly very different and apparently unrelated. The recessive allele *dl* in the tomato (*Lycopersicon esculentum*) reduces the development of hairs on the stems, peduncles, and stamens. The loss of hairs on the stamens causes the anthers, which are normally bound together into a tube, to spread apart. This divergence of the anthers now disrupts the normal mechanism of self-pollination that prevails in cultivated tomatoes. Normal tomato plants produce full sets of fruits by selfing supplemented by occasional (1–2%) outcrossing. But the *dl/dl* plants yield nearly 50% outcross progeny. Furthermore, the homozygous mutant plants (*dl/dl*) produce only 10% as many fruits as normal plants (+/+ and *dl/+*) under field conditions. An alteration in the breeding system and fruitfulness thus goes back to the loss of staminal hairs (Rick, 1947).

ON THE CAUSES OF CELL DIFFERENTIATION A basic aspect of development in multicellular organisms is cell differentiation. All of the different specialized cells of the body are derived by mitosis from a single unspecialized zygote cell. Do the various types of body cells differ in genetic constitution? If so, how do the genetic differences arise during mitotic cell division? Alternatively, if the body cells have the same genetic constitution, how do they come to differ structurally and functionally?

This problem was first formulated clearly by Weismann in 1883 (see Dunn, 1965). Weismann's solution of the problem assumed a genetic differentiation underlying the observed cell differentiation. There was a continual sorting out of genetic determinants during the course of somatic cell development and, as a consequence, the cells became differentiated into the various histological types.

The picture of cell and nuclear division which emerged from a period of active cytological investigation in the late nineteenth century put an end to Weismann's hypothesis. The multiplication of cells in the body was seen to proceed by a process of somatic mitosis. An essential feature of mitosis is the longitudinal division of the chromosomes and their equal distribution to the daughter nuclei. Mitosis yields two daughter nuclei equivalent to one another and to the mother nucleus. The genetic consequence of the mitotic mode of division is the duplication of the mother cells' genetic constitution in the daughter cells. Therefore, all the cells of the body from zygote to adult have, or normally should have, the same genetic constitution.

The facts of chromosome cytology led to the second main viewpoint regarding cell differentiation, a viewpoint which has held the center of the stage through most of the twentieth century. The different types of body cells derived from one zygote have the same genetic constitution and hence the same set of potentialities. The differences between them depend upon which genes act and which potentialities are realized. In one line of cells various internal and external influences may stimulate gene action on the part of certain genes of the complement so that the cells develop as typical members of one tissue system; in another parallel cell line exposed to different influences, other genes are released for gene action and the cells develop into another tissue system. In short, cell differentiation results from differential gene action.

Stern (1954) uses an analogy to clarify this concept of differentiation. A mountain contains gold, uranium, and water. One set of operations by one group of prospectors yields gold, but not uranium. Other operations release uranium, and still others are required to bring out water. The various potentialities are all together in the same mountain, but which ones come to expression at a given time and place depends upon specific responses to specific influences (Stern, 1954; see also Stern, 1968).

In recent years many cases have been found of chromosomal and genetic differences between somatic cells in the same individual. Such cases have come to light in both animals and plants. Genetic differentiation between cells and tissues of the body, which was formerly regarded as an exceptional event, is becoming recognized as a common and widespread phenomenon (Stern, 1968). The older view of cell differentiation can be reconsidered in a new light. Genetic differentiation does play a role in some cases of cell differentiation, and does provide a partial explanation of some aspects of development.

Genetic differentiation between cells and tissues is only part of the story, however, and will account for only some of the problems in development. Much of development takes place without genotypic changes in the cells involved. In plants this is shown by the processes of vegetative propagation and regeneration. Replication of a given genotype by vegetative propagation is a normal and very common mode of reproduction in vascular plants. Furthermore, specialized cells of the stem, root, or leaf may sometimes become dedifferentiated and then give rise to a new plant individual which is genotypically like the mother plant.

Phenomena of cellular and tissue specialization which can be related to genetic and chromosomal differentiation will be discussed in later chapters on mosaics and karyotypes. In the remainder of the present chapter we will consider the central and difficult problem of developmental changes which take place without genotypic changes in the cells involved.

PREPATTERNS   Cell specialization is regional. Different types of specialized cells regularly form in different parts of the developing body. The orderliness of tissue and organ development depends on this tendency of cell differentiation to occur in regional patterns. Underlying the regional patterns are embryonic fields or prepatterns (Stern, 1954, 1968).

Consider a developing body composed of genotypically identical cells. Different physical and chemical influences occur in different regions of the body. These influences evoke different genetic responses in the various regional groups of cells. Certain genes are active in one region of the body but not in others, while in other regions other genes will yield their characteristic products. The diverse responses of the genes or gene products to the specific influences in the various regions of the body may result in tissue and organ differentiation. The formation of different tissues and organs in their normal places may be said to be predetermined by the regional prepatterns in the developing body (Stern, 1954).

Examples of factors which may differ from region to region in an embryo or growing body are temperature, pH, and chemical growth regulators. The distribution of these or other factors throughout the developing body will establish the regional prepatterns.

Enzymes characteristically are active only within a certain range of temperatures or at certain levels of alkalinity or acidity. Let us say that an enzyme controlling a step in a metabolic process can act only at certain optimum temperatures. Assume further that temperatures range above and below the optimum in different regions of an embryo or developing body. Then the gene that produces the enzyme which yields the metabolic product will have a positive phenotypic expression in some parts of the body and no expression elsewhere in the same body.

Regional variations in the distribution of chemical growth regulators would evoke similar differential responses. The development of an organ may be under the control of a hormone. The hormone in turn is under the control of a gene, which is active only in certain regions of the developing body, and consequently the organ is formed in a particular place.

The foregoing mechanism is illustrated by the example of coat-color pattern in the Himalayan rabbit. At maturity under standard cold temperature conditions the Himalayan rabbit is white with black points. Genotypically this breed of rabbits carries the homozygous allele pair $c^h c^h$ of a gene C which governs the production of melanin pigment in the fur. The c^h allele is temperature sensitive, acting to produce melanin pigment at low temperatures (below 33 C), but not at higher ones.

There are variations in the skin temperature of the rabbit from one body part to another. These differences are due mainly to limitations in the efficiency of the blood circulation system, which does not bring enough warm blood to the extremities to maintain the normal body temperature in these parts. When the outside air is cold (below 25 C), the skin temperature falls below the critical level for action of the c^h allele in the extremities but not in the central part of the body. Consequently, the latter remains white, while black pigment forms in the ears, nose, and tail, and sometimes in the feet.

The temperature differences in the skin of the Himalayan rabbit constitute a prepattern. The black-and-white color pattern in the fur is superimposed on this prepattern (Stern, 1968).

Stomata in the epidermis of monocotyledons consist of specialized guard cells and subsidiary cells. The two types of cells develop from common mother cells in the young epidermis. The cytoplasm of the stomatal mother cells is polarized with respect to density and number of inclusions, and the mitotic figures are oriented across the cytoplasmic gradient, so that the daughter cells have different cytoplasmic densities. The asymmetrical mitosis thus yields daughter cells with identical nuclear genes but different cytoplasms and hence different nuclear-cytoplasmic interactions. The cytoplasmically different daughter cells develop into guard cells and subsidiary cells (Stebbins and Jain, 1960).

REGULATION Differentiation depends partly on the condition that only certain genes of the complement are active at any given time and place in the developing body. One gene or gene system is active in one sector of cells at one stage of development, and other genes are active in other regions at the same or other times. The *potentialities* are the same throughout the body, insofar as the cells are genotypically identical. The observed regional differences in tissues and organs then reflect differences in the *realization* of these potentialities. A basic problem of development, therefore, is the nature of gene regulation.

The question, how genes are turned on and off during development, can be answered fairly simply as long as we remain in the realm of generalities. Primary products of gene action such as enzymes are usually active only at certain optimum temperatures, pH levels, or other conditions, as noted earlier. Or the genes themselves may be active to produce the enzymes only under certain optimum conditions. In either case the whole series of developmental steps goes forward under one set of physical or chemical conditions only. Regulation here involves action or nonaction of particular genes or gene products.

Another possible mechanism of regulation is repression, in which the production of an enzyme ceases when the concentration of its metabolic products reaches a certain high level in the cell. Repression can be counteracted by derepression. The removal of the products of enzyme action from the sites of synthesis opens the way for a resumption of production. This is derepression.

Still another means of regulation is suppression, in which the action of one gene is blocked or sidetracked by the action of another suppressor or inhibitor gene. Finally, the state of a gene may be active or inactive depending on the nature of neighboring chromosome regions. We will call this activation and inactivation.

Although the basic principles involved in regulation may be simple, the details are complex in actual cases. Much of our current knowledge of regulation has been gained from studies of bacteria and may or may not be applicable to plants. The nature of regulators in higher plants is still poorly known. In later chapters of this book we will discuss gene systems in plants which contain suppressor or inhibitor components. In the present section we will briefly describe some modes of nonaction, repression, and inactivation.

The action or nonaction of a gene as controlled by environmental conditions is well illustrated by the Himalayan rabbit. As we have already seen, the C^h allele of the C gene for melanin pigment formation is active only at low temperatures. Rabbits with the genotype C^hC^h exposed to very cold air temperatures (below 1–2 C) become pigmented all over the body. At intermediate temperatures (15–24 C), the gene is active in the extremities of the body but not in the central parts, and the animal is white with black points. At outside temperatures above 29 C, the C^h allele is not active in any body part and the whole animal is white (Stern, 1968).

The diploid species of Phlox, *P. pilosa* and *P. drummondii,* have been compared with their allotetraploid derivatives, *P. villosissima* and *P. aspera,* with respect to flavonoid compounds. The flavonoids in the tetraploid species are not the same as those in the diploid species but are similar, being less highly glycosidated. Five novel types of flavonoids accumulate in the tetraploid species, and these novel types appear to be precursors of compounds found in the ancestral diploid species (Levy and Levin, 1971).

Levy and Levin suggest the following explanation for the non-additivity of biochemical products in Phlox and other similar cases. A biosynthesis proceeds to its normal end product in the parental species but not in the hybrid or hybrid derivative. Instead there is an accumulation of precursors in the latter. This accumulation of precursors could be a result of repression or suppression of gene action in the hybrid types (Levy and Levin, 1971).

HETEROCHROMATIN AND REGULATION The differences between heterochromatic and euchromatic regions of chromosomes as regards stainability have long been known. Euchromatic regions stain darkly during mitosis and meiosis, when the chromosomes are condensed, but do not stain when the cell is in the metabolic stage. Heterochromatic regions do not go through this same cycle, but remain condensed and therefore show up as stained blobs in the metabolic stage of the cell. These differences are due to differences in coiling. Euchromatin goes through a normal coiling cycle, becoming uncoiled at telophase and returning to a coiled state again in the next cell division, whereas heterochromatin remains coiled through the whole cell cycle. One and the same chromosome region may be euchromatic in some tissues and at some developmental stages and heterochromatic in others. It has been pointed out, therefore, that heterochromatin is not a substance, but a state (Brown, 1966; Ris and Kubai, 1970).

The old idea that heterochromatin is genetically inert has undergone important modifications in recent times. In fact, a number of unorthodox genetic properties are associated with heterochromatin (see Brown, 1966; Baker, 1968; and Ris and Kubai, 1970, for reviews). Here we are concerned with one of these properties which relates to gene regulation and more particularly to inactivation.

When a euchromatic segment is located close to heterochromatin, the action of the euchromatic genes tends to be depressed, and often appears in the form of a variegated phenotype. Variegation associated with proximity of a gene to heterochromatin has been observed in Drosophila and mice. Plant examples will be described in Chapter 7. The effect of the heterochromatin on gene expression is reversible, in some cases at least, inasmuch as the variegation disappears if the euchromatic gene is relocated away from heterochromatin. These facts suggest that heterochromatin acts in a generalized fashion as a nonspecific depressor of gene action (Brink, 1964).

Brink proposes a hypothesis to account for the correlation between type of gene expression and proximity to heterochromatin. The connection between the two features lies in the state of coiling of the chromosome region. He suggests that a euchromatic gene can act in its normal fashion only when the chromosome is uncoiled. The action of the same gene is depressed when the chromosome region is tightly coiled. This coiled state occurs in the metabolic stage of the cell when the gene is adjacent to heterochromatin. Heterochromatin thus plays a generalized role in the regulation of gene action (Brink, 1964).

A given gene may be active when located in a euchromatic segment but inactive when transposed to a position close to heterochromatin. Or the gene may remain in the same chromosome region but be active or inactive in different tissues and in different developmental stages, depending on the state of coiling in this region.

PHASE CHANGE In all seed plants the course of development passes through different stages from seedling to adult plant, and these stages differ both morphologically and physiologically. In a number of woody plants the change from the juvenile to the adult condition is striking and abrupt. Such abrupt switches from a juvenile to an adult type of growth are termed phase change (Brink, 1962).

A well-known example of phase change is found in English ivy (*Hedera helix*), in which the juvenile phase is a trailing vine with alternate palmate leaves, and the adult phase a semierect shrub with opposite oval leaves. In *Acacia melanoxylon* the juvenile phase is marked by bipinnate leaves and the adult phase by strap-shaped leafy petioles or phyllodes. The juvenile shoots of *Juniperus virginiana* have needle leaves whereas the adult branches bear scale leaves. The juvenile shoots of Citrus are thorny and vigorous, but the adult branches are thornless or mostly so and slow growing.

Many other examples could be cited in Eucalyptus, Pyrus, Fagus, Quercus, Acer, Thuja, and Araucaria (Brink, 1962; Hartmann and Kester, 1968).

The change of phase is not hereditary, inasmuch as the seedling progeny of adult fruiting branches invariably return to the juvenile condition. On the other hand, the phases, though not permanent, are sometimes quite persistent. Buds or cuttings of juvenile or adult shoots of the same plant, when propagated vegetatively, produce clonal derivatives which retain the characters of their respective phases, often for many years. The phases are usually reversible (Brink, 1962).

These facts rule out ordinary genotypic changes during development as an explanation of phase change. It is usually assumed, as Brink (1962) points out, that the only alternative is cytoplasmic differentiation in somatic cells becoming specialized for one phase or the other. But other intermediate possibilities should be considered, according to Brink. Certain accessory elements in the chromosomes, perhaps heterochromatic in nature, might induce so-called paramutable changes in particular unstable genes, that is, reversible changes of state in the affected or paramutagenic genes. (Paramutation is described in Chapter 6.) The reversible alteration in the condition and action of the paramutagenic genes in somatic cells during ontogeny could lead to developmental phase changes. Paramutation as an explanation of phase change warrants and requires further research (Brink, 1962).

ENVIRONMENTAL EFFECTS Gene action and its secondary effects in enzyme action are influenced by the physical and chemical conditions inside the cell; and the internal environment is influenced in turn by the innumerable variable environmental factors outside the body of the organism. Development follows one pathway and culminates in one exo-phenotype under one set of environmental conditions, and leads to a different phenotypic expression under other conditions. For any given genotype there is a range of phenotypic variation determined by the environmental component in development, as is well known.

In higher plants the range of normal phenotypic modifiability is quite great as compared with higher animals. The high phenotypic plasticity of plants is correlated with their sedentary habit, their open system of growth, and their relatively low degree of internal homeostatic buffering mechanisms.

The matter can be restated in terms of Waddington's (1957) concept of canalization. Developmental canalization is the reverse of phenotypic modifiability. The degree of canalization is under genetic control and varies in different kingdoms and phyla of organisms. The genotypes of higher plants, and perhaps most especially those of herbaceous plants, tend to be weakly canalized in comparison with those of animals. This difference may be a result of the course taken by canalizing selection in the different kingdoms.

Canalizing selection has evidently favored genotypes with a wide range of developmental pathways and phenotypic expressions in plants, with their sedentary habit which exposes them to a wide range of direct environmental influences, and their open system of growth which makes possible a series of separate developmental responses to these varying influences.

The wide phenotypic modifiability of plants is illustrated by innumerable experimental cases. Clonal divisions of a single individual of *Taraxacum officinale* grown in the European lowlands and in the Alps had the growth forms shown in Figure 6. In Figure 7 we see the products obtained by growing four clonal divisions of an individual of *Potentilla glandulosa* in four combinations of light and moisture conditions (Clausen, Keck and Hiesey, 1940). Many other cases are described in the classic monographs of Turesson (1922, 1925) and Clausen, Keck and Hiesey (1940, 1948).

Clausen and his associates studied the growth and development of many races of the *Achillea millefolium* group under controlled conditions in a phytotron. The environmental variables were length of daylight, day temperature, and night temperature. Members of the same clone were grown simultaneously in several different combinations of these light and temperature conditions. A typical result of this extensive series of experiments is shown in Figure 8. The graph shows growth curves for the Selma and San Gregorio races of *Achillea borealis,* from central and coastal California respectively. The variable factor here is night temperature. The curves represent

Figure 6. Clonal divisions of one individual of *Taraxacum officinale* grown in the lowland plains of Europe (*below*) and in the Alps (*above*). (From Baur, 1919, after Bonnier.)

stem growth and the dotted section alongside two of the curves indicates period of flowering. Marked differences between sister clone members in response to different night temperatures are evident from the graph. It is also evident that the two races respond differently

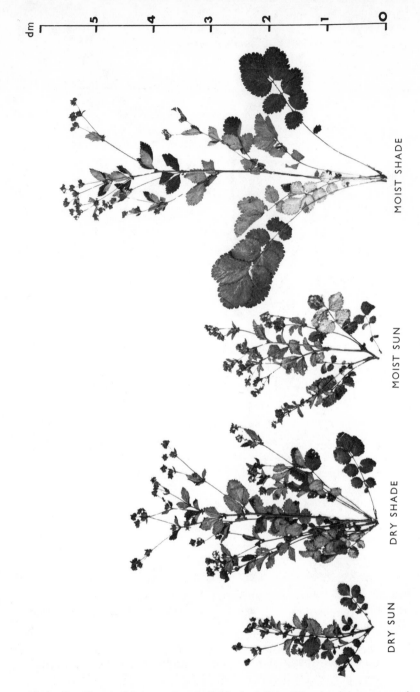

Figure 7. Clonal divisions of an individual of *Potentilla glandulosa* grown in four environments. (From Clausen, Keck and Hiesey, 1940.)

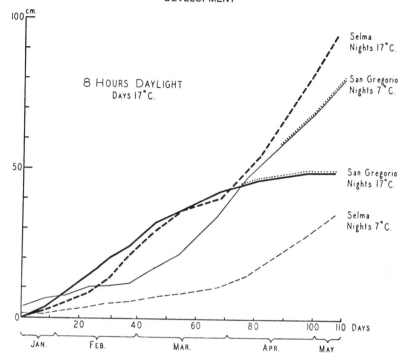

Figure 8. Growth curves for clonal divisions of two races of *Achillea borealis* at different night temperatures. (From Clausen, Keck and Hiesey, 1948.)

to night temperature. The interior valley race, Selma, produces its best growth in moderate cool nights, while the coastal race, San Gregorio, grows best in cold nights (Clausen, Keck and Hiesey, 1948; Hiesey, 1953a).

In a related study Hiesey compared the growth of clones of Poa at different temperatures in the phytotron. The plant materials consisted of five species and five interspecific F_1 hybrids. The environmental variables were nine combinations of various day temperatures and night temperatures. The growth and development of the clonal subdivisions in each environment were recorded in terms of dry weight of the plants at harvest time. A certain range was found in the amount of growth and yield of the species at different temperatures, as expected. The F_1 hybrid of *Poa scabrella* × *ampla* turned out to have a narrow tolerance range as compared with its parents. The hybrid genotypes derived from some other crosses, however, particularly *Poa arida × ampla* and *P. ampla × pratensis,* proved to have a wider range of responses than the parental species, producing good growth at different temperatures (Hiesey, 1953b).

ENVIRONMENTAL INDUCTION The old question of the induction of heritable changes by environmental influences has come to the fore again in recent years. Where the environmental influence involved is DNA, as in transformation and transduction in bacteria, the observations can be fitted into the framework of both classical and modern genetics. DNA-induced transformation also appears to be possible in Drosophila (Fox and Yoon, 1970) and presumably in higher plants. More puzzling are some cases of apparent induction of heritable changes in plants by inorganic chemical fertilizer elements.

Plants of *Linum usitatissimum* (flax) respond to varying dosage levels of the fertilizer elements nitrogen (N), phosphorus (P), and potassium (K) in their substrate in the well-known way, producing large forms on rich nutrient substrates and small forms on poor ones. Durrant (1962a, 1962b) found that, surprisingly, the changes apparently became hereditary. The progeny of induced large forms tended to remain large through seven generations, irrespective of the nutrient substrate; and similarly, the progeny of the original small forms remained small on different nutrient substrates for seven generations (Durrant, 1962a, 1962b). The size differences were later shown to hold up through the tenth generation (Evans, Durrant and Rees, 1966).

Similar results have been reported in *Nicotiana rustica*. Inbred varieties were subdivided into replicates and grown under different treatments of the nutrients N, P, and K. One variety responded to the different nutrient treatments by developing different characteristics of plant height and earliness. The differences in these characteristics were maintained in the progeny of the foundation stock through three subsequent generations in which the nutrient conditions were kept constant and alike (Hill, 1967).

Genetic analysis has proceeded farther in the Linum studies. Here reciprocal crosses between derived large and small types (L ♀ × S and S ♀ × L) yielded similar intermediate F_1s. These results rule out maternal effects and strengthen the case for hereditary changes. The same results also suggest that the hereditary changes are nuclear and not cytoplasmic (Durrant, 1962a, 1962b). Subsequent studies indicate that the amount of nuclear DNA differs in the derived large and small forms (Evans, Durrant and Rees, 1966).

It is not at all clear what is happening in the flax lines exposed to different nutrient treatments. One suggestion among others put forward by Durrant (1962a, 1962b) is that mutations affecting plant size were induced by chemical treatment in the foundation stock. The whole story is still developing at this writing.

INTEGRATION   There is a need for coordination among the various parts of a developing multicellular body. The body must develop as an internally integrated whole. This requirement is especially critical in higher animals, in which an extremely complex body must develop harmoniously by a closed system of growth. An array of embryonic organizers and endocrine hormones helps to bring about this result. The requirements are less stringent in higher plants, since the plant body is simpler and develops by an open system of growth in which the different parts are formed separately and in succession, but some means of integration is necessary here too.

Development in higher plants is controlled by various chemical growth regulators such as auxin, gibberellin, ethylene, cytokinin, abscisic acid, and phytochrome. These substances are produced as normal metabolic products in the cells and, in most cases, become diffused to other plant parts where they exercise their particular regulatory functions. This subject has been well reviewed recently by Galston and Davies (1970), and the reader may consult their book for more details.

The growth hormone auxin, or at least one of the auxins, has been identified chemically as indole-3-acetic acid (IAA). It is produced in the shoot tip, from the precursor tryptophan, and moves down the shoot axis from there, promoting cell elongation. Normally IAA is distributed equally on opposite sides of a vertical stem, with the result that stem growth in length is straight upward. An unequal distribution of auxin on opposite sides of a vertical stem can arise by light inactivation on one side. This leads to a growth movement or curvature which reorients the growing stem in the direction of the light source. Conversely, an unequal distribution of auxin may be brought about by gravity in an inclined stem, again producing a curvature toward the vertical position (see Galston and Davies, 1970, ch. 3).

Another hormone, gibberellin (GA), enhances stem and leaf growth, induces parthenocarpic fruit formation, and breaks dormancy. The differences between normal and dwarf plants in many species are due to deficient gibberellin formation in the dwarf types (Pelton, 1964; Galston and Davies, 1970).

Among the developmental processes controlled by other plant hormones are: cell division (by cytokinin), flower initiation (by ethylene and florigen), fruit ripening (ethylene), and induction of dormancy in buds and seeds (abscisic acid) (Galston and Davies, 1970).

Many developmental processes are regulated by interactions between different hormones or between a hormone and some other chemical growth substance. Thus the protein pigment phytochrome in the leaves exerts a control over flowering in other parts of the plant. Phytochrome absorbs light perceptions in the leaves, and transmits these to the hormone florigen, which then flows through the vascular system to other regions. Gibberellin and auxin have similar effects on growth and are synergistic. On the other hand, gibberellin may overcome the effects of abscisic acid so as to break dormancy. Auxin is inactivated by the enzyme peroxidase (Brian, 1959; Galston and Davies, 1970). Conversely, some hormones function by inducing enzyme synthesis (Markert and Ursprung, 1971).

We find interactions of an oppositional nature between plant hormones and we also find promoters and repressors of hormones at the level of enzymes. The obvious suggestion is that these relationships result from the action and interaction of different genes with oppositional effects.

Evidence for gene control of plant hormones is provided by the dwarf types in maize. Various dwarf types in *Zea mays* are inherited as single-gene recessives. Dwarfs are known to be determined at 20 gene loci. In the type designated as dwarf-one (with the recessive allele d_1), the coleoptile has reduced auxin production compared with the normal type. Nana-one (na_1) has greater destruction of auxin than the normal type. Other dwarf mutants such as anther-ear-one (an_1) and dwarf-8201 are deficient in gibberellin (Pelton, 1964).

Some dwarf mutants in maize, including anther-ear-one and dwarf-8201, respond to treatment with gibberellin by producing normal growth (Phinney, 1956). Also, grafts between dwarf seedlings and normal seedlings lead to nearly normal growth of the dwarfs shortly after grafting, indicating a diffusion of the growth regulator, probably gibberellin, from the normal to the dwarf plant. The dwarf mutation is evidently a block in the formation of gibberellin or a GA-like hormone. Furthermore, since there are different gene loci for dwarfism in maize, and since the various dwarf mutants react differently to GA treatment, it is probable that the various genes control different steps in the synthesis of gibberellin and related hormones (Pelton, 1964).

Single-gene dwarfs are known also in *Pisum sativum, Lycopersicon esculentum,* and many other species of angiosperms belonging to 17 families, according to a list compiled by Pelton (1964). The condition is widespread. Gibberellin deficiency is indicated in some dwarf mutants in the tomato, the pea, and other species.

Earlier in this chapter we tried to trace pathways from gene to specific character. Now we see evidence for pathways of a different sort from genes to growth regulators. The latter coordinate the developmental processes going on in different plant parts.

CONCLUSIONS The relation between gene action and character expression is highly complex. The complexity stems from the interactions among several sets of factors.

In the first place, many developmental steps usually intervene between primary gene action and ultimate phenotypic expression. A long stepwise process affords an opportunity for compounding of effects during development, that is, for pleiotropy. The gene has multiple phenotypic effects.

Furthermore, the character is apt to be the product not of one gene but of several or many. In *Zea mays* 20 gene loci are known to be concerned with normal or dwarf habit, and at least 50 gene loci are involved in chlorophyll production (Stern, 1949b; Pelton 1964).

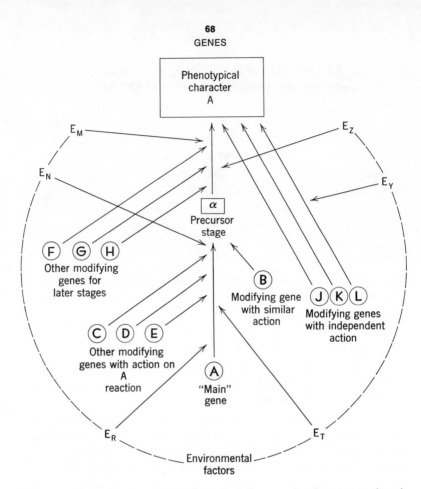

Figure 9. Generalized scheme to show the influence of various types of modifying genes and environmental factors on the developmental pathway leading from one main gene to a phenotypic character. (From Stern, 1960, after Timofeeff-Ressovsky, reproduced by permission of Freeman and Co.)

Some genes may contribute directly and primarily to the formation of a given character. Other genes produce chemical growth regulators with integrating functions. And still other genes with different primary functions of their own may affect the first character indirectly, as a pleiotropic effect or as a modifying factor (see Figure 9).

Finally, the effects of gene action are modified by environmental conditions. Enzymes are usually active only within a certain range of temperatures or pH values or other conditions. The products of enzyme action are therefore also manifested only at certain optimum conditions. And consequently the gene determining a given developmental sequence will have one phenotypic expression under one set of environmental conditions and different expression under other conditions (Fig. 9).

It is evident that if a character is determined directly by the joint action of several genes primarily concerned with the trait in question, and is affected indirectly by many other modifying genes, and if furthermore these genes all produce pleiotropic effects, the developmental pathway from gene to character will be a network of interwoven causes and effects. And if gene action, enzyme action, and hormone action are all subject to environmental influence, the number of possible developmental pathways and phenotypic expressions becomes very great indeed.

CHAPTER FOUR

THE NATURE OF
MENDELIAN GENES

INTRODUCTION   The purpose of this chapter is to consider the nature of chromosomal genes. We are interested here in the delimitation of genes, subgenes, and gene complexes. Needless to say, the problem is difficult and not completely solved, and furthermore, the search for a solution is beset with obstacles.

The evidence of classical genetics, as presented in Chapter 1, will take us far toward our goal of understanding the nature of the gene. Additional insights have come from the more recent studies in biochemical genetics, reviewed in Chapter 2. The nature of the genic units on the chromosomes, in other words, has been revealed to us gradually and in successive stages by breeding experiments, cytogenetic studies, detailed genetic analysis of small chromosome segments, and biochemical investigations. It would be convenient if the various lines of evidence regarding the delimitation of genes were in agreement with one another, but this happy state of affairs does not exist.

The results of breeding experiments in the nineteenth and early twentieth centuries were successfully integrated with the results of cytological investigations during the period from 1900 to the 1930s. These early studies supported the classical concept of the gene as a discrete hereditary particle. During the 1940s and since then, however, refined genetic analysis of chromosome segments and biochemical analysis of the genetic material have made the identity of the gene as a particulate unit questionable. It will be our difficult task in this chapter to consider the various lines of evidence on which conflicting concepts of the gene are based and then to attempt to reconcile, in some measure, the generalizations drawn from the different lines of evidence.

We cannot discuss the nature of the gene in terms of the plant evidence exclusively. Much of the information which is essential for a consideration of the problem has been obtained from studies of organisms belonging to other kingdoms, such as Drosophila, mammals, Neurospora, bacteria, and viruses. We must utilize this varied evidence and, moreover, utilize it properly, avoiding the pitfall of overgeneralizing from microbial genetics to the situation in higher multicellular organisms.

Our path is also strewn with some purely terminological problems. There are such synonymous or nearly synonymous terms as structural gene and Mendelian gene, or triplet and codon. There are misapplications of terms, as when the term chromosome is used to describe the genetic thread of bacteria and viruses. And there is an abundance of unnecessary terms. Our thinking tends to be channeled by the terminology we employ. Consequently, our search for a scientifically satisfactory concept of the gene must be accompanied by adoption of a suitable set of identifying terms.

THE CLASSICAL CONCEPT OF THE GENE   The existence of genes as particulate hereditary units was first inferred by Mendel from breeding data on peas. The genes had no known physical basis, other than their presence in the gametes, in Mendel's time. The next important step was showing that the genes are located in a linear order on the chromosomes. The location of the genes was next narrowed down to segments of a chromosome which become separated and recombined during crossing-over. The recombinable segments vary in length, however, and may be fairly broad, so that the criterion of crossing-over often leaves the exact delimitation of the gene unsettled.

The classical concept of the gene as elaborated by Morgan (1928) and other workers was a logical synthesis and development of the findings just described. According to the classical concept, the gene can be defined as a chromosomal region or locus that controls the development of some phenotypic characteristic, that can mutate to a new allelic form with altered phenotypic effects, and that is separable by crossing-over from adjacent loci governing other specific characters.

The discreteness of the gene was believed to be maintained by predetermined points of breakage along the length of the chromosome. The gene was bounded by zones of regular breakage and crossing-over. The physical boundaries of the gene as defined by breakage and crossing-over, moreover, were supposed to coincide with the boundaries defined by functional unity. The gene, then, was the structural and functional atom of genetics.

The chromosome was a string of these atomistic genes. A widely held model of genic organization in classical genetics was summed up in the figure of speech "beads on a string."

The chromosome does present a beaded appearance when seen in the stretched-out condition in an interphase or prophase nucleus. The beadlike chromomeres are connected by thin thread-like regions at this stage (Fig. 10). The chromomeres occur in fixed positions on the chromosomes in different cells, and evidently represent real entities (Belling, 1928; Lima-de-Faria, 1952; Beermann, 1964; Pelling, 1966). Electron-microscopic studies show that the chromomeres are locally tightly coiled regions in the otherwise uncoiled interphase chromosome (Buchholz, 1947; Ris, 1957; Ris and Kubai, 1970). Belling (1928), among other early workers, concluded that the chromomeres were the cytologically visible sites of the genes.

The correlation between genes and chromomeres could be established more clearly in the salivary-gland chromosomes of Drosophila. The giant salivary chromosomes are polytene, consisting of numerous parallel strands, and are extended in an uncoiled state in somatic tissue. Consequently, the chromomeres appear as

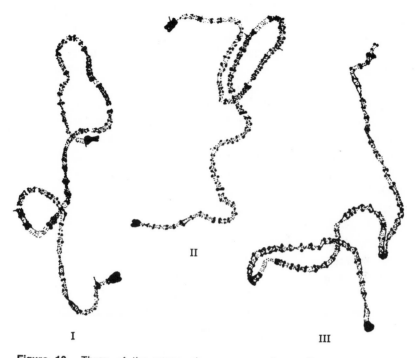

Figure 10. Three of the seven chromosomes of rye (*Secale cereale*) in meiotic prophase, showing the chromomeres. (From Lima-de-Faria, 1952, reproduced by permission of Springer-Verlag.)

recognizable bands. By the use of deletions and other cytogenetic techniques it is possible to pinpoint the location of various genes in or near specific bands.

The early Drosophila work did establish the location of various genes in particular bands. The second band in the X chromosome of *D. melanogaster* contains the genes for yellow body, achaete bristles, scute bristles, and a lethal (Muller and Prokofyeva, 1935). The gene roughest (*Rst*) is in the band designated 3C4, and facet (*Fa*) is in the 3C7 band of the X chromosome (Slizynska, 1938). A recent study places certain genes in bands in section 10 of the X chromosome (Lefevre, 1971).

The classical concept provided a satisfactory model of genic organization for the early modern period of genetics. It has since fallen into disuse. Several developments have contributed to its fall, some particular, others general. Two important lines of evidence which have called the classical view into question are those embodied in the phenomena of pseudoallelism and genetic fine structure, to be discussed in the sections that follow. More generally, the classical concept was just too simplistic.

However, the pendulum has perhaps swung too far from the classical concept toward other simplistic models, as we shall see in later sections. Some features of the classical concept appear to have permanent validity and should be incorporated into any final synthesis of the evidence.

PSEUDOALLELISM   In the early modern period of genetics, the gene was regarded as the smallest unit of recombination and as a functional unit, and the two criteria were supposed to coincide. A chromosome region was a string of different genes with different functions. But in the decade from 1945 to 1955 this concept was altered by the discovery of rare crossovers between genic components that had formerly been described as members of a single allelic series. The phenomenon of crossing-over within the limits of what had been considered a gene became known as pseudoallelism. Reviews of pseudoallelism are given by Komai (1950), Glass (1955), Symposium (1955), Wagner and Mitchell (1955, 1964), Pontecorvo (1958), and Carlson (1959).

The gene *W* controlling eye color in *Drosophila melanogaster* had long been known to occupy a specific locus on the X chromosome and to consist of a series of alleles determining white, apricot, eosin, and other colors of the compound eye. The different alleles were inherited as variants of a single gene. In 1952, however, Lewis found that crossing-over does occur rarely between two of these alleles, *w* for white eyes and w^a for apricot eyes, which must therefore exist at closely adjacent but separable loci. The distance between apricot and white on the genetic map was calculated to be between 0.02 and 0.005 crossover units (Lewis, 1952).

Let us pause to consider how small these map distances are. Recall that 1 crossover unit is a distance between two genes which leads to the formation of recombination types in the frequency of 1%. If A and B are 1 crossover unit apart, the double heterozygote $\dfrac{a_1 b_1}{a_2 b_2}$ will produce 1% recombination gametes $(a_1 b_2 + a_2 b_1)$. The estimates of 0.02 to 0.005 crossover units between pseudoalleles of the W locus in Drosophila therefore mean that a double heterozygote will yield recombinants in a frequency ranging from 2/10,000 to 5/100,000 gametes. In short, the pseudoalleles are so close on the genetic map that crossovers between them occur only very rarely, almost as rarely as mutational changes.

Other genes in Drosophila were also found to consist of closely linked but separable and recombinable subunits or pseudoalleles. Thus the pseudoallelic components, star and asteroid, lie 0.02 crossover units apart on the second chromosome (Lewis, 1945). The lozenge gene (Lz) on the X chromosome consists of three subgenes arranged in the order $lz^{Bs} - lz^{46} - lz^g$, with intervals between them of about 0.06 to 0.09 crossover units (Green and Green, 1949). Two recombinable components of the gene rosy, ry^{26} and ry^2, are calculated to be 0.00026 crossover units apart. This map distance can be expressed in molecular terms as about 40 nucleotide pairs (Chovnick, Schalet, Kernaghan and Talsma, 1962). Other cases of pseudoallelism in Drosophila are reviewed by Carlson (1959).

Parallel findings were made in plants. The A gene in *Zea mays*, governing color of kernels and other plant parts, consists of two recombinable subunits about 1 crossover unit apart (Laughnan, 1948, 1952). The R gene in maize is also compound (Rhoades, 1954). The gene Rpl in maize for rust resistance is a complex locus containing separable pseudoalleles (Hooker and Saxena, 1971). In flax (*Linum usitatissimum*), a gene for rust resistance also contains recombinable subunits (Shepherd and Mayo, 1972).

Pseudoallelism is found in the fungi (Pontecorvo, 1953, 1958; Winge, 1955; Jessop and Catcheside, 1965; Hayes, 1968). An example in yeast, Saccharomyces, concerns the genetic region governing sugar fermentation, which consists of closely linked but separable subunits controlling different stages in the process (Winge, 1955). We find comparable cases in the bacteria (see Hayes, 1968, for a review).

Thus pseudoallelism is a widespread phenomenon, as Pontecorvo (1953, 1958), Glass (1955), Carlson (1959), and others have pointed out.

CLUSTERS OF FUNCTIONALLY RELATED GENES　In the bacteriophage T4, numerous functionally similar mutations determining the ability of the virus to multiply in the cells of the host bacterium, *Escherichia coli,* are all located in one region of the genetic thread, designated *rII* (Benzer, 1955, 1957). The *rII* region lies between the genes *M* and *Tu* (Fig. 11).

On the basis of recombination data it has been possible to locate different types of *rII* mutations at particular points in the *rII* region, as shown in Figure 11. Benzer found cases of separation and recombination of *rII* mutations lying 0.02 crossover units apart on the genetic thread. In molecular terms this could represent recombination between nearly adjacent nucleotide pairs, perhaps separated by only one intervening nucleotide unit, on the DNA chain (Benzer, 1957).

The *rII* mutations fall into two classes, *A* and *B,* which have complementary effects. The *A* and *B* groups of mutations, moreover, are segregated into separate but adjacent parts of the *rII* region, as shown in Figure 11. Benzer (1957) refers to the *A* and *B* units as cistrons. The *rII* region as a whole is then a structural and functional unit composed of two cistrons with complementary effects.

Benzer (1957) proposed a test of the functional integrity of genic units like *A* and *B* based on the cis-trans comparison devised previously by Lewis (1951). Benzer developed his concept in relation to viruses, which are haploid, and expressed it in the terminology of recons and cistrons. The concept can be extended with appropriate modifications to dikaryotic fungi and diploid plants and animals (see Pontecorvo, 1958). Here we will state the cis-trans test of functional unity in terms of the diploid condition in higher plants and animals.

The cis-trans comparison in a diploid organism depends on the presence of two recessive, functionally similar mutations (m_1 and m_2) in the same nucleus, together with the corresponding dominant

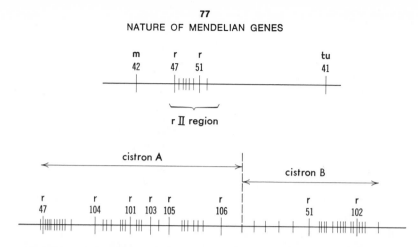

Figure 11. A portion of the genetic map of bacteriophage T4, showing the location of the *rII* mutations. *Above*, the *rII* region in relation to the genes *M* and *Tu*. *Below*, the *rII* region enlarged, showing the two cistrons *A* and *B*, and the location of various types of *rII* mutations. (Adapted from Benzer, 1955.)

normal alleles. The mutations m_1 and m_2 occur in the same chromosome region, and produce the same defective phenotype. The double heterozygote may have the two mutant alleles in either the cis or the trans position ($\frac{m_1 m_2}{+\,+}$ or $\frac{m_1 +}{+ m_2}$). The cis and trans arrangements may or may not produce the same phenotype.

Suppose that the cis arrangement $\frac{m_1 m_2}{+\,+}$ gives a normal phenotype, whereas the trans arrangement $\frac{m_1 +}{+ m_2}$ gives a recessive mutant phenotype. The recessive phenotype associated with the trans genotype indicates a lack of complementation, and hence a high degree of functional integration between m_1 and m_2. The functional unity of $+_{m1}$ and $+_{m2}$ in this case is shown by the fact that the cis genotype produces a normal phenotype, whereas the corresponding trans genotype does not. We could conclude that m_1 and m_2 are pseudo-alleles of the same gene.

An alternative possibility is that $\frac{m_1 +}{+ m_2}$ yields a normal phenotype, and the trans arrangement does not produce a mutant or recessive phenotype. Then one can conclude that m_1 and m_2 are not allelic or pseudoallelic. They are different adjacent complementary genes. Or, in Benzer's terminology (1957), m_1 and m_2 are separate cistrons.

In any event, m_1 and m_2 can be seen to be more separate and less integrated functional units in the case where the trans heterozygote engenders a complementary and normal phenotype than where it gives a recessive mutant phenotype.

The practical application of the cis-trans test for determining the degree of functional unity of a genetic segment is sometimes beset with difficulties and uncertainties, particularly in higher organisms. Yet it does provide a good approach to the problem.

In bacteria many cases are known of close linkage of functionally related genes. Well-analyzed examples are the clustered genes for tryptophan synthesis and histidine synthesis in *Salmonella typhimurium* (Demerec, Blomstrand and Demerec, 1955; Demerec, 1956). In *Escherichia coli* there is a similar clustering of genes for tryptophan synthesis and for arabinose fermentation. In each case, separate neighboring loci control different biochemical steps in a metabolic process (for reviews see Carlson, 1959; Hayes, 1968).

Clustering of functionally related genes is found again in the genomes of eukaryotic organisms from fungi to Drosophila and angiosperms, where, however, it does not appear to be as extensive as in bacteria. Examples in higher plants will be considered in Chapter 20.

The facts of pseudoallelism and gene clustering can be interpreted in different ways. Either adjacent genes, instead of possessing different functions, have similar and related functions; or else a single gene, defined as single by its function, consists of functional subdivisions. Either interpretation requires a modification of the classical theory of the gene. We have, in fact, not discrete atomistic genes, but a hierarchy of structural and functional genetic units.

THE MACROMOLECULAR CONCEPT OF THE GENE According to an alternative concept of the gene, heredity is not in fact particulate. Atomistic genes do not exist. Genic organization consists of a continuous chromosomal strand, and localized sites along its length control different biochemical and developmental reactions.

What we recognize as "genes" are these sites on the continuous genetic strand.

In the original version of this concept, the continuous genetic macromolecule was considered to be protein in nature, with nucleic acid serving as a scaffold around the genetically active protein core (Goldschmidt, 1938, 1955). The later discovery that the genetic material is DNA rather than protein does not require any essential change in the continuous-macromolecule concept of the gene, since DNA forms long polymeric chains as does protein.

The concept of the continuous macromolecule was explicitly stated by Goldschmidt (1938, 1955). The same concept is implicit in much of the literature of molecular genetics. We are given a continuous double-strand model of the DNA macromolecule. The problem of gene delimitation was rarely discussed by the early molecular geneticists, who seem to have tacitly accepted the continuous-macromolecule concept.

Molecular genetics has, at the same time, yielded new insights into gene action which provide one of the keys needed for a satisfactory definition of the gene. The important contribution of molecular genetics in this regard is the attempt to deduce the nature of genes from an analysis of gene products. A gene can be defined, to some extent anyway, in terms of its biochemical products. On this basis, a gene can be equated with the whole sequence of nucleotides required to specify a particular enzyme, polypeptide chain, or protein molecule (see Chapter 2).

The correlation of a gene with a protein product does not in itself settle the question of continuity or discontinuity in the genetic material. The decision to equate a gene with either a polypeptide chain or a protein macromolecule could be said to be arbitrary. And a linear series of such arbitrarily defined genes could form a continuum on the DNA chain. A fair statement of the problem was given by Beadle (1955, 1957a, 1962).

The main argument put forward by Goldschmidt (1938, 1955) for the continuous-macromolecule concept of the gene was the phenomenon of position effect with usually deleterious side effects. Position effect implies a functional and organizational pattern in a chromosomal strand taken as a whole. There would be no reason to expect position effects in rearrangements of a string of beadlike genes.

The force of this argument must be admitted. One must also recognize that the argument cuts two ways, for if the functional pattern resides in the chromosome as a whole, and if atomistic genes are lacking entirely, position effects should be a much more common result of chromosome breaks and rearrangements than they actually are.

Hundreds of chromosome rearrangements have been studied in both plants and animals, but position effects associated with the altered chromosome structures have been found in only a small fraction of the cases. In plants, where chromosomal rearrangements are common, position effects are known only in Oenothera and Zea, and these are special cases involving heterochromatin influences which do not necessarily support the continuous model of genic organization (see Chapter 8). In a sample of 64 translocations studied in corn, all were viable in the homozygous condition, which would not be the expected result with the continuous-macromolecule model (Anderson, 1935). In Drosophila, 134 out of 332 translocation heterozygotes were viable, which is again a result difficult to reconcile with the continuous-macromolecule concept (Patterson, Stone, Bedichek and Suche, 1934).

The existence of deleterious position effects is thus consistent with the hypothesis that the genes in at least *some* regions of the chromosomes are organized structurally to perform their collective function, and cannot continue to function well when their pattern of organization is disrupted. But the fact that position effects are not a universal or even a particularly common result of chromosomal rearrangements indicates that many other chromosomal regions in which breaks occur frequently do not contain integrated gene systems that are functionally inhibited when broken apart (Grant, 1964).

CONTINUITY VS. DISCONTINUITY IN THE GENETIC MATERIAL

The classical gene was a unit of function and a unit of crossing-over, and the two criteria of gene definition were supposed to coincide. The evidence of pseudoallelism, gene clustering, and position effect shows, however, that the two criteria do not coincide in many well-analyzed cases. The functional unit and the unit of crossing-over are not necessarily the same. The classical gene was also supposedly a unit of mutation. But there is no single genetic unit to which the criterion of mutability can be applied. Mutability is a property exhibited by all levels of organization of the genetic material, from nucleotide units through whole chromosomes to complete genomes. Consequently, as Beadle (1955, 1957a), Benzer (1957), Pontecorvo (1958), and others pointed out in the 1950s, the classical gene must be reexamined and perhaps broken down into different units with different definitions.

The atom of classical genetics has been all but lost in this process of reexamination. The phenomenon of pseudoallelism shows that the atom of genetics can be split. The clustering of functionally related genes and the phenomenon of position effect force us to view the gene as a subordinate component of a higher structural-functional unit. As an integral unit in itself, the gene appears to be vanishing into a continuous array of nucleotides in a DNA chain.

One response to these lines of evidence is to forget about genes as discrete units, and to regard them as successive but poorly demarcated regions in a continuous macromolecular chain. One can even discard the term gene altogether, as some authors (e.g., Benzer, 1957) have proposed to do.

The weakness of the continuous-macromolecule concept of the gene, however, is that it ignores the problem of regular zones of breakage and cross-over in the chromosomes. Such crossover points had been predicted from older cytological and cytogenetic evidence which is not necessarily invalidated by the more recent work at the molecular level. Furthermore, quite apart from the empirical evidence, the existence of predetermined zones for crossing-over in the chromosomes is a logical necessity.

An organism containing chromosomes composed of continuous genetic material, with potential crossover points distributed at random, would not be expected to be viable. Let it be granted that a certain linear sequence of nucleotides has a functional role of its own in specifying, say, a particular polypeptide or protein. Assume further that the polypeptide or protein is adaptively valuable to the organism possessing it. Then there is also adaptive value to the organism in protecting the functional genetic unit from disintegration by wholly random breakage and crossing-over.

Structural mechanisms that cause the breakage points to fall outside the functional unit would be expected to be built up by natural selection. If, in short, a particular nucleotide sequence is both functionally and adaptively valuable, mechanical devices will develop under selection to hold that sequence together. And structural units on a chromosome will come to coincide with functional units (Muller, 1959; Wright, 1959; Grant, 1964).

If the protein products of gene action are discrete units, then it follows that the genetic segments which produce them must also be discrete units. Actually there are special triplets, so-called nonsense triplets, which function to terminate the transcription and translation of a polypeptide chain (see Table 5 in Chapter 2). These nonsense triplets could be viewed as boundary markers between adjacent cistrons. Furthermore, we may be able to read back from protein structure to genetic units. A protein macromolecule composed of two or more polypeptide chains may be the product of a gene complex composed of two or more separable genes. The structural hierarchy in the protein molecule finds a corresponding structural-functional hierarchy in the organization of the genetic material.

We reach the conclusion that units of crossing-over corresponding to functional units are expected to exist in the chromosomes of higher organisms at least. This conclusion does not necessarily apply without modifications to prokaryotic organisms. It is quite conceivable that the genetic material of bacteria and viruses could consist largely of interchangeable nucleotide pairs, while that of plants and animals may be organized into large structural units (Dobzhansky, 1959; Muller, 1959; Wright, 1959; Grant, 1964).

THE SEARCH FOR EVIDENCE OF DISCONTINUITY The hypothesis of structural-functional genic units bounded by zones of regular breakage and crossing-over can be tested by evidence derived from studies of chromosome structure. The problem centers on the fine structure of true chromosomes as found in eukaryotic organisms.

Classical cytology revealed many cases in both plants and animals in which chiasmata, cytologically visible cross-connections between paired chromosomes, are localized in particular regions of the chromosomes. Examples of plants with localized chiasmata include Allium, Fritillaria, Oenothera, Veltheimia, Anemone, Lycopersicon, Plantago, Secale, and Scilla (Darlington, 1937; Japha, 1939; Coleman, 1940; Böcher, 1945; Barton, 1951; Hyde, 1953; Böcher, Larsen and Rahn, 1955; Rees and Thompson, 1955; Rees and Evans, 1966).

In Scilla, it has been found that chromosome regions which have a high chiasma frequency also have a correlated high rate of DNA synthesis (Rees and Evans, 1966).

If chiasmata are localized, the original crossover points are presumably localized too. To this extent the cytological evidence is consistent with the hypothesis of regular breakage zones. However, the chiasma localization which can be detected cytologically is a localization in relatively broad regions of a chromosome. A higher level of magnification is needed to demonstrate a localization of crossovers between genes and gene complexes.

Chromosome breaks induced by x-rays in Drosophila tend to fall preferentially in certain regions. The exposure of D. melanogaster to x-rays leads to much breakage throughout the genome, but a high proportion of the breaks occur in certain heterochromatic regions (Patterson, Stone, Bedichek and Suche, 1934; Lea, 1955). The distribution of induced breaks in a small region of the X chromosome of D. melanogaster containing the genes yellow and scute was studied by Muller and his co-workers. They found that 13 breaks in this small region fell into four positions so as to block out three very short segments (Muller and Prokofyeva, 1935; Raffel and Muller, 1940). In the lozenge region of the same chromosome, different x-ray-induced and spontaneous breaks likewise occur at definite points (Green and Green, 1956).

The evidence points to a nonrandom distribution of breaks in plant chromosomes as well. In Tradescantia, x-ray-induced breaks appear to be concentrated mainly in the proximal regions of the chromosomes (Sax and Mather, 1939; Sax, 1940; Lea, 1955). In inbred rye (Secale cereale), spontaneous breaks are localized in the distal regions (Rees and Thompson, 1955). Other plants in which breaks occur mainly in particular chromosome regions are Paris verticillata and Paeonia californica (Haga, 1953; Walters, 1956).

The basic morphological unit in a eukaryotic chromosome is a nucleohistone fiber about 100 Å in diameter, composed of DNA and firmly bound histone (Ris and Chandler, 1963; Ris and Kubai, 1970). Opinions differ in regard to the number and configuration of the fibers in a normal chromosome. One view is that each chromatid contains a bundle of 32 or 64 parallel fibers; 32 fibers before replication and 64 in prophase after replication (Kihlman, 1961, 1970; Rieger, 1965). An alternative view is that the chromosome contains a single long fiber which is folded and packed into a relatively short chromosome. The folds could come at points which are not DNA (Freese, 1958; Taylor, 1963; Uhl, 1965). Still another idea is that the strand is coiled in the form of a cycloid, with each complete loop corresponding to a chromosome (Whitehouse, 1967). Further work on chromosome ultrastructure may be able to reveal the presence or absence of regular crossover zones in the fibers.

Eukaryotic chromosomes contain various chemical components in addition to DNA: RNA, histone, so-called residual protein, lipids, calcium, and magnesium (Mazia, 1954; Ris, 1957; Steffensen, 1959; Kihlman, 1961; Ris and Chandler, 1963; Ris and Kubai, 1970; Dounce, 1971). It is of interest to know what role these accessory components are playing in the chromosome, and particularly whether any of them enter into the formation of crossover zones.

Several workers have considered and tested the possibility that calcium or magnesium may form links between sections of DNA and serve as sites for breakage and crossing-over.

The treatment of salivary-gland chromosomes of Drosophila with EDTA, a chelating agent which binds calcium and magnesium ions, was found to break the chromosomes up into short segments, some of which were about 4000 Å long (Mazia, 1954). Individual Drosophila flies reared on media containing this same chelating agent were found to have an increased frequency of crossing-over between certain marker genes (Levine, 1955; Colwell and Burdick, 1959). When an excess of calcium is fed to young adult female flies, however, the rate of crossing-over is decreased (Levine, 1955).

Similar results have been obtained in plants. Individuals of Tradescantia grown under conditions of magnesium or calcium deficiency have several times more breakage and fragmentation of the chromosomes than control plants (Steffensen, 1953, 1955). Tomato plants cultured under conditions of calcium deficiency have much chromosome fragmentation in comparison with control plants (Rüdenberg, 1965). The frequency of crossing-over in the alga Chlamydomonas is increased by treatment with a chelating agent. The effect of this agent can be reversed by subsequent incubation of the algal cells in a high concentration of calcium or magnesium ions (Eversole and Tatum, 1956).

Kaufman and McDonald (1956) repeated the Drosophila experiments using the digestive enzyme ribonuclease as well as the chelating agent EDTA. They found that chromosome breakage and recombination between certain marker genes can be increased by treatment with both types of agents. They concluded that widely different chemical agents are capable of disturbing the delicately balanced metabolism of the cell nucleus and bringing about chromosome breakage. The results obtained by other workers with chelating agents do not, according to Kaufman and McDonald (1956), necessarily point specifically to metallic ions as links between disjunct crossover units.

More recently Dounce (1971) has suggested that discontinuous segments of DNA may be linked by sulfur bonds belonging to adjacent chains of residual protein. Residual protein is a poorly known protein of high molecular weight which seems to be an important component of eukaryotic chromosomes. Evidence derived from studies of gels formed in isolated nuclei indicates that this residual protein is firmly bound to chromosomal DNA. Perhaps residual protein forms the connecting link between segments of DNA (Dounce, 1971).

We return finally to the question, What *are* the chromomeres? They are visible morphological units distributed at intervals along the length of the chromosome. Some modern workers have recently revived the old idea that they are also genetic units of one sort or another (Beermann, 1964; Pelling, 1966; Thomas, 1971).

The properties of chromomeres are best revealed in giant chromosomes of Drosophila, Chironomus, and other Diptera, where they are seen as bands. The sequence of bands is fixed. The bands are rich in DNA, whereas the intervening chromosome regions have a reduced DNA content, and may be heterochromatic. Consequently, there is a discontinuous linear arrangement of DNA-rich bands or chromomeres separated by DNA-poor zones (Beermann, 1964).

Later studies in Drosophila show that DNA-rich regions have the same discontinuous distribution in gametic chromosomes as in salivary chromosomes (Rudkin, 1965; Lefevre, 1971). The DNA content of a region in the gametic X chromosome can be correlated with the banding pattern in the corresponding region in the salivary X chromosome. Recombination studies utilizing marker genes show, moreover, that a high rate of crossing-over occurs in regions of the meiotic X chromosome that correspond with regions of prominent dark bands in the salivary X chromosome (Rudkin, 1965; Lefevre, 1971).

The bands of salivary chromosomes sometimes appear puffed out. This puffing seems to be a visible expression of a metabolically active state of the band or chromomere. The periodic and reversible condition of puffing fits in with the idea that the chromomere is a site of transcription (Beermann, 1964; Pavan, 1964). Autoradiographic evidence in Chironomus indicates that DNA replication takes place in each band more or less independently of other bands. The bands are thus also units of replication (Pelling, 1966).

In summary, the available evidence bearing on the problem of discrete genic units is suggestive but inconclusive. Classical cytology does not operate at a sufficiently high level of resolution to settle the question. The cytogenetic analysis of small chromosomal regions has revealed the phenomena of pseudoallelism, gene clustering, and position effect, and has thus exposed the problem as outlined earlier in this chapter but has not resolved it. More recent approaches utilize methods of electron microscopy and cytochemistry, which are potentially well suited for revealing the ultrastructure of chromosomes, but these methods also have not yielded wholly conclusive results because of the great complexity of the nucleus and the corresponding difficulties of interpreting the electron micrographs or cytochemical data (Ris, 1962). The search for approaches to the problem with a suitable "resolving power of genetic analysis," to use Pontecorvo's (1958) happy phrase, will have to continue.

THE ELEMENTARY GENETIC UNITS The recognition of a hierarchy of structural-functional units on a chromosome, ranging from small subdivisions of what was formerly considered one gene to larger but closely linked aggregations of these entities, requires a reappraisal of terminology. To which of the several structural-functional units does the term gene properly refer, and then how will the other units be designated?

Let us begin by equating the cistron with the polypeptide chain. A *cistron* is the string of triplets which is able to determine the complete sequence of amino acids in a polypeptide chain. This is the concept of one cistron–one polypeptide chain (Benzer, 1957; Rieger, Michaelis and Green, 1968).

A *gene cluster* or *gene complex* is then a functional unit consisting of two or more contiguous cistrons that produce different polypeptides in a complex protein macromolecule or different related proteins involved in a series of enzymatic steps (see Rieger, Michaelis and Green, 1968).

The unit of normal and regular crossing-over is designated a *Mendelian gene*. The Mendelian gene may embrace a single cistron or a gene cluster, depending on the complexity of the protein product and the adaptive value of constancy in its synthesis in each individual case. The term Mendelian gene is appropriate from the historical standpoint since it refers to the recombination units that Mendel and the early Mendelians worked with (Grant, 1964).

Pseudoalleles are subdivisions of a cistron which can separate and recombine rarely. Two pseudoalleles give different phenotypic effects in the cis versus trans positions in a double heterozygote, as described earlier. They are presumably subdivisions of a cistron concerned with the production of different components of a polypeptide (see Rieger, Michaelis and Green, 1968).

Needless to say, hard and fast distinctions cannot be drawn in every case between pseudoalleles and Mendelian genes or between cistrons and gene clusters. These units correspond to different groupings of the genetic material. The elementary genetic units thus stand in the good company of other biological units that likewise defy absolute definition.

At the lower molecular level of organization, the smallest structural-functional unit on the DNA chain is evidently the single *nucleotide unit,* and the next higher important unit is the *triplet* (see Chapter 2).

The term codon is simply another name for a triplet. The terms muton and recon refer to functions performed by neighboring nucleotides. A structural gene is approximately the same entity as a Mendelian gene. The assemblage which we previously (Grant, 1964) called a linked serial gene system is synonymous with gene cluster. Some terms in use for genic units are thus unnecessary and can be reduced to synonymy.

SIZE OF THE ELEMENTARY GENETIC UNITS   The dimensions of the DNA molecule are given by various authors (e.g., Hayes, 1968) and were quoted briefly in Chapter 2. The DNA double helix is 20 Å in diameter. One complete turn of the helix includes 10 base pairs and has a horizontal distance between gyres of 34 Å. It is apparent that nucleotide units lie 3.4 Å apart and that triplets are about 10 Å long.

If a cistron is assumed to contain the complete array of triplets for producing a polypeptide chain, and if we know the number of amino acids in a polypeptide chain, then it is a simple matter to calculate the estimated length of a cistron in triplets. An alpha chain in the human hemoglobin molecule contains 141 amino acids, as noted in Chapter 2. Such a polypeptide chain calls for a cistron with 141 triplets or 423 nucleotide pairs. A cistron of this size would have a linear length of about 1440 Å, in physical units. This length is probably close to the average size of many cistrons (see Hayes, 1968, pp. 147, 299).

A gene cluster controlling a complex protein or a series of metabolic steps would have a correspondingly larger size. The length of a Mendelian gene will vary from the size of a cistron to that of a gene cluster plus additional space at each end for crossing-over. An average crossover unit on the genetic map of Drosophila contains 3×10^5 base pairs, or 100,000 triplets, or enough DNA to make up numerous Mendelian genes (Hayes, 1968, p. 382).

In the preceding section, we quoted measurements of the basic nucleohistone fiber in eukaryotic chromosomes. This fiber is about

100 Å in diameter and is composed of a single DNA double helix and bound histone. The DNA helix is compacted in some way, perhaps by folding, in the nucleohistone fiber (Ris and Kubai, 1970). These characteristics of the chromosomal fiber will enter into estimates of the size of genic units.

A single large band in the salivary X chromosome of *Drosophila melanogaster* contains four genes—yellow, achaete, scute, and a lethal—and probably no others (Muller and Prokofyeva, 1935). This band is about 5000 Å wide. The average length of a chromosomal fiber occupied by each gene would then be about 1250 Å (Muller, 1935; Muller and Prokofyeva, 1935). The length of the DNA in each gene would be much greater than 1250 Å due to the folding and packing of the DNA strand in the chromosomal fiber.

Ris and Kubai (1970) quote estimates of the length of DNA in a chromomere in Drosophila. The length of DNA per chromomere varies from 10 to 160 μ (100,000 to 1,600,000 Å). A single chromomere thus contains a quantity of DNA which would make up at least a score and often several hundred cistrons of average size. In DNA content, a chromomere of Chironomus is in the same size range as the whole genetic thread in viruses (Pelling, 1966; Ris and Kubai, 1970).

The length of DNA in a chromosome of *Drosophila melanogaster* is about 10 mm (10,000 μ) for the short chromosome IV and 51 mm (51,000 μ) for chromosome III (Ris and Kubai, 1970).

CHAPTER FIVE

GENE MUTATION

SPONTANEOUS MUTATIONS
ADAPTIVELY VALUABLE MUTATIONS
MACROMUTATIONS

SPONTANEOUS MUTATIONS   A gene or point mutation is an alteration in the sequence of nucleotides which brings about a change in gene action. A simple change in the DNA code from CTT to GTT, for example, makes the difference between glutamic acid and glutamine in a polypeptide chain (see Chapter 2). Such molecular changes arise by the substitution of new nucleotides for pre-existing ones, as in the example, or by the addition or deletion of nucleotides in the DNA chain.

The important point for us is that gene mutations do arise spontaneously in frequencies which ensure a continuing source of mutant alleles in a population. The spontaneous mutation rates of seven genes in maize are given in Table 7. The mutation rates fall in the range of 10^{-4} to 10^{-6} (Stadler, 1942). Similar mutation rates for particular genes are found in Drosophila and in man.

It is probable that these estimates are somewhat too high, for the following reason: Any sample of mutant types is likely to include a fraction of rare recombinants, due to crossing-over within a compound gene as discussed in Chapter 4, in addition to true point mutations. The frequency of the latter would therefore be expected

90

TABLE 7

SPONTANEOUS MUTATION RATES OF SEVERAL ENDOSPERM GENES IN *ZEA MAYS* (STADLER, 1942)

Gene	Character	Mutation rate	Average no. of mutants per 1 million gametes	No. of gametes tested
R	aleurone color	0.00049	492	554,786
I	color inhibitor	0.00011	106	265,391
Pr	purple color	0.000011	11	647,102
Su	sugary endosperm	0.000002	2	1,678,736
C	aleurone color	0.000002	2	426,923
Y	yellow seeds	0.000002	2	1,745,280
Sh	shrunken seeds	0.000001	1	2,469,285
Wx	waxy starch	< 0.000001	0	1,503,744

to be somewhat lower than the overall mutation rates given in Table 7.

Even when we make allowance for rare intragenic recombination, however, it is apparent that a sample of one or two million gametes has a good chance of containing at least one new mutant allele of a specific gene, unless the gene in question is exceptionally stable.

The total mutation rate for all genes in the complement will naturally be much higher. In his pioneering studies of mutation in *Antirrhinum majus,* the snapdragon, Baur found rather large differences between strains in spontaneous mutation rates. Strain A2 had a high overall mutation rate of about 5% (Baur, 1924). The highly inbred strain 50, on the other hand, had a mutation rate of 0.74% in Baur's time (Baur, 1930b). The tests of strain 50 were continued on an extensive scale over the next 10 years by Stubbe (1966). Stubbe found year-to-year variation in mutation rates, ranging from 0.50 to 0.86%, and gives an overall average rate of 0.6% (Stubbe, 1966, p. 47).

A series of leaf mutants and one flower mutant in strain 50 of *A. majus* are illustrated in Figures 12 and 13.

The early Antirrhinum work furnished some preliminary indications concerning the causes of spontaneous gene mutations. Both genotypic and environmental factors appeared to be influencing mutation rates in Antirrhinum.

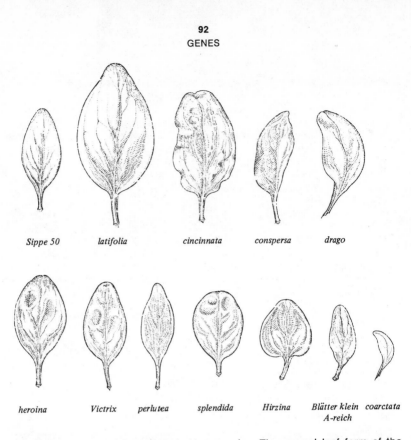

Sippe 50　　*latifolia*　　*cincinnata*　　*conspersa*　　*drago*

heroina　　*Victrix*　　*perlutea*　　*splendida*　　*Hirzina*　　*Blätter klein*　*coarctata*
　　　　　　　　　　　　　　　　　　　　　　　　　　A-reich

Figure 12. Leaf mutants in *Antirrhinum majus.* The normal leaf form of the parental strain 50 is shown at upper left. The leaves are all taken from the third leaf whorl. (From Stubbe, 1966, reproduced by permission of Veb Gustav Fischer Verlag.)

　　Genotypic control of mutations was later established more definitely in Drosophila and maize. A classical case in maize involves the gene *Dt* on chromosome 9 and elsewhere which induces mutations in the color gene A_1 on chromosome 3 (Rhoades, 1941). We will return to this case and related phenomena in Chapter 7.

　　The possible effects of natural environmental factors on gene mutation were investigated by Stubbe (1966, p. 49). Replicate lots of strain 50 of *A. majus* were grown in a greenhouse in Zurich, Switzerland, and in the high Alps at 3300 m elevation, and their progeny were tested for new mutations. The highland plants were, of course, exposed to higher levels of ultraviolet light and other forms of mutagenic radiation than the lowland plants.

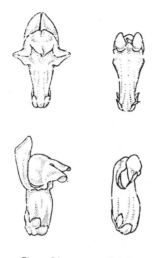

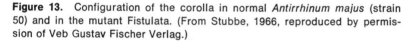

Sippe 50   *fistulata*

Figure 13. Configuration of the corolla in normal *Antirrhinum majus* (strain 50) and in the mutant Fistulata. (From Stubbe, 1966, reproduced by permission of Veb Gustav Fischer Verlag.)

The control population in Zurich yielded $0.54 \pm 0.31\%$ new mutations, which is close to the average frequency for this strain. A higher average mutation rate was found in the highland culture. The mountain-grown plants had a mutation rate of $0.89 \pm 0.33\%$ at the end of their first season of exposure, and a rate of $1.09 \pm 0.54\%$ after their second season in the mountains. The difference in means between the highland and lowland cultures, while suggestive, is not significant statistically (Stubbe, 1966, p. 49).

Populations of *Crepis capillaris* in the Alps also show an increase in spontaneous mutation rate with increasing altitude. The mutation rate reaches its maximum at 3000 m elevation (D'Amato and Hoffmann-Ostenhof, 1956).

Epilobium angustifolium is mostly uniform for flower color in Alaska and western Canada. However, a population growing on uranium ore at Port Radium on Great Bear Lake, Canada, contained a number of mutant individuals with pale pink flowers (Shacklette, 1964). Löve (1949) found an unusually high frequency of chromosomal mutations in plants grown from seeds placed near the crater of an erupting volcano in Iceland. Presumably gene mutation was also increased in these plants.

In the light of present evidence, the main causes of spontaneous gene mutations seem to be genotypic influences and accidents of DNA replication. Natural forms of radiation, both atmospheric and terrestrial, appear to be another cause of spontaneous mutations, concerning which more evidence is needed.

ADAPTIVELY VALUABLE MUTATIONS The great majority of new mutations are deleterious to some degree. In a sample of 2580 induced mutations in barley (*Hordeum sativum*), almost half (1250) were chlorophyll-deficient mutants, an equal number were recessive sterility mutants, and an additional complement of 40–45 had decreased viability. Thus 98.6% of the mutants in the sample were deleterious (Gustafsson, 1951). This is as expected, on the premise that any changes in a gene that has passed through many generations of natural selection for its functional usefulness are much more likely to be changes for the worse than for the better.

It is significant that a small minority of new mutations are superior to the wild type. In the same sample of 2580 mutations in barley, 5–10 mutants (0.2%) had a higher yield than the parental strain (Gustafsson, 1951).

In cultivated barley the so-called erectoides mutants are superior to the mother strain in a character of practical importance, namely, stiffness of straw, which is correlated with yield. About 60 erectoides mutants were observed in more than two decades of mutation breeding work on barley. Tests for allelism indicate that 14 loci are involved. The genes at 4 of these loci are more or less mutable, having mutated to the erectoides type repeatedly over the years, while the remaining 10 genes are relatively stable, having mutated only once (Hagberg, Nybom and Gustafsson, 1952).

A famous example of a mutation which was very successful agriculturally is hairlessness in the peach (*Prunus persica*), to give the nectarine. The condition of hairy fruit skin is dominant over glabrous skin in peaches. A number of instances have been recorded of hairy peaches mutating to smooth-skinned nectarines. In some of these instances the nectarines arose as bud sports and in others as seedling mutants. The mutant glabrous condition can be propagated sexually by seeds or vegetatively by grafts. The reverse mutation from glabrous nectarine to hairy peach has also been recorded. An interesting account of the origin of the nectarine is given by Darwin (1875, chs. 10, 11).

The adaptive value of a mutant allele is a function of the environment as well as the intrinsic properties of the new allele. An allele which is deleterious in the standard environment may well be normal or superior under other environmental conditions. This is clearly shown by Brücher's study of a series of mutations in *Antirrhinum majus.*

Brücher (1943) compared the growth and vigor of the standard strain 50 of snapdragon with those of six mutant types in a normal greenhouse environment and in four modified greenhouse environments. The conditions of light, temperature, and humidity in each environment are given at the bottom of Table 8. He scored the growth and vigor of the plants on a scale graded from 1 (normal growth) to 7 (poor growth).

The results are given in Table 8. The wild type is superior to all mutant types in the normal greenhouse environment. In abnormal greenhouse environments, however, the wild type is inferior, whereas certain mutant types have normal vigor. Furthermore, different mutants show normal growth in different environments (Brücher, 1943; Gustafsson, 1943).

Lotus corniculatus in England is polymorphic for a gene determining the presence or absence of a cyanogenic glucoside in the herbage. This cyanogenic glucoside can be converted to the poison cyanide, and is thought to be a defense mechanism against herbivorous animals. Feeding tests and natural population samples show that snails, slugs, and voles (*Microtus agrestis*) do feed preferentially on noncyanogenic plants in polymorphic populations and tend to avoid the cyanogenic plants. It seems probable that the allele for cyanogenic glucoside production is adaptively advantageous to the plants in relation to predation by herbivorous animals (Jones, 1966).

Herbivorous animals can acquire mutations enabling them to break down plant toxins and giving them immunity against the plant defenses. A case in point is the presence of an enzyme in the blood of some rabbits which hydrolizes the alkaloid atropine found

TABLE 8
RELATIVE VIGOR OF THE STANDARD STRAIN 50 AND SIX MUTANT TYPES OF
ANTIRRHINUM MAJUS UNDER DIFFERENT ENVIRONMENTAL CONDITIONS
(BRÜCHER, 1943; GUSTAFSSON, 1951)

Relative vigor	Environments (conditions specified below)				
	Normal greenhouse	R4	R3	R2	R1
1 (normal)	standard	argentea	matura	olive-green	delicate
2	argentea	—	olive-green	delicate	argentea
3	small-2	delicate	delicate	—	olive-green
4	delicate	small-2	small-2	—	small-2
5	olive-green	standard	—	small-2	—
6	salicifolia	matura	salicifolia	standard	standard
7 (weak)	matura	—	standard	argentea	—

Environment	Conditions		
	Light	Temperature	Humidity
Normal greenhouse	by day only	warm days and cool nights	moderate
R4	continuous light	5 C at start, then 12 C	low, 67%
R3	continuous light	constant 25 C	high, 98%
R2	continuous light	constant 12 C	high, 90–95%
R1	continuous light	constant 5 C	high, 100%

in plants such as Datura. The presence or absence of this enzyme
is inherited as a single gene difference (Wagner and Mitchell, 1955,
pp. 166–67). Here again a single-gene mutation could be useful to
its carriers by enabling them to exploit a new food plant.

This subject leads into the interesting relationships between
plants and animals with respect to toxic substances. For a recent
review of this fascinating problem see Rothschild (1972), and also
Harborne (1972).

MACROMUTATIONS In their morphological and physiological
effects, gene mutations form a spectrum from minor mutations to
macromutations. Mutations with minor and inconspicuous effects are
by far the most frequent, and undoubtedly play a very important
role in evolutionary changes in the genotype (Baur, 1924; Timo-

feeff-Ressovsky, 1940). A small fraction of new mutations, amounting to about 3% of all mutations in *Drosophila melanogaster,* have conspicuous phenotypic effects (Timofeeff-Ressovsky, 1940). Macromutations occur in this fraction and at the upper end of the spectrum of phenotypic effects.

Occasionally gene mutations arise that produce a phenotype completely outside the normal range of variation of the parental species. The mutant phenotype resembles or approaches the condition found in a distantly related species or genus. These are macromutations. Various authors have suggested that they may sometimes play a role in evolution out of proportion to their frequency of occurrence (Goldschmidt, 1940, 1955; Stubbe and Wettstein, 1941; Gustafsson, 1951; Schwanitz, 1956; Stubbe, 1959, 1966; Prazmo, 1965).

A number of macromutations have been isolated in *Antirrhinum majus* by Stubbe (1952, 1959, 1966). Some of these are listed below, together with the more or less distantly related genus of Scrophulariaceae which they simulate.

1. Transcendens: number of stamens reduced from the normal four to two; in this respect, like Mohavea
2. Hemiradialis: five stamens, approach to radial symmetry of corolla; as in Verbascum
3. Neohemiradialis: six anthers
4. Radialis: radially symmetrical flowers; as in Verbascum
5. Hirzina: floral spurs; approaching Linaria
6. Fistulata: reduction of corolla limb and a tubular corolla; as in Rhinanthus (Fig. 13)
7. Subconnata: one cotyledon

A macromutation in *Linaria maroccana* is the mutant type gratioloides, which behaves as a single-gene recessive (Schwanitz and Schwanitz, 1955). The configuration of the flowers is drastically altered in this mutant. The symmetry of the corolla is weakly bilateral or radial, the spur is absent, and the number of anthers is reduced from four to two or even zero.

The macromutations in Antirrhinum and Linaria generally show great variability in phenotypic expression. Thus the anther number varies in the mutant type Transcendens from four to three to two, and flowers in the same inflorescence often have different numbers of anthers. Plants of the mutant type Neohemiradialis have the mutant phenotypic condition of extra stamens expressed in about 77% of the flowers. The manifestation of the mutant condition is modified by environmental conditions and also by the genetic background. Stubbe found that when he transferred the mutant alleles transcendens and neohemiradialis to wild races of *A. majus* or to related species, and selected for the mutant phenotype in the early hybrid generations, the expressivity increased to 91–99% (Stubbe, 1952).

These results suggest that an adaptively valuable macromutation appearing in a natural population, and showing variable expressivity in the early generations, might mark the beginning but only the beginning of an important deviation from the ancestral stock. Selection for the mutant phenotype would necessarily involve selection not only for the macromutant gene but also for additional modifier genes which stabilize its expression. Evolution in such a case would proceed by the fixation of a macromutation together with an accompanying set of minor modifier genes (Stubbe and Wettstein, 1941; Stubbe, 1952, 1959, 1966).

The most prominent differences between the species of Aquilegia (the columbines) are characters of the flowers, such as erect or nodding position, color, and presence or absence of spurs. These characters adapt the species for pollination by different groups of insects and birds. Since the species are interfertile, it is possible to carry out a factorial analysis of the adaptive interspecific differences.

Prazmo has made a genetic analysis of 14 floral and vegetative character differences in numerous hybrid combinations in Aquilegia. Most of the characters show continuous variation in F_2, indicating multifactorial control. Some of the key floral characters, however, are controlled by allelic differences at one or a few gene loci (Prazmo, 1965).

Thus in crosses of spurless *A. ecalcarata* with spurred *A. vulgaris,* the floral spur is dominant in the F_1 hybrids, and segregates in a 3 : 1 ratio in F_2 as seen below:

obs.   373 spurs : 139 no spurs

exp.   384 spurs : 128 no spurs

Some F_3 families also show 3 : 1 ratios. Backcross hybrids to the *A. ecalcarata* parent give 1 : 1 ratios for presence or absence of spurs.

Similar monohybrid ratios for presence or absence of spur are found in crosses of *A. ecalcarata* × *A. canadensis*. The spur-bearing *A. alpina* and *A. chrysantha* differ from spurless *A. ecalcarata* by dominant duplicate factors for spur. The erect versus nodding position of the flowers in *A. chrysantha* and *A. ecalcarata* respectively is another single-gene difference (Prazmo, 1965).

On the basis of this evidence, Prazmo (1965) makes the plausible suggestion that macromutations for adaptive floral characters have played an important part in the evolution of Aquilegia.

CHAPTER SIX

ALLELE INTERACTION

INTRODUCTION   The two alleles of a Mendelian gene in a diploid cell may be alike, as in a homozygote, or unlike, as in the heterozygote. In either case the two alleles can interact in various ways. The ultimate phenotypic expression of the gene is affected in part by the mode of allele interaction. When the phenotypes of heterozygotes are compared with those of the corresponding homozygotes, the heterozygous phenotypes may be dominant, intermediate, codominant, or heterotic. Conversely, a homozygote, when compared with a corresponding heterozygote, may be similar in phenotype or may exhibit recessiveness, inbreeding depression, or some other difference.

In the present chapter we consider these and other modes of allele interaction and their phenotypic effects. The problem is discussed mainly in terms of the diploid condition, but brief reference is made to the situation in polyploids. Our subject leads into the interesting peripheral phenomena of heterosis, inbreeding depression, and homeostasis.

DOMINANCE   Mendel discovered in his hybridization experiments with peas that heterozygous individuals generally exhibit the same characteristics as one of the two contrasting parental types. Thus, in crosses between round-seeded plants and wrinkled-seeded plants, the heterozygotes have round seeds. As regards seed shape, round-ness is dominant and wrinkledness is recessive; in seed coat color, grayish-brown is dominant and white is recessive; and so on (see Table 1 in Chapter 1). Dominance in peas had been reported earlier by Thomas Andrew Knight, in 1823, and by several other nineteenth-century plant hybridizers (see Roberts, 1929).

When dominance in peas was reexamined in the early period of modern genetics, it was found that heterozygous smooth peas do differ from homozygous smooth peas. The difference, however, is histological rather than gross morphological, and can be reduced to the microscopic appearance of the starch grains in the pea seed (Sirks, 1956). The dominance of roundness over wrinkledness, or of R over r, in peas is incomplete.

Dominance is a common and widespread type of allele inter-action in diploid organisms generally. Where cases of dominance are investigated carefully at the endo-phenotypic level, incomplete dominance is perhaps the usual condition. But dominance, complete or incomplete, is by no means a universal feature of allele inter-action. Phenotypic intermediacy of heterozygotes is also common.

The classic case of intermediacy in plants involves flower color in *Mirabilis jalapa,* the four-o'clock. Correns crossed parental types of Mirabilis with white and with dark pink flowers. The F_1 plants had light pink flowers. The F_2 generation segregated in a $1 : 2 : 1$ ratio for flower color. The light pink F_2s continued to segregate in the same fashion in F_3 (Correns, 1905; Sirks, 1956).

At the fundamental level of gene action, dominance is a result of one allele, present singly in a heterozygote (Aa), bringing about the same phenotypic effect that arises when the same allele is present doubly in a homozygote (AA). Incomplete dominance is an approach to this condition.

Three possible modes of interaction between the alleles A and a in the heterozygote could lead to the end result of dominance (Stern, 1960): (1) The dominant allele A may be active and the recessive allele a inactive. (2) Both alleles may act, in the same direction, but A has a stronger action than a. (3) The two alleles may act in opposite directions, as when A induces one biochemical reaction and a an alternative reaction, but the effects of A prevail over those of a. The three modes of allele interaction leading to dominance may be represented diagrammatically, following Stern (1960):

$$(1) \quad A \longrightarrow$$
$$a \quad 0$$

$$(2) \quad A \longrightarrow$$
$$a \quad \rightarrow$$

$$(3) \quad A \longrightarrow$$
$$a \quad \leftarrow$$

Each of the three modes can be illustrated by actual examples. Mode 1, involving the inactivity of the recessive allele in the heterozygote, is exemplified by the combination of the alleles I^0 and I^A governing blood type in man. Mode 2, involving the cumulative action of two alleles with unequal strengths, is probably a common situation, a case in point being the Y alleles for vitamin A content in corn endosperm. Mode 3, entailing competition between two alleles of unequal strength, is illustrated by the alleles of the Ci gene for cubital wing vein in *Drosophila melanogaster* (Stern, 1943). See Grant (1964, pp. 55 ff) for review.

Dominance is not a result of the interaction between two homologous alleles exclusively. An allele which exhibits complete dominance in one genotype may, on being transferred by backcrossing to a different genetic background, sometimes show incomplete dominance or intermediacy in the new background. Dominance is determined in part by the influence of other genes of the complement, known in this context as dominance modifiers and discussed in Chapter 7.

CODOMINANCE In still another mode of allele interaction in a diploid heterozygote, each allele acts with equal or nearly equal strength but in a different direction. The action of both alleles leads to formation of a mixture of gene products at the endo-phenotypic level of expression. This situation is known as codominance.

Codominance is illustrated by blood types in man. The ABO blood groups are controlled by alleles of the gene I. I^A produces type A blood and I^B type B blood. The heterozygote I^A/I^B produces blood containing a mixture of both types A and B (see Stern, 1960). The type of hemoglobin is determined by the alleles of Hb. The homozygotes Hb^A/Hb^A and Hb^S/Hb^S have hemoglobins A and S, respectively, in the red blood cells, while the heterozygote Hb^A/Hb^S has blood cells with a mixture of both types of hemoglobin (see Chapter 3).

In plants codominance can be demonstrated by the methods of electrophoresis or chromatography for protein or flavonoid substances. Two parental types have different banding patterns for a protein or flavonoid. With codominance the hybrid exhibits a combination of both banding patterns. This result has been observed in some comparisons of parents and hybrids but not in others.

Gossypium hirsutum, G. sturtianum, their F$_1$ hybrid, and their amphiploid have been analyzed for seed proteins by electrophoresis. The parental species have different banding patterns for the seed proteins. The hybrids and hybrid derivatives show additive banding patterns, indicating the presence of both parental types of proteins (Cherry, Katterman and Endrizzi, 1971).

Similar results have been obtained in *Phlox pilosa, P. carolina,* and their naturally occurring hybrid derivatives *P. pilosa detonsa* and *P. pulcherrima.* Here again the seed proteins found in the hybrid derivatives are summations of those present in the ancestral species (Levin and Schaal, 1970). In such cases the opposite alleles of the parental species are apparently able to function uninhibitedly in the hybrids and hybrid derivatives.

The flavonoid compounds in related species and hybrids can be compared by paper chromatography. Some hybrids in Dicentra were found to have all the flavonoids present in their parental species. But other hybrid combinations in Dicentra lacked certain parental substances or showed novel substances not identified in the parents (Fahselt and Ownbey, 1968).

DOSAGE EFFECTS IN POLYPLOIDS The polyploid condition sets up the possibility of diverse combinations of alleles. A given dominant or recessive allele may be present in two or more doses in a heterozygote. The net gene action at a locus is the resultant of the different actions of the various alleles involved.

A particularly clearcut example of dosage effects in polyploid tissue is provided by the gene *Y* for vitamin A content in corn endosperm (Mangelsdorf and Fraps, 1931). Corn endosperm is triploid. The dominant and recessive alleles, *Y* and *y*, can be combined in four genotypes: *YYY, YYy, Yyy, yyy*. The vitamin A content of the endosperm was found to correlate very closely with the dosage of the dominant *Y* allele. This correlation is shown by the following data:

	vitamin A units per gram of endosperm
yyy	0.05
yyY	2.25
yYY	5.00
YYY	7.50

The *Y* allele is thus mainly responsible for the formation of vitamin A in endosperm tissue, and the effects of two or three *Y* alleles are additive (Mangelsdorf and Fraps, 1931).

The garden Dahlia, *Dahlia variabilis,* is an octoploid ($2n = 64$) of hybrid origin from tetraploid species with differently colored flowers. A great array of flower colors, ranging from white to yellow, purple, and red, is found within *D. variabilis.* The factorial analysis of Lawrence and co-workers shows that this variation is due basically to five genes (*I, Y, A, B, H*). The interactions among these genes determine the basic colors; we will return to the gene interactions involved, in Chapter 8. Of interest to us here is the fact that superimposed on the gene interactions are dosage effects of the alleles of a given gene, producing fine differences in shade within a basic color (Crane and Lawrence, 1934; Lawrence and Scott-Moncrieff, 1935).

The dominant allele of the *A* gene for anthocyanin has cumulative effects. The anthocyanin pigmentation in the flower becomes progressively deeper in the series of genotypes, *Aaaa, AAaa, AAAa,*

AAAA. The gene *I* for ivory flavone shows a similar dosage effect. The genotype *iiii* has no flavone, and *Iiii* has only a slight trace of pigment, while *IIii, IIIi* and *IIII* are fully pigmented with flavone (Crane and Lawrence, 1934; Lawrence and Scott-Moncrieff, 1935).

The eluta gene *El* in *Antirrhinum majus* causes bleaching of the red color in flowers. An induced tetraploid form of snapdragon showed a dosage effect for this gene. The incomplete dominant allele *El* in single dose makes the flower color one shade lighter. Flower color becomes progressively lighter with increasing doses of *El*, as indicated in Table 9. The anthocyanin concentration in floral extracts is correlated inversely with the dosage of *El*, as shown also in Table 9 (Stubbe, 1966, p. 38).

PARAMUTATION   It is a tenet of Mendelian genetics, supported by much empirical evidence, that the haploid gametes produced by a diploid heterozygote are themselves pure. The corollary of this tenet is that dissimilar alleles of a gene in a heterozygote may act in different ways biochemically but retain their individual identity unchanged in doing so. One allele does not permanently alter another allele in a heterozygous nucleus.

This conclusion is generally but not universally valid. Brink (1956, 1958) has discovered a significant exception in *Zea mays* which he describes under the term paramutation. The exception occurs in the interaction between certain alleles of the *R* gene, which controls anthocyanin pigmentation in the endosperm of corn kernels and in vegetative parts.

Our story involves the recessive allele *r* for colorless endosperm and two dominant alleles, R^r and R^{st}, for colored endosperm. The standard anthocyanin-producing allele, R^r (in single dose), gives dark, mottled endosperm, whereas R^{st} gives stippled endosperm (Brink, 1956).

The homozygote R^r/R^r and the heterozygote R^r/R^{st} were crossed as males to *rr* females to produce kernels with triploid endosperm. The R^rrr endosperm progeny of the two testcrosses should be alike phenotypically. In fact they are phenotypically different. The progeny of $R^r/R^r \times rr$ show normal dominant pigmentation in the endosperm. The R^rrr endosperm derived from

TABLE 9
DOSAGE EFFECT OF THE DOMINANT ALLELE *EL* ON FLOWER COLOR IN *ANTIR-RHINUM MAJUS* (STUBBE, 1966)

Genotype	Flower color	Colorimetric value for anthocyanin concentration
+ + + +	intense magenta	103.8
El + + +	dark magenta	60.2
El El + +	light magenta	43.9
El El El +	very light magenta	31.3
El El El El	nearly white	21.8

$R^r/R^{st} \times rr$, by contrast, have a lowered degree of pigmentation (Brink, 1956).

The R^r allele which has passed through an R^r/R^{st} heterozygote has a different, lowered pigment-producing activity than the R^r allele derived from an R^r/R^r homozygote. The R^r allele has been altered somehow in the R^r/R^{st} plants, and the alteration is hereditary, inasmuch as the modified R^r allele can be transmitted through the pollen to a subsequent generation. The modifiable R^r allele is said to be paramutable and the R^{st} allele paramutagenic (Brink, 1956, 1958; Brink, Styles and Axtell, 1968).

The paramutational change in R^r is reversible, at least partially. This is shown by putting altered R^r alleles from an R^r/R^{st} heterozygote into the homozygous condition R^r/R^r, and making the testcross $rr\,♀ \times R^r/R^r\,♂$. The progeny of the testcross show that the formerly altered R^r alleles have now regained nearly the same level of anthocyanin production as never-altered R^r alleles (Brink, 1956).

Brink and co-workers have offered some explanatory suggestions supported by evidence concerning the mechanism of paramutation. The R locus is compound. One of the components is assumed to be a heterochromatic segment which can repress the associated genic unit for anthocyanin formation. Further assumptions are that the heterochromatic segment can vary in length, that its repressive effect is directly proportional to its length, and that its length is susceptible to control during replication by certain homologous alleles. A paramutagenic allele, like R^{st}, could influence the extent of replication in the heterochromatic segment of a paramutable allele, like R^r, and in this way affect the strength of action of the latter (Brink, Styles and Axtell, 1968).

Paramutational changes in alleles in heterozygotes have also been reported for the Cr gene in Oenothera and $Sulf$ gene in Lycopersicon, and perhaps in some other plants such as Pisum and Malva (see Brink, Styles and Axtell, 1968).

SINGLE-GENE HETEROSIS Certain paired combinations of a gene may yield a greater productiveness than that of the corresponding homozygous combinations. The heterozygote then exceeds the mean of the corresponding homozygotes for some measurable phenotypic character like size, weight, vigor, or seed output. The general condition is: $Aa > AA$ or aa. This condition is known as single-gene heterosis or overdominance.

Single-gene heterosis as seen at the phenotypic level is a manifestation of a synergistic interaction between the dissimilar alleles in a heterozygote. The synergistic effects can be explained in terms of known properties of gene action. Haldane (1954a) has suggested that homozygotes may be able to produce only single kinds of enzymes for each gene, and may therefore be restricted to specialized reactions taking place within a narrow range of environmental conditions, whereas heterozygotes, by producing a mixture of enzymes, may be able to function in more diverse ways and over a wider range of environmental conditions. The superior yield or fitness of some heterozygotes is held to be a product of their greater biochemical versatility as compared with the corresponding homozygotes (Haldane, 1954a).

A case in Neurospora supports this hypothesis of the biochemical basis of single-gene heterosis. Certain qualifications must be made in this case, since Neurospora is a haploid organism and consequently cannot have a heterozygous constitution in the vegetative phase. However, the haploid nuclei of different parental strains can become associated in a single fungus body after cross-fertilization. Such a hybrid fungus is known as a heterokaryon. Genetically and physiologically, a heterokaryon in Neurospora is comparable to a heterozygote in diploid plants and animals.

Emerson (1952) compared the growth of a particular heterokaryotic strain of Neurospora with that of the two homokaryotic parental strains. The parental strains differed in the alleles of a single gene and not in other genes of the complement. The gene involved controls synthesis of a vitamin (para-aminobenzoic acid) necessary for growth. One homokaryotic strain, and hence one allele of this gene when present alone, apparently synthesizes too little of the vitamin. The other homokaryotic strain, and hence the alternative allele, produces a deleterious excess of the same growth substance. But the heterokaryon, in which the two alleles are present together, has a balance of vitamin production which promotes optimum growth (Emerson, 1952; Beadle, 1953).

We noted earlier in this chapter that the heterozygous genotype in man, Hb^A/Hb^S, produces both hemoglobin A and hemoglobin S. In this respect the two alleles, Hb^A and Hb^S, show codominance at the molecular level. We can add here that the Hb^A/Hb^S individuals have greater resistance to malaria than either homozygous type. The heterozygotes have superior viability in a malarial environment and thus exhibit heterosis at the organismic level (Allison, 1956).

Single-gene heterosis is exhibited by several mutant genes of *Antirrhinum majus*. The heterozygotes of the mutant types Spectabilis, Lurida, Vitellina, and Virescens are superior to the homozygotes in the size of various plant parts (Fig. 14). The Spectabilis heterozygote has in addition a higher chlorophyll content in the leaves than either homozygote (Stubbe, 1953, 1966).

Redei (1962) studied two leaf mutants in *Arabidopsis thaliana* (Cruciferae). This plant is autogamous and highly homozygous. Mutations were induced by x-rays in the genes *An* and *Er*. Since the x-rays could have also induced mutations in neighboring loci with no visible effects, the new mutants were interbred with normal plants for several generations in an effort to extract the mutant genes from the irradiated genetic background and transfer them as single genes to the background of normal Arabidopsis. The homozygotes and heterozygotes for the single genes were then compared with respect to size and yield. Some of the results are given in Table 10. The single-gene heterozygotes are seen to exceed the homozygotes in plant weight (Redei, 1962).

Cases of single-gene heterosis have also been reported in

Figure 14. Growth of plants heterozygous for a single gene compared with that of the corresponding homozygotes in *Antirrhinum majus. Left,* wild-type homozygote of strain 50. *Right,* mutant Vitellina homozygote. *Center,* F_1 heterozygote. (From Stubbe, 1966, reproduced by permission of Veb Gustav Fischer Verlag.)

Sorghum vulgare (Quinby and Karper, 1946), *Hordeum sativum* (Hagberg, 1953; Sarkissian and Huffaker, 1962), *Secale cereale* (Müntzing, 1963), *Dactylis glomerata* (Apirion and Zohary, 1961), *Phaseolus lunatus* (Allard and Workman, 1963), *Avena fatua* (Imam and Allard, 1965), *Zea mays,* (Crumpacker, 1966), and *Mimulus guttatus* (Kiang and Libby, 1972). Heterosis in intervarietal hybrids of *Nicotiana rustica* is due to a small number of genes for each heterotic character (Smith, 1952b).

In higher animals, apparent single-gene heterosis is reported in man, as noted earlier, and in mice (Dunn and Suckling, 1955) and Drosophila (Stern, Carson, Kinst, Novitski and Uphoff, 1952; Da Cunha, 1953; Mukai and Burdick, 1959; Schnick, Mikai and Burdick, 1960; Wallace, 1963; Richmond and Powell, 1970).

Single-gene heterosis has been the subject of controversy in the past. Is this term an accurate explanation of the observations when considered literally? Is the heterosis really a result of the interaction between a particular pair of alleles? Or, to pose an alternative objection, how is it possible to control the complicating effects of gene

TABLE 10
RELATIVE YIELD OF MUTANTS, NORMAL HOMOZYGOTES, AND HETEROZYGOTES
FOR TWO GENES IN *ARABIDOPSIS THALIANA* (REDEI, 1962)

Genotype		Plant weight, as percent of wild-type Fresh weight	Dry weight
An/An		100.0	100.0
an/an		90.8	83.4
An/an	(3 crosses)	111.3	116.7
		119.6	121.1
		123.6	122.6
Er/Er		100.0	100.0
er/er		90.3	77.1
Er/er	(3 crosses)	116.1	109.1
		126.3	125.3
		128.7	130.3

interaction on the particular allele pair under consideration (Schuler, 1954; Jinks, 1955; Mather, 1955)?

The objection to the hypothesis of single-gene heterosis has considerable force in cross-fertilizing and highly heterozygous organisms, like Mus, Drosophila, and Zea, where it is indeed difficult to separate the effects of allele interaction from those of gene interaction. The objection has force again in regard to observations of heterosis in natural populations, as in Drosophila, Dactylis, Avena, and Mimulus, where many genetic and selective processes are going on simultaneously. But it loses much of its force in the case of inbred and highly homozygous plants, like Antirrhinum, Hordeum, and Arabidopsis, in which the principal variable involved is homozygosity or heterozygosity at a single locus. In Arabidopsis a special effort was made to control the genetic background as well as the environment so that the phenotypic differences could be correlated with the genotypic condition at a single locus exclusively. Heterosis was observed here, too, as we have already seen. The objections to the hypothesis of single-gene selection appear to have been overstated.

HETEROSIS Heterosis or hybrid vigor is a phenomenon of widespread occurrence in both plants and animals. The vigor of the mule is legendary, and plant hybrids with similarly vigorous characteristics were often called mule plants by the early hybridizers. Heterotic hybrids have been observed in all major groups of higher plants. The phenomenon has been studied in depth in many cases and the literature on the subject is accordingly voluminous. Good reviews are given by Whaley (1944), Crow (1948), Gowen (1952), Rendel (1953), Allard (1960), Sehgal (1963), and Hiorth (1963).

The superiority of the hybrid compared with its parents may be manifested in a variety of ways: at the seedling or adult stage of development; in vegetative or reproductive parts; in visible features like size or invisible features like adaptive fitness. Visible heterosis is an ephemeral phenomenon, in that the superior vigor appears briefly in the F_1 generation and declines or disappears in the following sexual generations. The hybrid vigor can be perpetuated for many seasons of growth in vegetatively propagated crop plants, like potatoes and sugarcane, but not in sexually reproducing annual crops like maize.

The most widely known and economically important example of heterosis is hybrid corn. Parental inbred lines of corn with medium-sized plant bodies and ears consistently produce F_1 hybrids with very robust shoots and large ears. Photographic illustrations show these differences in all textbooks of genetics. The size differences are reflected in comparative yield in terms of bushels of corn per acre. Data on yield for 12 intervarietal hybrid combinations are given in Table 11. Every hybrid type is seen to exceed the mean yield of its two parents, and several of the hybrids exceed their parents by a wide margin.

The appearance of heterosis in the F_1 generation of any given cross depends upon a certain degree of genetic differentiation between the parent plants. Sister plants belonging to the same inbred family do not give rise to heterotic F_1s. Interracial and interspecific hybrids in plants frequently do exhibit heterosis. Interspecific hybrids also frequently exhibit lethality, semilethality, or reduced vigor. There is apparently an optimum range of genetic differentiation between parental populations which can engender heterosis in the hybrid combination.

This relationship between divergence and heterosis can be seen in the data of Table 11. Intervarietal hybrids within the group of dent corn show a moderate degree of heterosis for yield. By com-

TABLE 11

COMPARATIVE YIELD OF PARENTAL VARIETIES AND THEIR HYBRIDS IN 12
INTERVARIETAL COMBINATIONS IN CORN (DATA OF HAYES AND OLSON, AFTER
ALLARD, 1960)

Parental varieties		Yield, as percentage of P ♂		
P ♀	P ♂	P ♀	P ♂	F_1
Dents				
Minnesota 23	Minnesota 13 (Dent)	96.7	100.0	110.9
Northwestern	"	105.9	"	115.7
Chowen	"	99.0	"	114.9
Rustler	"	112.1	"	112.4
Silver King	"	100.9	"	106.7
Murdock	"	80.3	"	101.6
Flints				
Smutnose	"	110.8	"	127.6
King Phillip	"	100.0	"	119.9
Longfellow NR	"	100.5	"	119.6
Longfellow Bwls	"	104.9	"	114.1
Mercer	"	91.8	"	97.3
Flour				
Blue Soft	"	96.7	"	132.5

parison, the crosses of dent × flint corn and dent × flour corn usually produce marked increases in yield (Allard, 1960). The heterosis exploited agriculturally in the American corn belt is largely a product of northern flints × southern dents (Anderson and Brown, 1952).

These general relationships have been confirmed and quantified in a recent study (Moll, Lonnquist, Fortuno and Johnson, 1965). Moll and co-workers intercrossed populations of maize from different geographical regions in temperate and tropical North America in all possible combinations. Some hybrid combinations had geographically and genetically similar parents, but other combinations had very divergent parents. The amount of F_1 heterosis was compared with the degree of genetic divergence between the parental strains. Heterosis was found to rise with increasing divergence up to a certain level of divergence. But beyond this point the trend reversed and heterosis declined.

There is also a correlation between heterosis and breeding system. Heterosis is common in hybrids of outcrossing plants and uncommon in hybrids of autogamous plants. The strongest manifestations of heterosis are found in outcrossing and highly heterozygous plants such as maize, rye, cabbage, potatoes, and bananas. In many groups of inbreeding and highly homozygous plants the hybrids display no heterosis at all. A fair number of self-pollinating plants, such as tomatoes, the common bean, soybean, flax, barley, and wheat, do exhibit heterosis, but the degree of heterotic advantage is generally smaller than in cross-fertilizing plants (Whaley, 1944; Allard, 1960; Hiorth, 1963; Simmonds, 1969).

Attempts have been made to determine the developmental stage when hybrid vigor first appears. It is often visible in the seedling stage. In the embryo stage in maize the hybrid type does not have an advantage in size over inbred parental types. The hybrid seedlings gain their advantage over the inbred parents in root and shoot development during the first four weeks of growth, and maintain this advantage during the remainder of the growth period (Whaley, Heimsch and Rabideau, 1950).

The question then narrows down to the nature of the physiological advantage of the hybrid type in the early embryonic stages. Hybrid embryos of maize have a greater amount of nutrient material available to them, and may be more efficient in utilizing these nutrients (Whaley, Heimsch and Rabideau, 1950). The hybrid soon develops a larger root system than the inbred parents and absorbs a greater amount of minerals, as indicated by measurements of radioactive phosphorus absorption (Rabideau, Whaley and Heimsch, 1950). The hybrid embryos may also have a superior ability to synthesize or utilize growth substances such as auxins (Whaley, 1952).

Part of the nutritional advantage of the hybrid embryo may result from association with a hybrid endosperm. In alfalfa (*Medicago sativa*), an outcrossing species, the endosperm develops more rapidly after double fertilization in outcrossed seeds than in selfed seeds. A hybrid embryo thus has the advantage of being able to draw nutrients from a vigorous endosperm (Brink, 1952).

Robbins (1941) compared the growth of excised roots of *Lycopersicon esculentum, L. pimpinellifolium,* and their F_1 hybrid in nutrient solutions containing various growth substances. The F_1 hybrid was superior to the parents in rate and amount of root growth. By varying the composition of the nutrient medium, Robbins showed that one parental type was defective in the ability to synthesize the vitamin pyridoxin, while the other parent was deficient in nicotinamide. The hybrid roots grew well on substrates lacking these substances. The hybrid genotype evidently has a complementary combination of factors covering up the separate nutritional deficiencies of the two parents (Robbins, 1941). A similar nutritional complementation was found in hybrids between different inbred lines of *Lycopersicon esculentum* (Whaley, 1952).

Other aspects of metabolism are implicated in heterosis. Evidence in rye (*Secale cereale*) suggests that heterotic hybrid seedlings are more efficient in the utilization of phosphorus, potassium, and iron than inbreds (Rees and Hassouna, 1964). In barley (*Hordeum sativum*), the heterotic F_1 hybrid between two inbred lines proved to be superior to the parents in photosynthetic fixation of CO_2 (Sarkissian and Huffaker, 1962). Heterotic hybrids of maize have a higher oxygen uptake and respiration rate than the parental inbreds (Sarkissian and Srivastava, 1967). This last finding leads into the cytoplasmic aspects of heterosis, which are discussed in Chapter 11.

The expression of heterosis is influenced by environmental conditions. A particular heterotic hybrid combination of *Antirrhinum majus* was grown in a greenhouse in three different months of winter and spring. The main difference between the replicate cultures was the amount of sunlight received. The degree of heterosis was much greater in the spring cultures receiving more sunlight (Haney, Gartner and Wilson, 1953). Later experiments with Arabidopsis, in which the only variable environmental factor was temperature, confirmed and extended these results. The expression of heterosis in Arabidopsis was found to be most marked at high temperatures, but was slight or absent at lower temperatures near or below the growth optimum (Langridge, 1962).

CAUSES OF HETEROSIS The genetic mechanisms responsible for heterosis have been a subject of much interest since the early period of Mendelian genetics. Speculations on the subject date back to pre-Mendelian times.

Two classical hypotheses were proposed early in this century. These were the dominance hypothesis of Davenport (1908), Jones

(1917), and others, and the concurrent overdominance hypothesis proposed independently by East and Shull in 1908 (see East, 1936; Shull, 1952). These hypotheses have undergone much discussion and testing in the years since their formulation. They have withstood the test of time and are still valid today. Meanwhile some additional hypotheses have been proposed, two of which seem to be necessary. One of these is the complementary gene hypothesis, or epistasis hypothesis, which is a modification and extension of the dominance hypothesis (Powers, 1944; Jinks, 1955). The other important theory is the cytoplasmic hypothesis of Michaelis (1951).

The dominance hypothesis assumes that the quantitative characters expressed in heterosis are controlled by many genes linked in various linkage groups. Some of these genes will normally be represented by deleterious alleles in a population. The deleterious alleles may be recessive or incompletely dominant. The F_1 hybrid between two such populations will have a genotype in which the unfavorable alleles of each parent are covered up by favorable dominant alleles of the other parent. The F_1 hybrid will therefore exceed both parents in size or vigor. The perpetuation of the heterosis into later sexual generations would require the formation of a recombination type containing the favorable alleles of both parents in homozygous condition. But the formation of such a recombination type is opposed by the large number and linkage relations of the genes involved. Consequently, the heterosis is lost as a result of segregation in the progeny of the F_1 generation.

The complementary gene hypothesis starts with essentially the same setup. The premise is changed in one respect, however: it is not necessary to assume complete dominance of the favorable alleles. The main emphasis in this hypothesis is on the synergistic complementary effects between different genes in the hybrid as a cause of heterosis, rather than on the dominance relations between alleles of the same genes (Powers, 1944; Jinks, 1955).

The overdominance hypothesis postulates a synergistic interaction between certain alleles or their gene products at each of a series of loci. Thus $Aa > AA$ or aa, $Bb > BB$ or bb, and so on. The superiority of the heterozygote could be due to a favorable complementarity in enzyme action or gene products of the two divergent alleles at a locus, as we noted in our earlier discussion of single-gene heterosis, and it would be appropriate therefore to speak of allele complementation, as some authors have done.

The more recent cytoplasmic hypothesis of Michaelis (1951) suggests that heterosis may arise from favorable interactions between nuclear genes and cytoplasmic genes or from favorable combinations of different cytoplasmic genes. Still more recently some of the heterosis-affecting cytoplasmic genes have been identified as mitochondria. Certain mixtures of dissimilar mitochondria from different parents show complementation in respiration, and this is correlated with the manifestation of heterosis in the hybrid seedlings (Sarkissian and Srivastava, 1967; see also Chapter 11). In such cases mitochondrial complementation appears to be a cause of heterosis (Sarkissian and Srivastava, 1967).

Any attempt to assess the cogency of the various hypotheses of heterosis must take into account the heterogeneous nature of the phenomena involved. Heterosis does indeed comprise a diverse array of phenomena, including various vegetative and reproductive features, ranging from sheer size or weight to adaptively important traits.

Dobzhansky (1952) has classified these diverse phenomena in the following manner: (1) Hybrid luxuriance; the hybrids are larger or faster growing but are not necessarily superior adaptively to the smaller parental types. (2) Euheterosis or adaptive heterosis; the hybrids have superior adaptive fitness. (2a) Mutational euheterosis; deleterious recessive alleles are covered up by dominant wild-type alleles in hybrids. (2b) Balanced euheterosis; heterozygotes are adaptively superior to either homozygote.

Gustafsson (1951) proposed a similar classification which is especially applicable to the situation in plants: (1) Somatic heterosis, in which the vegetative system is affected. (2) Reproductive heterosis, involving the reproductive capacity of the hybrid. (3) Adaptive heterosis, in which the hybrid genotype produces a greater than average number of surviving new individuals. Somatic and reproductive heterosis in Gustafsson's classification correspond to hybrid vigor in the traditional sense.

A great difference exists between heterosis in the strict sense, or hybrid vigor, and adaptive heterosis or balanced euheterosis. Heterosis in the original sense of the term is a visible but ephemeral superiority of heterozygotes as compared with the corresponding homozygotes. The heterosis concept has been extended by Dobzhansky, Gustafsson, and others to include the population-genetics phenomenon of heterozygote advantage in adaptive fitness. Heterozygote superiority in this respect is real but usually invisible and often persists from generation to generation.

The diversity of heterotic phenomena has another dimension. Heterosis occurs in plants with different breeding systems, as we have already seen. Many differences in the genetic system exist between an outcrossing and highly heterozygous plant group, like maize, and an inbreeding and fairly homozygous group, like the cultivated tomato. Heterosis in the contrasting plant groups is not likely to be the same in terms of its underlying genetic mechanisms.

Students of heterosis have been divided between those advocating one particular hypothesis more or less exclusively and those who regard the different hypotheses as complementary. The wide variety of phenomena subsumed under the common term of heterosis suggests that the latter group of workers is on the right track. It is unlikely that any single genetic mechanism is responsible for all types of heterotic expression. The case has been well stated by Allard (1960, p. 227):

> In most situations, the dominance and overdominance hypotheses lead to exactly the same expectations. . . . Despite extensive experimentation over more than forty years, conclusive evidence favoring one or the other of these hypotheses has not been forthcoming. There are, in fact, no good reasons for believing that both systems cannot operate simultaneously in producing heterotic effects. If these two hypotheses are not mutually exclusive, neither do they together exhaust all of the possibilities. There is every reason to believe that the genes concerned with heterotic effects may be fully as complex in their interactions as the qualitative genes of classical genetics. If this is indeed true, then we must expect all sorts of complex interactions in heterosis and must entertain the possibility that both hypotheses, at least in their simplest form, are gross oversimplifications of the actual situation.

The problem of heterosis, considered in this light, becomes one of estimating the relative roles of the different genetic mechanisms in a series of cases.

The evidence for single-gene heterosis presented earlier is critical evidence in favor of the overdominance hypothesis. This evidence was not available when the hypothesis was first proposed in 1908. Later it was resisted by some students. But, as the evidence for single-gene heterosis has grown in quantity and in quality, it can no longer be ignored. Single gene heterosis, or allele complementation, is almost certainly a component mechanism in many of the more complex cases of heterosis, as Redei (1962), Grant (1964), Stubbe (1966), and others have argued.

In autogamous plants, heterosis occurs in two rather different situations: as a continuing feature in natural populations, and as a sporadic incident in artificial crosses. Different genetic mechanisms may often be involved in the two situations.

Allard and his co-workers have discovered adaptive heterosis in natural populations of *Avena fatua* and some other autogamous species. The heterozygotes for certain genes leave more offspring than the homozygotes. Consequently, the heterozygous allele combination persists from generation to generation in the population despite self-pollination (Imam and Allard, 1965; Allard, Jain and Workman, 1968).

Heterosis appearing in artificial hybrids between different strains of autogamous plants may be due to two or three genetic mechanisms. Allele complementation is a definite possibility, and is the actual cause in the verified cases of single-gene heterosis, but it cannot be automatically assumed. Favorable complementary gene interaction is also a definite possibility here. At least two instances of heterosis in hybrids of inbred tomatoes are attributed to complementary-gene interaction (Powers, 1944; Burdick, 1954). Complementary factors are also implicated in heterosis in the autogamous *Phaseolus vulgaris* (Malinowski, 1955).

Dominance is not likely to play an important role in heterosis in autogamous plants, as several authors have pointed out. Deleterious recessive alleles arising by mutation in the autogamous populations are quickly put into homozygous condition by the breeding system, where they are exposed to selection and can be eliminated. By the same token, the dominance factor may contribute importantly to heterotic expressions in outcrossing plants, whose populations do carry hidden deleterious genes.

The various theories of heterosis converge in the case of corn. Overdominance contributes to heterosis here (Brieger, 1950; Crumpacker, 1966). Dominance is also a factor in this outcrossing plant. A statistical analysis indicates that dominance is more important than overdominance as a source of heterosis in corn (Moll, Lindsey and Robinson, 1964). Introgressive segments of Tripsacum and teosinte in corn chromosomes contribute to heterosis (Sehgal, 1963). Finally, the cytoplasmic factor of mitochondrial complementation plays a role in heterosis in corn (Sarkissian and Srivastava, 1967).

HOMEOSTASIS   Organisms generally exhibit some degree of stability or constancy in the face of changing environmental conditions. An obvious example is the nearly constant body temperature of warm-blooded animals over a wide range of air temperatures. This buffering of the individual against fluctuating external conditions is called homeostasis (Cannon, 1932). More exactly, it is physiological homeostasis (Cannon, 1932) or developmental homeostasis (Lerner, 1954). The two forms are closely related, differing as indicated by the self-explanatory terminology, and they can be discussed together. Developmental homeostasis is also closely related to developmental canalization (see Chapter 3). A third form, genetic homeostasis, is a population phenomenon which falls outside the scope of the present discussion (but see Lerner, 1954).

Much evidence shows that physiological and developmental homeostasis are associated with heterozygosity in Drosophila, mice, chickens, and other organisms. This evidence consists of comparisons between related heterozygotes and homozygotes with respect to the degree of constancy of phenotypic characters under varying environmental conditions (Lerner, 1954; Falconer, 1960). The most conclusive evidence is derived from studies of animals, and we will consider this briefly before turning to the situation in plants.

In *Drosophila melanogaster* the number of sternopleural bristles sometimes differs markedly between the two sides of an individual fly. Furthermore, the difference between opposite sides varies with genotype and culture conditions. Mather (1953) compared the difference in bristle number on opposite sides of individual flies in two inbred lines and in their F_1 and F_2 hybrids. The difference between sides was found to be more variable in the inbred lines than in their hybrids. The more highly heterozygous types thus show greater stability in the development of this quantitative characteristic (Mather, 1953; but see also Thoday, 1958a).

Similarly, in the house mouse the variation in time of vaginal opening is correlated with the degree of homozygosity. The variation in this character is high in strains that have long been inbred. It becomes low in their F_1, B_1, and B_2 hybrids. The variation then becomes high again in the fairly homozygous B_{11} and B_{22} generations (Yoon, 1955).

The homeostasis of heterozygotes affects not only various morphological characters which may or may not enhance the fitness of the individuals manifesting them, but also physiological characters related to fitness. In *Drosophila pseudoobscura, D. persimilis, D. prosaltans,* and *D. melanogaster,* the rates of survival of flies that were heterozygous or homozygous for particular chromosomes were determined in replicated crowded cultures. Environmental conditions varied somewhat from culture to culture. The homozygotes showed significantly different survival rates in the different cultures, whereas the heterozygotes were generally uniform among the replicates. The viability of the homozygous flies, in other words, was modified by environmental conditions to a much greater extent than was the viability of the corresponding heterozygotes (Dobzhansky and Wallace, 1953; Dobzhansky and Levene, 1955).

Apparently in Drosophila some physiological processes related to viability are better buffered against environmental fluctuations in heterozygous than in homozygous flies. This is shown by comparisons of the inversion heterozygote ST/CH and the homozygous types ST/ST and CH/CH in *Drosophila pseudoobscura.* The inversion heterozygote is heterozygous not only for the inverted chromosome segments themselves, but also for blocks of genes affecting viability and fitness that are borne on these segments. Egg-laying rate, larval viability, growth rate, and adult longevity all showed greater uniformity in response to environmental diversity in heterozygous flies ST/CH than in the homozygous types ST/ST and CH/CH (Moos, 1955).

Individuals that possess internal self-regulating mechanisms buffering them against changing environmental conditions can perform successfully in a variety of environments. Less homeostatic individuals are likely to perform well only in a narrow range of conditions. The homeostatic types will thus exhibit a higher average fitness in various environments and hence a higher degree of adaptive heterosis than their less homeostatic relatives. Adaptive heterosis is associated with homeostasis, and both phenomena are based on heterozygosity. These correlations suggest that adaptive heterosis may be due to the superior homeostatic properties of some heterozygous combinations of alleles (Dobzhansky and Wallace, 1953; Lerner, 1954).

Lerner (1954) and Haldane (1954a, 1954b) have suggested that the property of a heterozygote providing the basis for both its homeostasis and its heterosis could be its relatively great biochemical complexity compared with the homozygotes. The two identical alleles of any gene in a homozygote may produce a single type of enzyme. The dissimilar alleles of the same gene in a heterozygote may determine the formation of two slightly different enzymes. Whereas a biochemically homogeneous enzyme may have a sharp optimum for pH or temperature, a pair of related enzymes is likely to be active over a broader zone of conditions. A mixture of enzymes resulting from heterozygosity would, therefore, be advantageous in an organism exposed to a wide range of environmental conditions (Haldane, 1954a, 1954b; Lerner, 1954). Single-gene heterosis thus provides a partial explanation for homeostasis as well as for general heterosis.

HOMEOSTASIS IN PLANTS In higher animals, physiological and developmental homeostasis play important roles in the self-regulation of individual organisms and in the maintenance of populations. The situation in higher plants is much less clear-cut. Evidence of homeostasis has been found in some plants. In other cases homeostasis has been sought but not found.

One of the first reported cases of homeostasis in a higher plant was described by Mather (1950) in *Primula sinensis,* the primrose. Mather measured the variation in style length between flowers of the same plant in long inbred lines of the primrose and in their F_1 hybrids. The phenotypic variation due to differences in the development of separate homologous organs on an individual plant turned out to be much less in the heterozygotes than in the related homozygotes produced by inbreeding. As shown in Table 12, the variance of style lengths on a single plant was consistently and significantly lower for the F_1s and F_2s than for the inbreds. Evidently the heterozygous genotypes are more stable than the homozygotes in the development of styles.

A lower variance and hence a greater phenotypic stability in heterozygotes than in the corresponding homozygotes have been found also for corolla length in petunias (Mather, 1949), rate of growth in tomatoes (Mertens, Burdick and Gomes, 1956), plant height and ear weight in maize (Adams and Shank, 1959), and number of chiasmata in rye (Rees and Thompson, 1956).

Hiesey (1953b) measured the growth and development of clones of five species and five hybrids of Poa under different environmental conditions. Subdivisions of each clone were grown in nine different combinations of day and night temperatures. The dry weights of the plants were recorded at the end of the experiment. Hiesey found that F_1 hybrids of *Poa ampla* (Washington) $\times$ *Poa pratensis* (Sierra Nevada) have a wide tolerance range for temperature, a range wider than that of the parental species, so that they produced good growth in different environments. Conversely, F_1 hybrids of *Poa scabrella* $\times$ *P. ampla* have a rather narrow tolerance range (Hiesey, 1953b).

Levin and co-workers used gel electrophoresis to assay populations of the permanent translocation heterozygote *Oenothera biennis* in Connecticut. An open field encompassing a variety of habitats was found to be occupied by sister plants with an identical heterozygous genotype. This hybrid genotype is apparently well adapted to a wide range of environmental conditions in its area (Levin, Howland and Steiner, 1972).

A case in which heterozygosity does not enhance homeostasis occurs in the cotton *Gossypium hirsutum*. Nine homozygous lines and their F_1 hybrids were measured for five quantitative characters. No clear distinction could be found between heterozygotes and homozygotes in plant-to-plant phenotypic variability (Kohel, 1969).

TABLE 12

VARIATION IN STYLE LENGTH BETWEEN FLOWERS OF THE SAME PLANT IN
PRIMULA SINENSIS (MATHER, 1950)

Year	Variance for style length in:		
	Inbreds	F_1s	F_2s
1946	0.36	0.12	0.29
1947	0.60	—	0.18
1948	0.41	0.21	—

Likewise in *Nicotiana rustica* some inbred lines and F_1 hybrids do not differ significantly in environmental variance (Jinks and Mather, 1955). The allotetraploid species *Gossypium hirsutum*, which contains internal hybridity, is not more stable phenotypically in different experimental environments than the related diploid, *G. arboreum* (Quisenberry and Kohel, 1971).

The evidence in plants is not as extensive or conclusive as could be desired. Certainly physiological and developmental homeostasis occur in higher plants. Berg (1959) pointed out that floral parts are less variable than vegetative parts in 19 plant species. And the pollen deposition mechanism of flowers pollinated by specific insects is the most stable feature of all. Selection has apparently favored a high degree of developmental homeostasis in plant organs which require precision for their proper functioning (Berg, 1959).

But the homeostasis is not always connected with heterozygosity in plants. The two conditions are associated in some outcrossing plants. In some self-pollinating plants like *Gossypium hirsutum* and *Nicotiana rustica*, on the other hand, heterozygosity does not enhance homeostasis, as Kohel (1969) and others have pointed out.

One may ask, finally, whether homeostasis is as important for plants as it is for animals in any case. Plants, with their open system of growth, often respond to changes in environmental conditions by dropping one set of leaves or other organs and initiating a new and phenotypically different set. Plants adjust to their external environment by phenotypic stability in some cases and by phenotypic modifiability in others.

INBREEDING DEPRESSION   Inbreeding depression is the reverse aspect of hybrid vigor. In many organisms, both plants and animals, the progeny derived from inbreeding show a decrease in vigor and fecundity compared with the progeny of normal outcrosses. The decline in vigor often continues through successive generations of inbreeding and may end finally in the extinction of the inbred line.

Many studies have been made of inbreeding depression in *Zea mays,* one of which will serve for purposes of illustration. As shown in Table 13, two low-yielding inbred lines of maize produce a heterotic F_1 hybrid. The yield then declines rapidly in the early inbred generations derived from this hybrid, levelling off at a low figure in the I_4 generation (Jones and Mangelsdorf, 1925).

The effects of inbreeding have also been studied extensively in rye (*Secale cereale*). This outcrossing and incompletely self-incompatible species can be selfed. The curves for the performance of the inbreds then show a decline in plant height and weight through six generations of inbreeding (Jain, 1960). Inbreeding effects in rye also include reduced pollen and seed fertility and various meiotic abnormalities such as low chiasma frequency, sticky chromosomes, and poor pairing (Lamm, 1936; Müntzing and Akdik, 1948; Putt, 1954).

The manifestation of inbreeding depression is closely correlated with the breeding system in plants. Outcrossers generally show much inbreeding depression. Autogamous plants, at the opposite extreme, show none. The many species with intermediate breeding systems, permitting a mixture of natural outcrossing and natural selfing, have intermediate degrees of tolerance of forced inbreeding.

This correlation favors the classical idea that inbreeding depression is a result mainly of the exposure to expression of deleterious recessive alleles. Such alleles occur as hidden recessives in the normally outcrossing population, but they are put into homozygous condition by inbreeding and are then expressed phenotypically. In autogamous plants, as a general rule and apart from some special exceptions, the deleterious recessives have been eliminated from the gene pool of the population by previous generations of selection. The intermediate situation in a facultative outcrosser-inbreeder pre-

TABLE 13

YIELD IN TWO INBRED LINES OF MAIZE, THEIR F_1 HYBRID, AND INBRED DE-
RIVATIVES OF THE HYBRID (JONES AND MANGELSDORF, 1925, AFTER HIORTH,
1963)

Generation	Bushels per acre	Generation	Bushels per acre
P_1 (inbred line)	20	I_3	44
P_2 (inbred line)	20	I_4	23
F_1	101	I_5	27
I_1 (F_2 generation)	69	I_6	25
I_2	43	I_7	27

sumably involves an intermediate quantity of hidden deleterious recessives.

An example of the intermediate situation is provided by *Gilia achilleaefolia* (Polemoniaceae). This annual herb of California is self-compatible. Many local races are predominantly outcrossing, being cross-pollinated by various bees, but engage in some self-pollination, especially at the end of the flowering season. Another series of local races is predominantly autogamous. The inbred progeny of the autogamous races are fully vigorous, as expected (Grant, 1954; Grant and Grant, 1965).

The I_1 progeny derived from artificially selfed plants of a predominantly outcrossing race, on the other hand, consisted largely of weak or inviable plants, some of which died in the seedling stage. A small proportion of the I_1 plants had normal vigor. Some of these were self-pollinated to continue the inbred lines. The I_2 and I_3 progeny derived from the exceptional normal I_1 plants also included some inviable and semiviable individuals, but in relatively low frequencies (Grant, 1954; Grant and Grant, 1965).

Progeny tests of open-pollinated plants in a natural population of the same race reveal the presence of a few weak individuals intermixed with a majority of normal, vigorous individuals. Apparently this population carries a moderate number of deleterious recessive alleles, which usually remain hidden but are sometimes exposed by self-pollination; this load of deleterious recessives can be tolerated by the population under normal conditions of predominant but not exclusive outcrossing (Grant, 1954; Grant and Grant, 1965).

The degree of inbreeding depression and the tolerance of inbreeding vary widely in different species of outcrossing plants. Among cultivated plants, alfalfa and carrots are very intolerant of inbreeding, maize is fairly intolerant though not as much so as alfalfa, while onions and rye are fairly tolerant of inbreeding (see Allard, 1960).

PHENODEVIANTS Cross-fertilizing organisms have evolved genotypes which produce developmental patterns with much self-regulation so that relatively uniform phenotypes arise. The uniform phenotypes can arise despite underlying genetic variability; this is developmental homeostasis. The uniform phenotypes can arise and persist in spite of environmental fluctuations; this is developmental and physiological homeostasis. The homeostasis of cross-fertilizing organisms is based in a large measure on the superior buffering properties of heterozygotes, as we saw in an earlier section (Lerner, 1954).

Loss of heterozygosity in such organisms reduces this buffering, with various results. One result is an increased environmental variance, as noted earlier. A second result is reduced viability, as described in the section on inbreeding depression. A third result is the appearance, sporadically, of morphologically abnormal types, which Lerner (1954) refers to as phenodeviants.

Lerner (1954) lists a number of features of phenodeviants. A particular kind of abnormality, such as crooked toes in chickens or extra wing venation in Drosophila, is characteristic of any given type of phenodeviant. Phenodeviants of the same type appear on rare occasion in various populations or strains of a species, and the genetic potential for them is therefore widely distributed (Lerner, 1954).

The frequency of phenodeviants is increased in inbred lines and decreases in outcross progeny. Different nonallelic genes and gene complexes may give rise to the same phenodeviant trait. Moreover, the genes involved usually cannot be located readily on the genetic map. Inheritance of the phenodeviant character does not follow any simple Mendelian pattern (Lerner, 1954).

All of these features taken together support the hypothesis that phenodeviants are a special form of inbreeding depression. In species that segregate occasional phenodeviants, the normal phenotype requires a certain level of heterozygosity, involving perhaps many different genes. Inbreeding lowers the heterozygosity below the threshold for normalcy and upsets the homeostatic properties of the genotype. Loss of heterozygosity in any one of various parts of the genotype can then result in the same type of phenodeviant (Lerner, 1954).

Three of the best examples of phenodeviants are provided by Lerner's own study of crooked toes in chickens, Dubinin's study of extra wing venation in *Drosophila melanogaster,* and Goldschmidt's study of the podoptera type of wing in *D. melanogaster.* Other cases occur in Drosophila and rodents (see Lerner, 1954, for a review).

Dubinin's (1948) study of extra wing venation in *Drosophila melanogaster* revealed that aberrant individuals with extra veinlets occur in low frequencies (0.05–1.13%) in many populations throughout Eurasia. A substantial proportion (60/348) of normal females carry genes for the trait and give rise to phenodeviant progeny. Inbreeding of normal wild females from an Alma-Ata population brought out extra wing veinlets in 68% of the lines tested. Other populations of *D. melanogaster,* and other species including *D. obscura,* when tested in the same way also yielded phenodeviants with extra veinlets.

Phenodeviants, as opposed to the ordinary forms of inbreeding depression, have received little attention in plants. For this reason we will describe a case in the *Linanthus parviflorus* group investigated first by K. A. Grant (unpublished data, summary in Grant and Grant, 1965) and later by Huether (1968, 1969). The story is inconclusive but strongly suggestive of phenodeviants.

Linanthus parviflorus (Polemoniaceae) is a group of annual herbs of the California foothills. The plants are self-compatible but predominantly outcrossed by long-tongued flies. The I_1 and I_2 progeny of artificial self-pollinations mostly exhibit normal vigor. But among the inbred progeny of one population tested there were some individuals with abnormal flowers (Grant and Grant, 1965).

Normal flowers of *L. parviflorus* have a tubular corolla with five petal lobes, five stamens, and a long style with three stigma lobes. Floral abnormalities found in some individuals in the I_1 and I_2 generations included deviant numbers of petal lobes (six, seven, four, etc.), odd numbers of stamens, extra long or short styles, and four stigma lobes. The phenodeviant individuals appeared in low frequencies in both the I_1 and I_2 generations. The abnormal floral characters showed variable expressivity, being expressed most completely at high temperatures (K. Grant, unpublished data; Grant and Grant, 1965).

The genetic condition underlying the floral abnormalities is widely distributed throughout the species, as shown by the occasional appearance of abnormal flowers in various natural populations. This conclusion is confirmed by the segregation of a few phenodeviants in progeny of artificial outcrosses between different normal parent plants belonging to the one natural population tested experimentally (K. Grant, unpublished data). These findings have been extended to the related species, *L. androsaceus,* and quantified by Huether (1968, 1969).

We have evidence, then, for the existence of a widespread genetic condition for abnormal flower development in the outcrossing members of the *Linanthus parviflorus* group. The abnormal phenotypic expression is enhanced by inbreeding, when the normal heterozygous buffering breaks down partially, and is enhanced still more by stressful environmental factors in the inbred lines.

PART II : GENE SYSTEMS

GENE INTERACTION

INTRODUCTION   Classical genetics was focused primarily on the entities which the early workers called simply "genes" and which we now propose to call Mendelian genes. The continued investigation of inheritance soon revealed, however, that the genotype is not composed exclusively of these Mendelian genes. In the early modern period, some additional classes of genetic elements were inferred from the evidence of formal genetics, quantitative genetics, and cytogenetics.

Our first task in this chapter is to recognize the main classes of genetic elements that we have to deal with. We discuss this problem and suggest a useful classification in the next section. We can then go on to consider the main types of interaction between these classes of genetic units. The modes of gene interaction are discussed in a general way in the third section. Some important modes are introduced here and taken up more fully in later chapters. Others are described in the present chapter.

Purely terminological problems will continue to occupy a part of our attention in this chapter and those that follow. The terminology is now in such a state of confusion that it hinders about as much as it facilitates an understanding of the underlying genetic phenomena.

Not only must we search for the principles of gene interaction, therefore, but also we must select, somewhat arbitrarily, a suitable set of labeling terms.

THE GENETIC ELEMENTS   The progress of genetics has brought in its train a great enrichment of our vocabulary. With reference to the types of genes we have such terms as major gene, Mendelian gene, inhibitor gene, oppositional gene, modifier gene, multiple gene, oligogene, polygene, cistron, structural gene, architectural gene, regulator gene, controlling gene, and plasmagene. Some of these entities are covered by overlapping terms like heterochromatin, repeat, and redundant DNA.

It is of interest to note in passing how this semantic multiplicity developed from several sources. In the first place, the progress of genetic research has yielded evidence for several real classes of genetic elements. Second, the same genetic elements may be studied by various methods, such as the classical and the biochemical, which reveal different aspects; then we often get different names applied to aspects of the same genetic unit. Third, a terminology derived from studies of, say, bacteria is sometimes extended to higher multicellular organisms for which it may or may not be suited.

We can reduce the confusion to some degree of order by recognizing four main classes of genetic elements on the basis of their nature, size, and position in the cell. The classes are intergrading or overlapping in certain cases, but they nevertheless provide a convenient grouping for our purposes in this book. The classes are listed below under the terms which the author regards as preferable. Near synonyms, partial synonyms, and names of related entities are given in parentheses. Some of these synonymous terms are also useful in the proper context.

1. Mendelian gene (classical gene; major gene; modifier gene, in part; multiple factor; cistron; structural gene)
2. Polygene (modifier gene, in part; redundant DNA; repetitive DNA, in part)
3. Controlling element (heterochromatin block, in part)
4. Cytoplasmic gene (plasmagene, organelle gene)

Mendelian genes are the "genes" of classical genetics. Their mutant alleles produce marked phenotypic effects. Heterozygotes for individual Mendelian genes segregate into more or less clear-cut phenotypic classes. They are discussed in Chapters 1, 2, 4, and 7 to 10. Polygenes, taken individually, have slight phenotypic effects, but they normally exist in sets, and an organism segregating for a set of them shows quantitative inheritance (see Chapter 11). Controlling elements are genic entities with regulatory functions in the chromosomes. They do not have fixed positions in the chromosomes, but become transposed to different sites (see Chapter 8). Cytoplasmic genes include plastids, mitochondria, and other sorts of self-reproducing particles in the cytoplasm. They bring about maternal inheritance (see Chapter 12).

The chromosomal genes can be subdivided further according to their direction of action. Many chromosomal genes have what can be called a positive action. Their primary product is an enzyme or other protein which plays a positive role in biosynthesis and development. Our discussion of gene action in Chapter 2 dealt with genes of this sort. Other chromosomal genes, known collectively as controlling genes, have a regulatory function. They regulate positive chromosomal genes by suppressing or releasing their action. A partial synonymy of controlling genes is: inhibitor gene, suppressor gene, minus modifier, minus polygene, regulator gene, controlling element, heterochromatin block.

Several of these types of controlling genes are Mendelian genes or polygenes with a regulatory function. As such they do not form a separate class of genetic elements. Certain other types of controlling genes, particularly the so-called controlling elements, seem to form a truly distinct class of genetic elements.

MODES OF GENE INTERACTION The simplified classification of genetic elements just given serves as a convenient starting point for a discussion of gene interaction. Different modes of interaction result from different combinations of the genetic elements. We will merely list the various modes here as a prelude to discussions in this and following chapters.

1. Two or more Mendelian genes controlling successive steps in a metabolic or developmental sequence, forming a so-called serial gene system (Chapter 9)

2. Two or more Mendelian genes with similar and usually cumulative actions, forming a multiple gene system (Chapter 10)

3. Numerous polygenes, forming a polygenic system (Chapter 11)

4. A combination of a Mendelian or major gene with modifiers, forming a modifier system (Chapter 7)

5. A combination of controlling genes with positive chromosomal genes, forming an oppositional gene system (Chapters 8–10)

6. Combinations of chromosomal and cytoplasmic genes (Chapter 12)

GENE NUMBER   The number of Mendelian or classical genes in the genotype of a multicellular organism can be estimated by several methods, all of which contain uncertain assumptions. Perhaps the most reliable of these methods is to count the number of chromomeres (which are visible), correlate them with genes in some numerical ratio, and derive an estimate of the total number of gene loci.

The best-analyzed organism from this standpoint is *Drosophila melanogaster*. This fly has about 5000 bands of varying size in the whole complement of four salivary chromosomes. One region in the X chromosome contains 4 known genes per large band (Muller and Prokofyeva, 1935). Another region of the X chromosome has 12 known genes in 12 bands (Shannon, Kaufman and Judd, 1970).

The estimated number of gene loci depends on one's assumption concerning the gene content of an average band. Drosophila geneticists have assumed average figures of one or slightly more Mendelian genes per band. This figure leads to suggestions of 5,000–10,000 genes in *D. melanogaster* (Muller and Prokofyeva, 1935), 5,000–15,000 genes (Stern, 1960), and about 6,000 genes (Crow, 1972).

A second method for estimating gene number utilizes mutation rates. Extrapolation from the average mutation rate for single genes to the collective mutation rate for the genotype as a whole suggests the total number of genes. Estimates obtained by this method in Drosophila are of the same order as estimates based on salivary-band counts.

Similar estimates, namely, about 10,000 Mendelian genes, have been given for the genotypes of mammals including man (Stern, 1960; Frota-Pessoa, 1961). The evidence is less direct here than in Drosophila.

Belling (1928) counted about 2,200 chromomeres in the haploid chromosome complement of *Lilium pardalinum,* about 1600 in *Agapanthus umbellatus,* and over 1400 in *Aloe striata.* Belling himself argued that the chromomeres are genes (1928). Our present knowledge of the size relations between chromomeres and genes rules out this older assumption; each chromomere might contain several or many genes (see Chapter 4). Therefore the number of gene loci in Lilium, Agapanthus, and Aloe could be of the order of several thousand.

Higher plants have a much simpler body organization than higher animals. It is therefore quite possible that plants also have a simpler genotype containing fewer Mendelian genes. An alternative possibility is that the difference in complexity of body organization between the two kingdoms is due primarily to disparities at the developmental level rather than at the genic level of control. We have no way of deciding between these two possibilities at present. My own hunch is that the number of Mendelian genes has the same order of magnitude in diploid plants and in insects, but is substantially smaller in plants, being perhaps one-half the number in the insects.

Whatever the correct figure for gene number, we can readily see that interaction between genes is inevitable and complex, and must play an important role in the formation of phenotypic characteristics.

MODIFIER GENES The action and expression of a Mendelian gene is influenced by other genes of the complement. Such a gene combination is referred to as a system of major genes and modifiers.

The existence of modifier genes was first demonstrated in *Drosophila melanogaster* by Bridges (1919) and Sturtevant (1918). Bridges found eight genic factors modifying the mutant gene for eosine eye color; seven of these factors diluted the color and one darkened it. Sturtevant found plus and minus modifiers for bristle number in the bristle character dichaete.

The effects of modifiers on a Mendelian or major gene are

well illustrated by the classic case of dichaete in *Drosophila melano-gaster*. Normal flies have four bristles in each group on the back of the thorax, whereas the dichaete mutants possess fewer bristles per group. The difference between the normal and the reduced number of bristles is due primarily to allelic differences in a single gene, D, the normal flies being homozygous recessive (d/d) and the dichaete flies heterozygous for a dominant allele (D/d). The genotype D/D is lethal.

Among a series of dichaete flies with the uniform constitution D/d, all of which were reared in a uniform environment, the average number of bristles varied significantly in different lines. Thus some D/d flies had two bristles, and other lines with the same constitution for the D gene had one, zero, three, or even four bristles. It has been shown that the D/d genotypes with different bristle numbers also differ with respect to other genes of the complement, which have relatively slight effects on the bristle character and do not modify the expression of this character at all in normal d/d flies. These modifier genes do, however, affect the expression of the D allele in the D/d flies (Sturtevant, 1918).

Several modifier genes affect the dichaete character. Some of these modifiers enhance the expression of the D allele; other modifier alleles act to suppress its effects. The enhancers, or plus modifiers, account for the phenotypes of D/d with extremely reduced numbers of bristles, whereas the minus modifiers lead to phenotypes that are normal or nearly so in D/d flies. The number of bristles in dichaete flies is controlled, therefore, not by the major gene D alone but by the combination of this gene with a system of modifiers (Sturtevant, 1918).

The dominant allele D for dichaete bristles in Drosophila has incomplete penetrance, in that not all the individuals carrying this allele exhibit a mutant phenotype, and this incomplete penetrance is due to the effects of the modifier genes. The latter, therefore, are called penetrance modifiers. Modifier genes have been found corresponding to other aspects of major gene expression. In the next section we will examine dominance modifiers in Gossypium, the cottons.

What are modifier genes in terms of the types of genetic elements presented earlier? Are they polygenes? Or are they Mendelian genes with important primary functions of their own and secondary, pleiotropic side effects on other Mendelian genes? In certain cases it has been possible to identify modifier genes as Mendelian genes, for example, in Gossypium. In other cases modifiers may be polygenes with slight individual effects. There is not sufficient evidence in most studies of modifier genes to distinguish between these two possibilities.

MODIFIER GENES IN GOSSYPIUM The concept of modifying genes was taken up by Harland in a series of studies from 1929 to 1936 to explain some of the variations observed in F_2 backcross generations derived from hybrids in cotton. The two closely related and interfertile species of New World cotton, *Gossypium hirsutum* and *G. barbadense*, possess a number of homologous genes in common. In certain cases identical alleles of the same genes are found in both species. Thus the gene S controlling petal spot is represented by the dominant allele S for spot and the recessive allele s for spotless in the two species. Similarly the dominant and recessive alleles, R and r, which cause the plant body to become red or green, respectively, are present in both species of cotton (Harland, 1936). [The modern nomenclature for these genes is R_2 ($= S$) and R_1 ($= R$ sensu Harland).]

These and many other character differences segregate as single-gene differences in crosses within either species. For example, if a form of *G. barbadense* with purple spots at the base of the petals is crossed with a spotless form of the same species, the F_1 hybrid is spotted, and the F_2 segregates into two clearly distinguishable phenotypic classes, spotted and spotless, in a 3 : 1 ratio. Similar results are obtained when spotted and spotless forms of *G. hirsutum* are hybridized (Harland, 1936).

But the same character differences, which segregate in a simple Mendelian ratio in the F_2 generations derived from intraspecific crosses, exhibit continuous variation in the progeny of interspecific crosses. Thus if a spotted plant of *G. barbadense* is crossed with a spotless plant of *G. hirsutum,* the F_1 is intermediate in the intensity of the petal spots, and the F_2 generation contains 22 degrees of spotting, varying by minute stages from complete absence of spots to the presence of spots more intense than those of *G. barbadense*. The same graded series of petal spots is found in the F_3 generation (Harland, 1936).

Other character differences, such as the red or green color of the plant body, and the crinkled or even margins of the leaf, exhibit the same contrasting behavior in intraspecific and interspecific crosses.

Since the alleles of the major genes concerned with petal spots, plant color, leaf margin, and other characters are the same in *G. barbadense* and *G. hirsutum,* the continuous variation in the F_2 generation of interspecific crosses is probably a result of allelic differences in other parts of the two genotypes. Harland suggested that the simple and clear-cut segregation of a character difference in an intraspecific cross is due to the fact that the rest of the genotype provides a relatively constant genetic background against which the dominant and recessive phases of a major gene can express themselves clearly. By contrast, the genotypes of the two species differ in many other modifying genes besides the one major gene under consideration, and the segregation of these modifiers in the interspecific crosses produces many different genetic backgrounds upon which the major gene must manifest itself. The continuous variation in the F_2 and F_3 generations of the interspecific crosses results from segregation of the different alleles of numerous modifier genes affecting the expression of the dominant and recessive alleles of the major genes. The species of cotton thus differ in their assemblages of modifier genes (Harland, 1936).

This hypothesis can be tested by transferring an allele of a major gene, controlling a particular character, from one species to the other by means of repeated backcrossing. If the continuous variation in this character in the interspecific crosses is due, as postulated, to the influence of segregating modifier genes, the inheritance of the character should cease to be multifactorial and should become simple Mendelian as soon as the genetic background is made uniform. The genetic background can be made uniform by backcrossing the hybrids repeatedly for several generations to one parental species. For practical reasons the parent carrying the dominant allele of the major gene under investigation is used as the recurrent parent (Harland, 1936).

The hybrid of spotted *barbadense* × spotless *hirsutum* was back-crossed generation after generation to the spotted *barbadense* parent. In this way the recessive allele for spotless (*s*), derived originally from *G. hirsutum,* was transferred to the genetic background of *G. barbadense.* After several generations of backcrossing, the variation in the degree of spottedness became discontinuous. Heterozygous individuals in later backcross generations, on selfing, yielded progeny that segregated into distinct phenotypic classes, normal spots versus weak spots, in a simple 3 : 1 ratio (Harland, 1936). These and similar results obtained for other character differences between the two species of cotton confirm Harland's hypothesis that the two species differ with respect to many modifier genes.

The *s* alleles of *G. hirsutum* in the genetic background of *G. barbadense* are associated with weak petal spots instead of no spots. This fact suggests that the modifier system of *G. barbadense* enhances the gene for petal spot. When, however, the reciprocal backcross is made between the hybrid and *G. hirsutum,* the spotting becomes gradually weaker in successive generations and eventually disappears. This result indicates that the modifier complex of *G. hirsutum* tends to suppress the formation of petal spots (Harland, 1936).

Several other genes give parallel differences of expression in the two genetic backgrounds. Table 14 lists five allele pairs which occur normally in *G. barbadense* and gives their normal character expressions in the *barbadense* genetic background. The dominant alleles of these genes were transferred to *G. hirsutum* by repeated backcrossing. In each case the dominant allele of *G. barbadense* has a weaker expression when in a *hirsutum* genetic background (Table 14). In general, therefore, *G. barbadense* carries a series of plus modifiers, and *G. hirsutum* a series of minus modifiers with suppressing effects on the major genes (Harland, 1936).

Furthermore, since the dominant member of an allele pair exhibits complete dominance only in certain genetic backgrounds and not in others, and since this dominance or nondominance is due to the modifier genes, the latter can be regarded as dominance modifiers.

One of the modifiers of the gene *S* for petal spot in *Gossypium barbadense* turned out, on further analysis, to be another Mendelian

TABLE 14
EXPRESSION OF GENES OF *GOSSYPIUM BARBADENSE* IN THE GENETIC BACK-
GROUNDS OF *G. BARBADENSE* AND *G. HIRSUTUM* (HARLAND, 1936)

Gene	Character	Allele pair	Expression in barbadense	Expression in hirsutum
R	plant body color	R	red	weak red
		r	green	green
Y	corolla color	Y	yellow	pale yellow
		y	cream	near white
P	pollen color	P	strong yellow	pale yellow
		p	dark cream	near white
S	petal spot	S	strong spot	weak spot
		s	weak spot	no spot
K	lint color	K	khaki	pale khaki
		k	cream	white

gene *Y*. The primary function of this gene *Y* is to control the production of yellow pigment in the corolla. The dominant allele *S* produces much more intensely colored petal spots in combination with the allele *Y* than it does with *y* (Harland, 1936).

Later studies elucidated the influence of species-specific modifier systems on the expression of the gene *L* for leaf shape in both the New World and Old World cultivated cottons (Harland, 1936; Silow, 1939; Stephens, 1945a, 1945b). Some of the modifiers of leaf shape are Mendelian genes with other primary functions of their own. Thus one of the modifiers affecting leaf shape is the gene *H* for hairiness (Harland, 1936); other modifiers of leaf shape are genes controlling early or late flowering (Stephens, 1945b).

The linkage relations of the modifier genes in cotton form another interesting story in cotton genetics, which we will consider in Chapter 19.

MACRORECOMBINATIONS   An interesting aspect of gene interaction is the phenomenon of macrorecombination. Morphological characteristics unlike those found in either parent occasionally appear in the progeny of interspecific hybrids. Such novel forms have been called macrorecombinations (Grant, 1956a). The phenomenon itself has long been known, and has been reported in a wide variety of plant groups.

An early example of a macrorecombination is the Rhinanthoides form in Antirrhinum (Lotsy, 1916). The parental species, *Antirrhinum majus* and *A. glutinosum,* both have the typical snapdragon flower with a prominent two-lipped corolla limb. Some individuals in the F_2 generation of the interspecific cross had tubular corollas with reduced limbs, as in the related genus Rhinanthus. The Rhinanthoides type, which occurred sporadically in the F_2 generation, became frequent in some F_4 families selected for this trait and was predominant in other F_4 families (Lotsy, 1916).

There is now some doubt whether the Rhinanthoides type is a macrorecombination or a macromutation. Stubbe (1966) repeated the cross without obtaining any Rhinanthoides types. A similar morphological deviation in *Antirrhinum majus* is caused by the mutation Fistulata. It is possible that Rhinanthoides and Fistulata are identical. If so, the parental stock of *A. majus* in Lotsy's experiment could have carried the mutation Fistulata in heterozygous condition (Stubbe, 1966, pp. 243–44, 311).

The flowers of Digitalis (Scrophulariaceae) lack spurs. A complex cross between the spurless *D. purpurea* and a spurless amphiploid yielded numerous F_1 hybrids. Most of the F_1 plants were also spurless, but a few aberrant F_1 individuals had spurred flowers (Schwanitz, 1957).

A case in Gilia involves the number of parts in each whorl of the flowers. *Gilia achilleaefolia* and *G. millefoliata* have the normal number of 5 calyx lobes, 5 petal lobes, and 5 stamens that is found generally in the family Polemoniaceae. Their F_1 hybrid also has normal 5-merous flowers. The F_1 hybrids, though highly sterile themselves, gave rise spontaneously to fertile allotetraploid plants in F_2, which made it possible to continue the experiment in later generations. One F_3 family had a high proportion of plants with aberrant 4-merous and 7-merous flowers. Another F_3 family contained some 6-merous and 3-merous types. The proportion of individuals with abnormal flowers became greater in F_4 families derived from single-plant selections in F_3 (Grant, 1956a).

Some plants in the F_2 and later-generation progeny of *Gossypium anomalum* × *G. barbadense* have abnormal flowers in which the ovules are borne on the margins of the petals (Meyer, 1970). This is technically a gymnospermous condition, as Meyer points out. In the F_1 hybrid of *Geum montanum* × *G. macrophyllum* (Rosaceae), the sepals become transformed into leaves and the stamens into petals (Gajewski, 1953). The F_2 populations of *Lycopersicon esculentum* × *L. chilense* contain variants with entire leaves and compound inflorescences (Rick and Smith, 1953). Hybrids of *Layia platyglossa* × *Madia elegans* (Compositae) have pappus bristles on the achenes unlike either parent (Clausen and Hiesey, 1958).

The explanation of macrorecombinations is not hard to find. The combining of well-differentiated gene systems in interspecific hybridization upsets preexisting gene interactions and gives rise to novel gene expressions. The new gene interactions appear at the phenotypic level as radical and unpredictable departures from the normal morphological condition in either parental population.

GENIC BALANCE IN NICOTIANA Genes and gene systems are components of whole genotypes. The organism develops and functions as an integrated whole, and the genotype which determines this development and functional unity must also possess an internal integration. The several thousand genes constituting the genotype of a complex organism must be able to work together harmoniously if that organism is to be successful in life.

Evidence of coordination among the genes of a genotype is found repeatedly. By far the majority of the mutational changes in any genotype are changes for the worse, the mutant individuals usually being physiologically deficient, physically deformed, inviable, or sterile. The loss of a chromosome segment of moderate length, a long deletion, is apt to be lethal even in the heterozygous condition. Small deficiencies of chromosomal material, resulting from the deletion of a short segment, may or may not reduce viability in heterozygotes, but are nearly always lethal as homozygotes.

The wild diploid tobacco species, *Nicotiana sylvestris,* has 12 pairs of chromosomes in the normal complement. Goodspeed and Avery (1939, 1941) obtained a series of trisomic types in which one or another of the 12 chromosomes was represented in triplex condition. If the different individual chromosomes constituting the haploid set are designated A, B, C, D, . . . , L, one trisomic type would have an extra A chromosome, and would thus have the chromosomal constitution AAABBCCDD . . . LL. Trisomic type B would have the constitution AABBBCCDD . . . LL, and so on. All of the 12 possible trisomic types were produced, maintained as separate lines, and studied morphologically. For convenience they were designated by names referring to some conspicuous feature, such as Compact or Stubby (Goodspeed and Avery, 1939, 1941).

Each trisomic type deviates from the normal diploid condition in various morphological characters. Thus the average length of the corolla tube in diploid plants of *Nicotiana sylvestris* is 8.46 cm. The trisomic type Compact has corollas only 6.65 cm long, but the trisomic Enlarged has corollas 10.83 cm long. The various trisomics deviate from the norm in gross features of the growth form, the leaves, and the capsules as well as in minute details of the flowers. Two accurately measurable dimensions of the corolla, the length of the tube and the diameter of the petallike part or limb, are given for 10 trisomic types in Table 15. Some trisomic types differ strongly and others slightly from the diploid in the corolla characters measured, but all the trisomics differ phenotypically from the diploid in some degree (Goodspeed and Avery, 1939).

The table reveals another significant fact. The average length or breadth of the corolla for all trisomic types combined is close to the condition in the diploid plants. Not only do the different trisomic types deviate from the norm, therefore, but their individual deviations tend to balance out and give a statistical condition approaching the norm (Goodspeed and Avery, 1939).

The leaves show a similar pattern of deviation from the norm in the various trisomic types. The leaf characters, however, cannot be expressed by simple measurements. They are described in a quotation from Goodspeed and Avery (1939, p. 452), and illustrated in Figure 15:

> A character which cannot be subjected to statistical analysis but in which the diploid represents a balance between the variations occurring in trisomic types, concerns distinctions in growth rates of different parts of the leaf expressed in terms of contour and fullness at maturity. Thus the elliptic-ovate rosette

leaf of the diploid is relatively flat with the margin slightly erect or reflexed according to age and environmental conditions. In Enlarged, however, the midrib elongates so much more rapidly than the tissue of the leaf blade that the margin and the tips of the leaves are early turned under. In Pointed, on the other hand, it is not the midrib but the leaf-blade tissue bordering the midrib which apparently grows more rapidly so that the leaf blade becomes "fulled in" along the midrib. With Puckered, extra fullness of leaf tissue accumulates not along the midrib as in Pointed but in "puckers" between the veins. In Bent, "blisters" at the outer edge of the veins mark the position of tissue expanding more rapidly than the rest of the leaf. In Stubby, the extra fullness is in still a different location, the leaf tissue being "fulled in" along the veins, instead of along the midrib as in Pointed or between the veins as in Puckered. In this way one can follow the distribution of the control of the rate of development of different parts of the leaf to particular chromosomes of the complement. Many more such comparisons could be made but in all of them the diploid is the centre of the circle of variation.

TABLE 15

DIMENSIONS OF THE COROLLA IN DIPLOID AND TRISOMIC FORMS OF *NICOTIANA SYLVESTRIS* (GOODSPEED AND AVERY, 1939)

	Average corolla tube length, mm	*Average corolla limb diameter, mm*
Diploids	84.6	37.8
Trisomics		
Compact	66.5	35.3
Stubby	67.6	33.8
Pointed	71.0	33.2
Late	74.5	29.3
Inflated	75.2	40.4
Puckered	77.2	37.1
Narrow	81.1	35.1
Bent	88.1	39.9
Recurved	89.3	37.5
Enlarged	108.3	40.4
Average for all trisomic types combined	79.9	35.9

Figure 15. Leaves of diploid and five trisomic types of *Nicotiana sylvestris*. (*A*) Diploid. (*B*) Enlarged. (*C*) Pointed. (*D*) Puckered. (*E*) Bent. (*F*) Stubby. (From Goodspeed and Avery, 1939; photograph courtesy of the late T. H. Goodspeed.)

In short, the leaf form of diploid *N. sylvestris* is the result of a balance between the growth rates of the midrib, veins, and blade. This balance is upset in compensating directions in different trisomic types. Thus in Enlarged the midrib has a growth rate faster than normal, while in Pointed the blade has the faster growth rate (Fig. 15B,C).

CONCLUSIONS Gene interaction starts at the elementary functional level. Consideration of the modes of interaction at this level leads to the recognition of various types of gene systems: serial gene systems, multiple gene systems, modifier systems, polygenic systems, oppositional gene systems, and cytoplasmic gene systems. Gene interaction can also be considered at a level in the developmental process far removed from primary action. The phenotypic expression of some allele combinations is favorable for the life and reproduction of the organism, while that of other combinations is unfavorable. Gene interaction has adaptive aspects as well as functional aspects.

Timofeeff-Ressovsky (1934a, 1934b) compared the viability of two different mutant types of *Drosophila funebris* with the viability of the double mutant. The viabilities of the mutants were expressed as percentages of the viability of normal nonmutant flies under constant environmental conditions. The single-mutant type Abnormes has a moderately reduced viability of 89%, and the mutant Lozenge has 74% viability. But the interaction between the mutant alleles in the double-mutant type Abnormes-Lozenge reduces the viability to 59%.

Another comparison, with effects in the opposite direction, involves the mutants Miniature and Bobbed. The viabilities of Miniature and Bobbed genotypes are 69% and 85%, respectively. The viability of the allele combination Miniature-Bobbed, however, is almost normal (97%), and is thus significantly higher than that engendered by either mutant allele singly (Timofeeff-Ressovsky, 1934a, 1934b).

Whether or not gene interaction in any given case brings about synergistic effects depends on the relation between the phenotypic characteristics produced by the particular gene combination and the environment in which the phenotype lives. The luxuriance and adaptedness of a phenotype and of the genotype which determines that phenotype are in the last analysis a function of the organism-environment interrelationship. The forms of primary gene action and interaction are somewhat beside the point in regard to the luxuriance or adaptedness of the final phenotypic product. A synergistic interaction between genes, like a heterotic interaction between alleles, can arise as an outcome of different modes of gene action.

CHAPTER EIGHT

CONTROLLING GENES

INTRODUCTION Controlling genes form a highly varied assemblage of genetic units. The main types of controlling genes which can be recognized at present are inhibitor genes, minus modifiers, minus polygenes, regulator genes, controlling elements, and heterochromatin blocks. These entities belong to different classes of genetic elements. Nevertheless, they all have a regulatory function in common. They all block or depress the normal action of other chromosomal genes. The types of controlling genes listed above are overlapping in certain cases, and a better classification is needed.

Minus modifiers and minus polygenes are discussed in Chapters 7 and 11. This chapter introduces the other types of controlling genes.

INHIBITOR GENES Inhibitor genes are Mendelian genes which suppress the action and expression of other particular genes in the complement. Such genes were found in various plant groups in the early modern period of Mendelian genetics and were designated by the gene symbol *I*. The synonymous term suppressor gene and the symbol *S* are also used.

One of the earliest examples of an inhibitor gene was found in corn (East and Hayes, 1911; Emerson, 1912). Purple and red endosperm in corn is determined by a series of complementary factors—*C, R,* and others. An inhibitor gene *I* is also involved in the control of endosperm pigmentation. The recessive allele *i* of this gene permits the expression of the color genes. But the dominant allele *I*, carried by some varieties of popcorn, blocks the expression of *C* and *R*; consequently, genotypes containing *I* in combination with the color genes have white kernels (East and Hayes, 1911; Emerson, 1912).

The flowers of *Viola arvensis* usually lack spots, but spotted plants occur sporadically. Segregations in the progeny of the cross spotted × spotless indicate that *Viola arvensis* possesses two positive genes for spots, *S* and *K*, like the related *V. tricolor* which normally has spotted flowers, and in addition two inhibitor genes, *I* and *H,* which suppress spot formation. The spotless condition in *V. arvensis* is due to the overruling effect of these inhibitor genes (Clausen, 1926, 1951; Clausen and Hiesey, 1958).

Layia platyglossa (Compositae) has pappus bristles on the achenes which vary with respect to hairiness. A southern race of this California species has floccose hairs on the pappus. The floccosity is determined by the dominant allele of a pappus gene *F* which occurs in this race and in other races as well. The nonfloccose races carry in addition one or several inhibitors of pappus hair formation. There is one strong inhibitor gene (*I*) in a southern California race, while a central California race has a set of two or three weaker inhibitor genes (I_1–I_3) (Clausen, 1951; Clausen and Hiesey, 1958).

In *Potentilla glandulosa* (Rosaceae), inhibitor genes occur as components of the gene systems governing interracial differences in petal shape, petal color, and stem length (Clausen and Hiesey, 1958).

CONTROLLING ELEMENTS Controlling elements are a type of chromosomal gene found by McClintock in maize. Unlike ordinary genes, they do not have fixed positions in the chromosome complement, but may change their position by means of transpositions.

They function as inhibitors, modifiers, or mutators, and thus affect the action and phenotypic expression of other ordinary genes (McClintock, 1953, 1956, 1961).

The best-known controlling elements in maize are activator (Ac) and dissociation (Ds). Ac activates Ds, which in turn induces chromosome breaks and the formation of dicentric chromatid bridges (McClintock, 1953, 1956, 1961).

The gene dotted (Dt), discovered by Rhoades, induces mutability in the A_1 gene for anthocyanin in maize. The recessive allele of this gene, a_1, does not produce anthocyanin in the kernels or plant body, and is usually stable. But in the presence of Dt, a_1 mutates to A_1, which does produce anthocyanin. The phenotypic effect of the combination of Dt and a_1 is dots of deep anthocyanin pigment in the kernels and streaks of anthocyanin in the leaves and stems. The A_1 gene has a fixed position in chromosome 3. Dt has been found in chromosomes 9, 7, and 6, and thus seems to be subject to transpositions (Rhoades, 1941, 1945, 1954; Neuffer, 1955). Dt is similar to Ac in that it is transposable and induces mutability in another gene; Dt could therefore be regarded as another controlling element (Rhoades, 1954; McClintock, 1956).

Blocks of heterochromatin tend to depress the action of neighboring euchromatic genes and to give a variable phenotypic expression of these genes. This effect of heterochromatin was introduced in Chapter 3, and is discussed in this chapter under the heading of Variegated Position Effect. It has been suggested but not established that the controlling elements in maize are heterochromatin blocks (Rhoades, 1954; Brink, 1964).

REGULATOR GENES   Regulator genes are a type of controlling gene discovered in studies of enzyme synthesis in bacteria, particularly *Escherichia coli* (Jacob and Monod, 1959, 1961). Regulator genes differ from and interact with structural genes (i.e., cistrons). The latter determine the molecular structure of enzymes by genetic transcription (see Chapter 2). The action of these genes is subject to regulation by feedback from the gene products. Accumulation of gene products may repress gene action, while removal of these products may bring about derepression and the resumption of primary gene activity (see Chapter 3). Repression can be induced by chemical agents known as repressors. Regulator genes produce such repressors and thereby control the state of activity of structural genes (Jacob and Monod, 1959, 1961). Regulator genes are perhaps a bacterial counterpart of some types of controlling genes found in higher plants.

OPPOSITIONAL GENE SYSTEMS   Many gene systems in plants and animals are composed of genes with opposing positive and negative effects. The net gene action and phenotypic expression is a resultant of the balance between the opposing factors. A modifier system consisting of plus and minus modifiers is a good example. Another example is a combination of positive Mendelian genes and inhibitor genes.

Clausen and Hiesey studied the inheritance of 19 characters in interracial hybrids of *Potentilla glandulosa*. The character differences include vegetative, floral, fruit, and physiological features. Sixteen of the 19 characters studied are controlled by allelic differences in three or more genes. It is significant that oppositional gene systems control 7 or perhaps 8 of the 16 multifactorial characters (Clausen and Hiesey, 1958, pp. 108–9).

These authors go on to stress the general importance of oppositional gene systems in the control of morphological and physiological characters in higher plants. Such systems, composed of balanced plus and minus factors, permit a fine regulation of developmental processes and phenotypic expressions. Oppositional gene systems have been found widely in plants whenever a thorough factorial analysis has been made (Clausen, 1951, ch. 5; Clausen and Hiesey, 1958, chs. 4, 5).

An analogous system in bacteria is the combination of operon and regulator gene. Operons and regulator genes have been discovered as a result of experimental work on enzyme synthesis in *Escherichia coli*. They are defined in biochemical-genetical terms. Nevertheless, the operon–regulator gene system requires some mention here because of the prominence this system has achieved in recent discussions of genetic control mechanisms in higher organisms.

The operon is defined as a unit of genetic transcription consisting of two or more contiguous structural genes and a closely linked operator gene. The operator gene reacts with chemical repressors, and then, depending on its state, switches the structural genes on or off (Jacob and Monod, 1961). The regulator gene occurs at a different locus but interacts with the operon through the chemical repressors. The regulator gene produces a repressor that determines the state of the operator gene in the operon (Jacob and Monod, 1961).

It seems to me that the operon is essentially a special type of gene cluster or complex locus. The combination of operon and regulator gene is analogous to some kinds of oppositional gene systems in higher plants.

Jacob and Monod and McClintock have drawn attention to similarities in the genetic control mechanisms of bacteria and higher plants, particularly maize. The question posed is whether some phenomena of gene expression and repression in higher plants can be explained by an extension of the operon-regulator concept. Jacob and Monod (1961) suggest tentatively and McClintock (1961) argues persuasively that the operon-regulator concept is applicable in plants and other higher organisms.

Brink (1964), however, in a penetrating review of the problem, emphasizes the enormous difference between bacteria and higher organisms in the complexity of developmental processes and the gene systems which control them. A control mechanism of primary importance in bacteria is apt to play only a minor role in multicellular organisms. As regards regulation of gene action in higher organisms, Brink attaches much greater importance to the generalized effects of heterochromatin blocks on euchromatic genes (see Chapter 3 and also the next section of this chapter).

The original question posed by Jacob and Monod (1961) and McClintock (1961), whether operon-regulator systems play an important role in higher plants, is misleading. These authors do not refer to oppositional gene systems. The concept of oppositional gene systems has grown out of experimental work in plant genetics in the same way that the more recent operon-regulator concept has grown out of experimental work on bacteria. We certainly need to learn more about genetic control mechanisms in higher plants. We can pursue this goal quite well, however, within the framework of the indigenous concept of oppositional gene systems.

VARIEGATED POSITION EFFECT A gene lying in a euchromatic region of a chromosome, and having a constant mode of action in its normal site, often exhibits depressed activity when transposed to a new site close to heterochromatin. The variable and depressed action of the gene in its position near heterochromatin results in a variegated phenotype. This is the variegated type of position effect, which is well known in Drosophila and also in mice.

The clearest example in plants is in Oenothera. It is also the first case of position effect to have been demonstrated in the plant kingdom (Catcheside, 1939, 1947). The plant is a strain of the *Oenothera lamarckiana* complex known as *Oe. blandina.* Translocations were induced in *Oe. blandina* by x-ray treatment. Two translocation types were associated with various alterations in the phenotypic characters of the plants. The phenotypic changes associated with one of these translocations are probably but not certainly position effects. Position effect can be established experimentally for the other translocation, owing to the presence of a marker gene (Catcheside, 1939).

The gene P in the *Oe. lamarckiana* group comprises a series of multiple alleles governing pigmentation of the flower buds and other parts. The standard type of *Oe. blandina* carries the allele P_s for red-and-green-striped flower buds in homozygous condition. P_s is located on arm 3 of chromosome 3–4 in standard *Oe. blandina.*

In the translocation type of *Oe. blandina,* an interchange of segments has taken place between the 3–4 and the 11–12 chromosomes; a portion of the 3 arm bearing the P_s gene has been transferred to a new position on the 11 arm. The P locus is close to the translocation breakage point. Catcheside believes that it is also close to heterochromatin in its new position (1939, 1947).

The phenotype of the flower buds changes along with the chromosome translocation, from broad red stripes on a green background to uneven light red striations. The pigmentation becomes reduced and variable in the new translocation type. P_s thus produces a regular striped phenotype when on the 3–4 chromosome and a variegated phenotype when in the altered 3–11 chromosome (Catcheside, 1939, 1947).

Catcheside then transferred the P_s gene in the translocation type back to its original position in the 3–4 chromosome by backcrossing the translocation type to standard *blandina* and obtaining the appropriate crossovers. When the P_s allele is restored to its original site it resumes its normal gene action and produces the normal amount of pigmentation on the flower buds. The phenotypic change associated with change in gene location is thus reversible. The reversibility is strong evidence against point mutation and equally strong evidence for position effect as the cause of the observed phenotypic changes (Catcheside, 1939, 1947).

This conclusion was confirmed by putting another allele of the series, P_r, into the same alternative chromosome positions. P_r in its normal position on the 3–4 chromosome produces a solid red flower bud with no variegation. When P_r is in a translocated 3–11 chromosome, the flower bud is variegated red and green. The return of the P_r allele to the 3–4 chromosome restores its normal phenotypic expression (Catcheside, 1947).

The pigmentation in the pericarp of corn kernels (*Zea mays*) is governed by various genes whose normal action yields solid yellow or solid colored kernels. Variegated kernels also occur. In at least some cases this variegation appears to be a type of variegated position effect associated with transposable controlling elements.

Controlling elements in maize are suspected of being heterochromatic. When a controlling element is transposed to a new region in a chromosome, it may depress gene action in that region. When the controller is transposed away again, the gene resumes its normal function. Controllers thus act as nonspecific regulators of gene action. They can be involved in many instances of unstable and apparently mutable loci (McClintock, 1953, 1956, 1961; Rhoades, 1954; Brink, 1964). By the same token, the controlling elements can produce variegated tissues. Suppose a maize plant is heterozygous for a gene affecting color of the pericarp, and has in addition a transposable controlling element. Then colored and yellow sectors will appear in the pericarp, depending on the inactivation and reactivation of the pericarp color alleles by the controller (McClintock, 1953, 1956; Rhoades, 1954; Greenblatt, 1966).

Variegated flowers in hybrids of *Nicotiana sanderae* and *N. langsdorffii* may also be due to position effect. But other possible causes, particularly gene mutability, cannot be excluded in this case (Smith and Sand, 1957; see also Chapter 15).

CHAPTER NINE

SERIAL
GENE SYSTEMS

INTRODUCTION Gene systems composed of Mendelian genes can be divided into two main classes on the basis of functional relationships between the component genes. We have previously referred to these two types as multiple gene systems and serial gene systems (Chapter 7, and Grant, 1964).

The individual members of a multiple gene system produce the same growth substance or similar ones. Their combined actions are cumulative, and their phenotypic end product is a quantitative character. The progeny of a cross between parents that differ in a multiple gene system usually show continuous variation. We will discuss multiple gene systems at greater length in Chapter 10.

In a serial gene system, by contrast, the component genes control different steps in a developmental sequence. The progeny of parents differing in the alleles of such genes show discontinuous variation. The present chapter is concerned with gene systems of this sort.

A well-known example of a developmental sequence which takes place in a series of steps is the synthesis of arginine in

Neurospora. The synthesis of this amino acid involves at least seven steps, and each step is controlled by a different gene (the sequence is described in Chapter 2).

On the basis of this and similar examples, Beadle (1955) and others have suggested that the relation between genes and steps in a biochemical reaction chain can be described by the model shown in Figure 16. The formation of the final product (product C in the figure) depends on the cooperation of three genes, *A, B,* and *C,* which work on different steps in the developmental process. These three genes accordingly form a serial gene system.

If any single critical gene in a serial gene system is represented by a mutant allele which does not properly activate an essential step, product C will not develop and the individual will appear as a mutant type deficient in C. There are, moreover, different C-deficient mutants which on a first classification by phenotypic characteristics would be regarded as the same, but which on further genetic analysis could be proved to be nonallelic. The interaction between the separate genes comprising a serial gene system may take various forms, which we will now examine.

COMPLEMENTARY FACTOR INHERITANCE In the arrangement shown in Figure 16, character C is the end product of a stepwise development requiring the successive action of three genes, *A, B,* and *C.* Assume that the normal genotype is homozygous for the dominant alleles of these genes, being *AABBCC,* and that this genotype yields a normal phenotype exhibiting product C.

Assume further that two mutant types exist, one of which carries mutant alleles for the *A* gene and the other mutant alleles for the *B* gene. The former has the constitution *aaBBCC,* the latter the constitution *AAbbCC.* Both are alike phenotypically in that the character C is lacking. Intercrossing of two such C-deficient individuals will lead, however, to the production of normal individuals in the F_1 generation; the gametes of the two mutant types (*aBC* and *AbC*) produce an F_1 zygote (*AaBbCC*) containing all the dominant alleles needed to carry normal development through to completion.

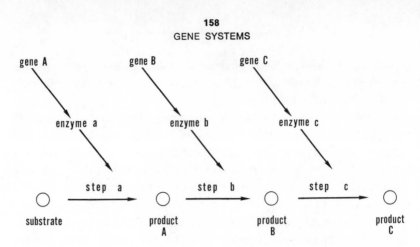

Figure 16. General model to show the relation between different genes and different steps in a biochemical reaction chain. (From Beadle, 1955.)

If the genes *A* and *B* are on separate chromosomes, they will segregate independently, so that the F_1 hybrid produces four classes of gametes (*AB, Ab, aB, ab*) in equal frequencies. The random union of these gametes will then give normal individuals and C-deficient mutants in a modified dihybrid ratio, a 9 : 7 ratio, in the F_2 generation.

The classic case of complementary factor inheritance in *Lathyrus odoratus,* the sweet pea, was reported by Bateson, Saunders and Punnett in 1906. The two parental types were both true-breeding for white flowers. Their F_1 hybrid was blue-flowered. The F_2 progeny segregated into 2132 individuals with colored flowers and 1593 with white flowers, which is very close to a ratio of 9 : 7. The authors reasoned from these results that flower color in the sweet pea requires the complementary action of two (or more) independent genes (Bateson, Saunders and Punnett, 1906; review in Clausen and Hiesey, 1958, pp. 160–61).

INHIBITORS   Although we already discussed inhibitor genes in Chapter 7, it is worthwhile to reconsider the subject briefly here, since inhibitors are often components of serial gene systems.

A gene can acquire the status of an inhibitor gene simply by blocking a precursor step in a developmental sequence. Consider the genotype *AAbbCC* for a serial gene system conforming to the model shown in Figure 16. In this genotype the allele *b* blocks the formation of product C, and can therefore be regarded as an inhibitor of gene C. A single gene *B* may therefore be a complementary factor in one allelic form (*B*) and an inhibitor gene in another (*b*).

Some inhibitor genes are normal components of serial gene systems. In the model in Figure 16 it was assumed that the presence of product C is normal for an organism. Let us alter this assumption; suppose that the absence of product C is normal. The normal C-deficient phenotype can be produced by a genotype which contains gene C and, in addition, an inhibitor of C such as *bb*. If the normal allele is *b*, and the normal genotype *AAbbCC*, then a C-deficient phenotype will also be normal.

The example of *Viola tricolor* and *V. arvensis,* introduced in Chapter 7, serves to illustrate this point. *Viola tricolor* usually has a dark spot or nectar guide on the pistil in the flowers, though mutant types without spots also occur; *V. arvensis* is normally spotless, but has occasional spotted individuals. Factorial analysis indicates that the formation of spots is determined by two complementary factors (*S* and *K*). These genes are present in both species. The gene system governing flower spots contains in addition two inhibitor genes (*I* and *H*). The active alleles of these inhibitors are present in *V. arvensis* where they suppress spot formation, probably by blocking an early stage of development (Clausen, 1926, 1951).

A similar situation is found in Primula. *Primula elatior* is yellow-flowered and does not have red or blue anthocyanin pigments in its flowers. But it does carry an inhibitor gene (*I*) which suppresses the formation of these pigments in hybrids with other species, *P. juliae* and *P. vulgaris,* which normally have red or blue flowers. Some English populations of *P. elatior* are polymorphic for the gene *I,* containing both the active dominant allele *I* and a recessive noninhibiting allele *i* (Valentine, 1970).

EPISTASIS   A form of gene interaction discovered in the early period of Mendelian genetics was that wherein one gene masks the effects of another, so that the phenotype is determined effectively by the former and not the latter when both genes are present. The phenotypic effects are comparable to those produced by the combination of a dominant and a recessive allele of one gene, but the genotypic arrangement responsible for these phenotypic effects differs—the dominant character results from the interaction between separate genes. This relation of dominance and recessiveness between separate genes is known as epistasis. The term was introduced by Bateson in 1907 (see Rieger, Michaelis and Green, 1968).

The term has been given a much broader meaning in recent usage in population genetics and quantitative genetics. The literature in these fields often employs the term, "epistatic interaction" to refer to various types of nonadditive gene interaction in general. In this book epistasis is used in its original and more specialized sense.

It is possible to relate epistasis to the model diagrammed in Figure 16 by making a simple modification in the developmental sequence. Assume that the developmental pathway is not linear but branching. At step b the course of development can be sidetracked toward the formation of an alternative product, beta. The gene B controlling step b in the reaction now plays a critical role, for different alleles of B might: (1) prepare the way for the gene C, (2) simply block the action of C, or (3) switch the metabolic pathway in a new direction toward the formation of beta. The functional relation of the two genes in these cases is then: (1) B is complementary to C, (2) B is an inhibitor of C, or (3) B is epistatic to C.

With epistasis, therefore, an allele of gene B (say, B_b) not only inhibits the action of gene C but also has a separate and divergent action of its own. The phenotype resulting from the action of the genotype AAB_bB_bCC consequently lacks product C and possesses the alternative product beta instead. The genotype $BBCC$ exhibits character C, whereas B_bB_bCC exhibits character beta.

Nilsson-Ehle (1909) studied the inheritance of glume color in oats (*Avena sativa*). The parental strains had yellow glumes and black glumes. Their F_2 and F_3 progeny segregated in ratios suggesting an epistatic series of at least two genes for glume color

(S and Y). Gene Y considered by itself determines yellow pigmentation. But its effects can be masked by the dominant epistatic allele of S. Consequently, genotypes which are recessive for S ($ssYY$, $ssYy$) give yellow spikelets, while genotypes containing the dominant allele of S in combination with Y ($SSYY$, $SSYy$, SsY_-), have black spikelets. Apparently the formation of glume pigment can be switched to either yellow or black at a point controlled by gene S (Nilsson-Ehle, 1909; review in Sirks, 1956, pp. 149–52).

Clausen and Hiesey (1958) have reanalyzed this study, on the basis of the combined data for several different crosses in oats, and have concluded that the results indicate a longer and more complex epistatic series than that described above. Actually there are several color classes of glumes: black, gray, white, and yellow. The epistatic series includes at least four genes. S_1 and S_2 are the top epistatic members of this series; their dominant alleles predominate over the other genes in the series to produce black glumes. The next lower genes in the system are Gr for gray and then W for white. Gene Y for yellow glumes is the bottom hypostatic gene of the series (Clausen and Hiesey, 1958, pp. 162–64).

The more complex epistatic gene system can be taken to mean that the developmental pathway leading to spikelet color in oats is equally complex. The alternative routes in pigment formation are not merely twofold, black versus yellow, but are multiple and probably follow a series of branchings.

The flowers of normal wild plants of *Viola tricolor* are violet. A garden form of the pansy has jet black flowers. Several intermediate shades (referred to as velvet) connect these two extremes. The series of shades is determined by an epistatic series of five genes which predominate over one another in the following order: $M_1 > M_2 > M_3 > M_4 > M_5$ (Clausen, 1926; Clausen and Hiesey, 1958).

Genotypes containing the top epistatic allele M_1 ($M_1 \ldots$) have the normal violet color. Successively darker shades of velvet are produced by the alleles M_2, M_3, and M_4. Thus genotypes $m_1m_1M_2$ . . . give light velvet flowers, and genotypes $m_1m_1m_2m_2m_3m_3M_4$ . . . have deep velvet flowers. Plants that are homozygous recessive for the top four genes (M_1 to M_4) but carry a dominant allele of the hypostatic member M_5 ($m_1m_1m_2m_2 \ldots M_5$) have velvety black flowers. Finally, plants that are homozygous recessive for all five M genes ($m_1m_1m_2m_2 \ldots m_5m_5$) have jet black flowers (Clausen, 1926; Clausen and Hiesey, 1958).

FLOWER COLOR INHERITANCE IN ANTIRRHINUM AND DAHLIA

The flowers of *Antirrhinum majus,* the snapdragon, exhibit a wide range of variations with respect to both form and color. As regards the configuration of the corolla, the limb is normally two-lipped, but sometimes radial; the lobes are normally large, but sometimes reduced; and the tube is normally stout, but sometimes long and slender. Seven color classes can be recognized in the corolla, including magenta, pink, ivory, yellow, and white.

The classic studies of Baur on inheritance of floral characters in the snapdragon were continued and amplified by his followers, including Stubbe. The early work is summarized by Baur (1919, 1924), and an up-to-date account is given by Stubbe (1966).

The first stage of the investigations showed that certain floral character differences are inherited as simple Mendelian differences. Thus the cross of red × ivory-flowered plants gives an F_1 with pink flowers and an F_2 that segregates into red, pink, and ivory-flowered individuals in a 1 : 2 : 1 ratio. The F_3 progeny of pink F_2s segregate in a similar manner. Red flowers differ from ivory flowers by a single gene (F) and from yellow flowers by two independent genes (F and C). The difference between normal two-lipped and peloric radial corollas is due to one main gene (Baur, 1919, 1924).

The next stage of the Antirrhinum work demonstrated the independent assortment of some of the flower genes. The cross of red peloric × ivory normal gives a modified dihybrid ratio in the F_2. And the cross of red peloric × yellow normal yields an F_2 with a more complex segregation, as expected (Baur, 1919, 1924).

Some crosses between different color forms gave complex segregations indicating a larger number of gene differences. Baur identified the following separate genes: C for ivory or yellow; F for red; G modifying red; and D controlling color in the tube. Later

TABLE 16

GENOTYPES FOR DIFFERENT FLOWER COLORS IN *ANTIRRHINUM MAJUS* (STUBBE, 1966)

Color	Genotype for four genes			
white	*niv/niv*			
yellow	*niv⁺*	*sulf/sulf*	*inc/inc*	
ivory	*niv⁺*	*sulf⁺*	*inc/inc*	
bronze	*niv⁺*	*sulf/sulf*	*inc⁺*	*eos/eos*
pink	*niv⁺*	*sulf⁺*	*inc⁺*	*eos/eos*
crimson	*niv⁺*	*sulf/sulf*	*inc⁺*	*eos⁺*
magenta	*niv⁺*	*sulf⁺*	*inc⁺*	*eos⁺*

Note: the plus superscript, as in *niv⁺*, indicates the dominant wild-type allele.

work added other genes to the list and changed the gene nomenclature. The main flower color genes are nivea (*Niv*), sulfurea (*Sulf = C*), incolorata (*Inc = F*), eosinea (*Eos*), pallida (*Pal*), diluta (*Dil*), and eluta (*El*) (Stubbe, 1966).

Factorial analysis reveals that these genes determine different flower colors, and biochemical analysis reveals that they make different pigments. The seven main phenotypic color classes are: white, yellow, ivory, bronze, pink, crimson, and magenta. These form a more or less graded series of pigmentation. The genotypes corresponding to the different phenotypes are shown in Table 16 (Stubbe, 1966).

The four genes listed in Table 16 appear to have a graduated effect on pigmentation. Homozygous recessive *niv/niv* blocks pigment formation and results in white flowers, whereas dominant *niv⁺* permits pigment formation to take place. The color is yellow in the presence of *niv⁺* and the homozygous recessive condition for the next two genes (*Sulf* and *Inc*). The gene *Inc* exercises a switching action between yellowish anthoxanthins and reddish anthocyanins, a dominant *inc⁺* allele being necessary for anthocyanin formation. The color becomes pink when the first three genes are represented by dominant alleles, and magenta with dominant alleles of all four genes (See Table 16). The four genes seem to form a serial gene system.

Three additional color genes are *Pal, Dil,* and *El. Pal* is complementary to *Inc;* both dominant alleles, *inc*[+] and *pal*[+] must be present for anthocyanin production. *Dil* is a modifier which intensifies the coloration. Different allele combinations of *Inc, Pal,* and *Dil* determine different fine shades of bronze or pink. *El* causes bleaching of the reddish color, and, depending on the dosage of the dominant *El* allele, determines different shades of magenta (see Table 9 in Chapter 6) (Stubbe, 1966).

The garden dahlia, *Dahlia variabilis,* is highly variable in flower color. There are white, ivory, yellow, magenta, purple, orange, and red forms, and various tones in some of these color groups. The complex genetics of flower color in Dahlia has been unraveled by Lawrence and Scott-Moncrieff (1935). The story is reviewed by Crane and Lawrence (1934, 1938) and Lawrence and Price (1940). A simplified account will be presented here.

Five main genes determine flower color in *Dahlia variabilis—I, Y, A, B, H.* They fall into three categories. *I* and *Y* produce flavone pigments, *Y* being epistatic to *I. A* and *B* produce anthocyanins (particularly cyanidin and pelargonidin). *H* is an inhibitor. The actions of the dominant alleles of these genes are as follows (Crane and Lawrence, 1938):

I produces ivory flavone
Y produces yellow flavone
A produces pale anthocyanin
B produces deep anthocyanin
H inhibits yellow flavone

Color variation in the garden dahlia has two basic components: the flavone pigments, producing ivory or yellow colors; and the anthocyanin pigments, giving reddish to purplish colors. The total array of colors is a result of various mixtures of these components. Some examples are:

Ivory background + pale anthocyanin = magenta
Ivory background + deep anthocyanin = purple
Cream background + either anthocyanin = crimson
Yellow background + pale anthocyanin = apricot
Yellow background + deep anthocyanin = scarlet

TABLE 17

MAIN GENOTYPES FOR DIFFERENT FLOWER COLORS IN *DAHLIA VARIABILIS*
(LAWRENCE AND SCOTT-MONCRIEFF, 1935)

Genotype	Flower color	Genotype	Flower color
i y a b	white	i Y a b	yellow
I y a b	ivory	I Y a b	yellow
i y A b	rosy-magenta	i Y a b H	cream to primrose
i y a B	rosy-purple	i Y A b h	apricot
I y A b	bluish-magenta	i Y a B h	scarlet
I y a B	bluish-purple	i Y a B H	crimson

When both the flavone and anthocyanin pigments are absent, the resulting color is white (Crane and Lawrence, 1938).

The interaction of the two flavone-producing genes (*I, Y*) and the two anthocyanin-producing genes (*A, B*) determines the mixtures of pigments and type of color expression. Additional effects are caused by the color inhibitor gene *H*. The main genotypes and the corresponding color phenotypes are listed in Table 17.

Each color gene is present in four doses in the polyploid garden dahlia, and consequently dosage effects of these genes can come into play. Examples of dosage effects of the *I* and *A* genes were described in Chapter 6. It suffices to say here that a genotype which is constant for four color genes but varies in the number of dominant alleles of the fifth gene, where the fifth gene is *I, A,* or *H,* will give a graded series of shades within a given color class. Thus anthocyanin pigmentation becomes progressively deeper in the series *Aaaa, AAaa, AAAa,* and *AAAA.* Dosage effects are superimposed on gene interactions to produce the great array of flower colors and shades in the garden dahlia (Crane and Lawrence, 1938).

GENERAL GENETICS OF FLOWER COLOR   Flower color inheritance has been studied in scores of plant species. *Lathyrus odoratus, Viola tricolor, Antirrhinum majus,* and *Dahlia variabilis* have been discussed in this chapter. *Primula sinensis, Matthiola incana,* Streptocarpus, Delphinium, Aquilegia, Geum, and Canna are among the other examples. The subject has been reviewed by Crane and Lawrence (1938), Lawrence and Price (1940), and Paris, Haney and Wilson (1960).

Certain common features are found in the numerous different cases. Complementary factors for flower color are one such feature, and epistasis is another. Different workers have independently observed that fine shades within a given color class are determined by intensifier genes or bleaching genes. The latter grade into suppressors and inhibitors of pigment formation. All of these features indicate serial gene systems.

Paris, Haney and Wilson (1960) surveyed 75 plant species in which flower color had been analyzed genetically. Their objective was to search for common denominators in the different specific gene systems for flower color. They found that most cases of flower color inheritance can be described in terms of six genes with epistatic interactions.

The six genes as designated by a uniform gene nomenclature are W, Iv, Y, B, P, Dil. Their actions are as follows: W switches color production on or off. Iv and Y control ivory and yellow colors, respectively. Genes B and P produce anthocyanin pigments which have purple, blue, or red colors. Gene Dil in one allelic form or another acts as an intensifier or bleaching factor (Paris, Haney and Wilson, 1960).

The dominant and recessive alleles of these genes have the following phenotypic effects:

W	colored	ww	white
Iv	nonivory	$iv\,iv$	ivory
Y	nonyellow	yy	yellow
B	purple or violet	bb	blue
P	purple or violet	pp	pink, rose, red
Dil	intense color	$dil\,dil$	dilute color

The six color genes form an epistatic series running from W, the top epistatic member, to Dil, the bottom hypostatic member. The series is usually $W > Iv > Y > B > P > Dil$ (Paris, Haney and Wilson, 1960).

We wish to call attention, finally, to a remarkable parallelism between flower-color inheritance in angiosperms and coat-color inheritance in mammals. Castle (1954) showed that coat color in horses, rodents, and other mammals is determined by the interaction of four main color genes: *C, B, A,* and *E.* The colors of mammals are due to two different groups of pigments, black-brown and red-yellow, in the coat. Gene *C* controls the initial step in pigment production. In the presence of the dominant wild-type allele *C,* gene *B* gives black pigments, while *A* gives yellow-red pigments. Genes *E* and *A* are pattern genes controlling the distribution of black and red or yellow areas on the body. When *C* is in the homozygous recessive condition *cc*, on the other hand, the coat is white (Castle, 1954).

CONCLUSIONS Several previous attempts to classify gene systems have recognized complementary factors and epistatic series as different major types of systems (Darlington and Mather, 1949; Sirks, 1956; Clausen and Hiesey, 1958). Here we have brought them together, along with some cases of inhibitor genes, under the common heading of serial gene systems, following an earlier treatment (Grant, 1964). This way of grouping these gene systems emphasizes the relationships between them, and helps to set them off collectively from multiple gene systems.

Serial gene systems consist of two or more Mendelian genes controlling different steps in a developmental sequence. The mode of action of the normal allele of a gene controlling a step in the early or middle stages of the sequence determines the type of serial gene system.

Two or more components of a serial gene system are complementary factors if one gene provides the necessary conditions for the action of another gene controlling a subsequent step in the reaction chain. The first gene is an inhibitor if it blocks the action of genes controlling later stages in the reaction chain. The genes form an epistatic series if the first gene switches the reaction onto an alternate pathway.

In certain cases, different alleles of the same gene may be

a complementary factor, an inhibitor, and an epistatic gene. Such cases show the basic similarity between these three modes of gene interaction. Complementary factors, some inhibitors, and epistatic series are different aspects of serial gene interaction.

CHAPTER TEN

MULTIPLE
GENE SYSTEMS

INTRODUCTION The early successes of Mendelian genetics depended upon the deliberate selection of discontinuous variations for factorial analysis. Much of hereditary variation, however, is continuous and blending. Blending inheritance was well known to nineteenth-century students of heredity, including Mendel, and was a stumbling block to the acceptance of Mendelism as a general theory of heredity in the early years of the twentieth century.

The impasse was resolved, and the observed facts of blending inheritance were satisfactorily explained in terms of Mendelian genes, by the multiple factor theory of Nilsson-Ehle, East, and others. This theory was formulated, and critical evidence for it was obtained, in the period from 1909 to 1916.

The multiple factor theory holds that hereditary differences in quantitative characters are usually determined by two or more independent genes with similar and cumulative effects. The segregation of such multiple factors in the progeny of hybrids then results in a continuous or semicontinuous series of phenotypes for the quantitative character in question.

The topic of quantitative inheritance goes well beyond multiple factor inheritance in the strict sense. Two or three types of gene systems are involved. The present chapter is concerned with one of these, multiple gene systems as defined in a rather narrow sense. Other gene systems which bring about similar results are discussed in Chapter 11.

Multiple gene systems are circumscribed as follows for our purposes in this chapter. The gene system controls quantitative characters such as growth rate, size, proportions, and density of pigmentation or hairiness. The genes that make up the system elaborate the same or similar growth substances or growth regulators and thus cooperate to produce the quantitative character. The number of independent genes may be two or more but is usually not large. The effects of the individual genes are equal or nearly so, and additive or mainly so, and the phenotypic expression is the result of the combined effects of the individual genes.

There is good experimental evidence for the existence of gene systems with the properties just described. To this extent our discussion of multiple gene systems deals with real entities in the genotype. What is not so clear is whether such systems can be distinguished in a nonarbitrary manner from other gene systems controlling quantitative characters, particularly polygene systems and modifier systems. This problem is deferred to Chapter 11.

HISTORICAL CASES   The concept of multiple factor inheritance was foreseen by Mendel and stated in his 1866 paper. A cross between different flower-color forms of the *Phaseolus multiflorus* group, one white-flowered and the other purple-flowered, yielded F_1s with purple flowers and an F_2 generation ranging in a graded series from purple to white. The recessive white-flowered type appeared in a low frequency (1 in 31 plants) in F_2. The recessive type was constant in F_2 and F_3, but some of the colored F_2s continued to segregate in later generations. Mendel suggested that these results could be explained by a modification of a basic dihybrid or trihybrid ratio in which two or three independent genes were concerned with a single color character rather than with different characters. But the experimental evidence was not sufficient to take the suggestion any further at that time (Mendel, 1866, 1965).

Definite evidence for multifactorial control of quantitative characters was brought forward by several workers in the years immediately following the rediscovery of Mendel's principles.

The first important study was Nilsson-Ehle's (1909) investigation of kernel color in wheat, *Triticum aestivum*. The parental varieties had red or white kernels. The F_1s were red. Some F_2 families segregated in a ratio of 3 red to 1 white, but one large F_2 family was all red. A series of F_3 families derived from different red F_2 plants then gave various red/white ratios including 3 : 1, 15 : 1, 63 : 1, and all red. Nilsson-Ehle concluded that the differences in grain color are due to three independent multiple factors (R_1, R_2, R_3). The alleles for red are dominant over those for white at each locus. Consequently only the triple recessive ($r_1r_1r_2r_2r_3r_3$) is white. It would be expected to occur in a frequency of $1/64$ in F_2 and in some F_3 families. This expectation is partly realized in the experiment (Nilsson-Ehle, 1909).

Clausen and Hiesey (1958, p. 162) have reanalyzed Nilsson-Ehle's data, bringing out some new points, and believe that four rather than three R genes may be involved.

Shull (1914) reported a similar case of duplicate factor inheritance of capsule shape in *Capsella bursa-pastoris* (Cruciferae). One parental type with triangular capsules and another parent with oval capsules yielded F_1 hybrids with triangular capsules. The F_2 generation segregated in a ratio approximating 15 triangular to 1 oval. Some of the triangular-fruited F_2 individuals bred true to type in the F_3 and F_4 generations; others gave segregating F_3 and F_4 progenies. Some of the segregating F_3 and F_4 families gave 3: 1 ratios and others 15 : 1 ratios for the shape of the capsule. These results are explained by the hypothesis that the differences in capsule shape are controlled by duplicate factors, with dominance in the alleles for triangular capsules.

Multiple factors were first related to continuous phenotypic variation in the work of East and Emerson on ear size in corn. A variety of popcorn has short ears, which are usually 6–7 cm long, while the long ears of a variety of sweet corn are usually 14–19 cm long. Their F_1 hybrids mostly have ears 10–13 cm long and are thus intermediate. The ears of the F_2 progeny range continuously from short (7 cm) to long (21 cm), with most of the F_2 ears being medium long (10–14 cm). The range and distribution of the F_1 and F_2 generations indicate that the differences in ear length between the parental varieties of corn are due to allelic differences in multiple genes (Emerson and East, 1913). The number of seed rows per ear varied in a similar continuous pattern in hybrid progenies and also appeared to be under multifactorial control (East, 1910; Emerson and East, 1913).

Baur (1914) observed complex and intergrading variations in the progeny of interracial hybrids in *Antirrhinum majus* and interspecific hybrids of *A. majus* × *A. molle*. Characters such as plant stature, habit of branching, and corolla form showed blending inheritance. Baur (1914, 1924) concluded that many of the character differences between races and species of plants are controlled by multiple factors with small individual effects.

The breeding experiment which furnished conclusive verification of the multiple factor hypothesis was that of East (1916) on flower size in *Nicotiana longiflora*. We will review this case in the next section.

Many other cases of multifactorial inheritance have been worked out in a wide range of plant groups. A valuable comprehensive review is given by Clausen and Heisey (1958, ch. 4). See also Hiorth (1963, pp. 60 ff.).

INHERITANCE OF FLOWER SIZE IN *NICOTIANA LONGIFLORA*

East put the multiple factor hypothesis, which was still disputed in many quarters in 1916, to a decisive test. He reasoned that if size characters are in fact determined by multiple factors with generally cumulative effects and no dominance, then several predictions could be made and tested experimentally.

The expectations on this hypothesis are as follows (East, 1916, slightly paraphrased): (1) Crosses between inbred parental strains differing in a size character should give F_1 populations which are as uniform phenotypically as the parental strains. (2) The variability of the F_2 population should be far greater than that of the

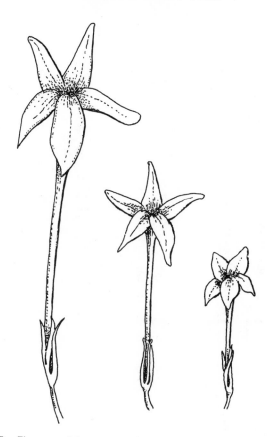

Figure 17. Flowers of two parental types of *Nicotiana longiflora* (*left* and *right*) and of a modal intermediate type in the F$_2$ generation (*center*). (Redrawn from East, 1916.)

F$_1$ generation. (3) The parental types should reappear in F$_2$ if the F$_2$ population is sufficiently large, or otherwise in subsequent generations. (4) "Individuals from various points on the frequency curve of an F$_2$ population should give F$_3$ populations differing markedly in their modes and means." (5) Different F$_2$ individuals should yield F$_3$ populations with different ranges of variation. (6) The range of variation in later-generation families should center on the quantitative value of the F$_2$ parent. (7) "In generations succeeding the F$_2$, the variability of any family may be less but

TABLE 18

INHERITANCE OF COROLLA LENGTH IN AN INTERVARIETAL CROSS IN *NICOTIANA LONGIFLORA* (EAST, 1916, DATA SIMPLIFIED)

Generation	Corolla size of parent plant, mm	Frequency distribution of individuals with corollas of given lengths																						
		34	37	40	43	46	49	52	55	58	61	64	67	70	73	76	79	82	85	88	91	94	97	100 mm
P_1		13	80	32																				
P_2																				6	22	49	11	
F_1									4	10	41	75	40	3										
F_2	61							1	5	16	23	18	62	37	25	16	4	2	2					
F_3, fam. 12	46				1	4	26	44	38	22	7	1												
F_4, fam. 121	44			8	42	95	38	1																
F_5, fam. 1311	41	3	6	48	90	14																		
F_3, fam. 26	82												3	5	12	20	40	41	30	9	2			
F_4, fam. 262	87												4	5	6	11	21	33	41	29	8	5	1	
F_5, fam. 2621	90														2	3	8	14	20	25	25	20	8	

TABLE 19

STATISTICS OF COROLLA LENGTH VARIATION IN HYBRIDIZATION EXPERIMENT
WITH *NICOTIANA LONGIFLORA* (EAST, 1916, DATA SIMPLIFIED)

Generation	Parental plant, corolla length, mm	No. of individuals in progeny	Mean corolla length in progeny, mm	Coefficient of variation
P_1		125	40.46	4.33
P_2		88	93.22	2.46
F_1		173	63.53	4.60
F_2	61	211	67.51	8.75
F_3, fam. 12	46	143	53.47	6.99
F_4, fam. 121	44	184	45.71	5.18
F_5, fam. 1311	41	161	41.98	5.49
F_3, fam. 26	82	162	80.20	5.93
F_4, fam. 262	87	164	82.86	7.04
F_5, fam. 2621	90	125	87.88	6.28

never greater than the variability of the population from which it came."

To test the hypothesis, East used as parental types two inbred varieties of *Nicotiana longiflora* that differ in flower size. One parental form has a long-tubed corolla and the other a short-tubed corolla, as shown in Figure 17. The character of corolla size has a relatively high degree of heritability in *Nicotiana longiflora*. Populations were measured for corolla tube length in 1911, and the measurements were grouped in classes on class centers 3 mm apart, with the differences in mode and range shown in Table 18. Replicate cultures in subsequent years (omitted from the table) gave similar results.

The large- and small-flowered types were crossed to produce large F_1 and F_2 generations. Various F_3 families were derived from phenotypically different F_2 plants, and some of these lines were continued to F_4 and F_5. As before, the individuals in each family were scored for corolla length, the measurements were grouped into size classes, and the frequency distribution of the classes was plotted.

The main results are summarized in Table 18. Some replicate cultures and some advanced families are omitted from the table for simplicity. The pattern of variation is shown graphically in

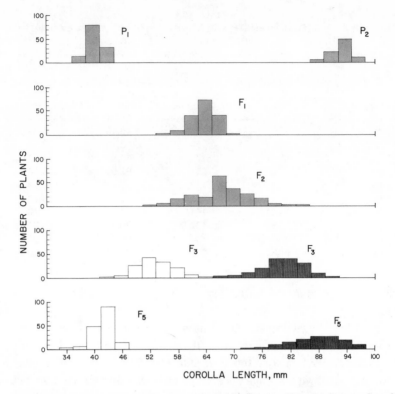

Figure 18. Frequency distribution of individuals with different flower sizes in an intervarietal cross in *Nicotiana longiflora*. (Drawn from data of East, 1916; from Grant, 1971.)

Figure 18. Table 19 gives the statistical analysis of the data in Table 18.

The tables and graph show that nearly all of the hypothetical expectations are fulfilled by the experimental data in the *Nicotiana longiflora* cross. The F_1 and F_2 generations are intermediate in mode and mean. The F_1 is uniform and the F_2 highly variable. Although the parental types are not recovered in F_2, they do segregate out in later generations. Different ranges and means are exhibited by the nine different F_3 families grown (only two of these are presented in Tables 18 and 19). The variability of these nine F_3 families, as indicated by the coefficient of variation, is consistently less than that of the F_2, and in at least one line the variability continues to decline in the succeeding F_4 and F_5 generations. All of these features support the hypothesis that the intervarietal hybrid is segregating for a series of multiple factors determining the differences in corolla size between the parental forms (East, 1916).

THE CONSTITUTION OF MULTIPLE GENE SYSTEMS Multiple gene systems can and do vary in their constitution. Six important variable factors in the organization and workings of multiple gene systems are discussed: the number of multiple genes in the gene system; their linkage relations; heritability; the mode of allele interaction; the relative strength of action; and direction of action of the component genes.

The number of multiple genes governing a quantitative character may be two or more, up to some undefined point at which the multiple gene system grades into a polygene system. If these genes are unlinked, are equal in strength, and do not exert dominance, they will segregate in F_2 to produce a frequency distribution of phenotypes according to the ratios given in Table 20.

The number of unlinked multiple genes for which a hybrid is segregating can be inferred from the frequency with which the parental types are recovered in the F_2 generation. One parental type is recovered in the following proportions in an F_2 generation segregating for g independent multiple factors (Anderson, 1949, p. 29; Clausen and Hiesey, 1958, p. 44):

$$g = 2, \quad 1/16$$
$$g = 3, \quad 1/64$$
$$g = 4, \quad 1/256$$
$$g = 5, \quad 1/1024$$
$$g = n, \quad (1/4)^n$$

The numerator in these fractions may represent either parent if the multiple genes have equal and additive effects and no dominance.

TABLE 20

THEORETICAL RATIOS IN AN F_2 POPULATION SEGREGATING FOR VARIOUS
NUMBERS OF INDEPENDENT GENES WITH EQUAL AND ADDITIVE EFFECTS AND
NO DOMINANCE (MODIFIED FROM CLAUSEN AND HIESEY, 1958)

No. of allele pairs	Ratio	n
1	1 : 2 : 1	4
2	1 : 4 : 6 : 4 : 1	16
3	1 : 6 : 15 : 20 : 15 : 6 : 1	64
4	1 : 8 : 28 : 56 : 70 : 56 : 28 : 8 : 1	256
5	1 : 10 : 45 : 120 : 210 : 252 : 210 : 120 : 45 : 10 : 1	1024

In more complex multiple factor systems that are complicated by the existence of dominance or antagonism between the component genes, the numerator represents the bottom recessive type only.

If the multiple genes are linked, the ratios for a given number of genes will approach those resulting from the segregation of fewer genes. The general effect of linkage is to cause a complex multiple gene system to simulate a simpler gene system in its breeding behavior.

Heritability is the fraction of the observed variation in a character which is due to genotypic differences, as contrasted with the fraction caused by environmental factors. The heritability of many quantitative characters is often relatively low. It is high in some cases, including corolla tube length in *Nicotiana longiflora*. The effect of decreasing heritability is to make the frequency distribution of observed phenotypic variations in F_2 more continuous for any given number of segregating multiple genes under any condition of dominance (Allard, 1960, p. 84). In other words, a simple multiple gene system with low heritability will produce a frequency curve of phenotypes in F_2 like a more complex multiple gene system with higher heritability. For further discussion of heritability see Lerner (1958), Allard (1960), and Falconer (1960).

Interallele relations of dominance and recessiveness may exist in all genes of a multiple gene system, in none of them, or in some but not others. Dominance causes the F_2 frequency curve to be skewed toward the condition of the dominant parent. This aspect is discussed in more detail later.

The individual genes in a multiple gene system may have equal and additive effects (polymery), unequal and additive effects (anisomery), or total effects which are the summation of additive and subtractive components (oppositional systems, in part).

These three possibilities can be illustrated by hypothetical examples. Suppose that a quantitative character, a certain growth rate or mature size, requires the production of a growth substance at an appropriate quantitative level, for example, 12 units. Suppose further that there are three multiple genes in the genotype to produce this growth substance. Then it is clearly possible to achieve the same end result of 12 units of growth substance by several alternative pathways. A polymeric multiple gene system would be one in which each of three allele pairs produces 4 units of growth substance. An anisomeric multiple gene system might produce the 12 units by the summation of unequal gene products (e.g., $6 + 3 + 3$). An oppositional multiple gene system could produce 12 units by a combination of additive and subtractive gene effects (e.g., $7 + 7 - 2$).

The characteristics of polymeric, anisomeric, and oppositional systems are discussed further in the sections that follow. To facilitate the discussion we will hold other variable factors constant. In particular, we will assume that we are dealing in all cases with three independent multiple genes with complete heritability. The first set of variable factors is then the relative strength and direction of action of the three component genes. The second factor which is made to vary within each type of multiple gene system is the presence or absence of dominance. (See also Grant, 1964, ch. 5.)

POLYMERIC MODELS   Polymeric genes are multiple genes with additive and equal effects (Nilsson-Ehle, 1909). Here we consider the inheritance of a hypothetical quantitative character controlled by three independent polymeric genes (A, B, C) in a diploid system. We assume that the quantitative character has a value of 18 in one parental type and a value of 6 in the other. These character expressions have high heritability. We can then proceed to examine the expected distribution of phenotypes in F_1 and F_2 under several conditions of dominance.

In the first case, no dominance is present between homologous alleles at any gene locus. The genotypes of the parents and F_1 and the phenotypic score values determined by these genotypes will be:

$$P_1 \quad AABBCC \quad 6+6+6 = 18$$
$$P_2 \quad aabbcc \quad 2+2+2 = 6$$
$$F_1 \quad AaBbCc \quad 4+4+4 = 12$$

The hybrid is exactly intermediate between the parents. The frequency distribution of phenotypes in F_2 now forms a symmetrical bell-shaped curve with a peak at 12 and extremes at 18 and 6, as shown in Figure 19A.

In the next case, dominance is present in the A gene, but not in B and C, so that Aa has the same score value as AA, while Bb and Cc have intermediate values as before. The phenotypic values of parents and F_1 are now:

$$P_1 \quad AABBCC \quad 6+6+6 = 18$$
$$P_2 \quad aabbcc \quad 2+2+2 = 6$$
$$F_1 \quad AaBbCc \quad 6+4+4 = 14$$

The F_1 hybrid, though intermediate, is skewed in the direction of the parental type carrying the one dominant allele pair (see Fig. 19B). The F_2 generation now forms a frequency curve which is also skewed toward the P_1 type (Fig. 19B).

If two genes, A and B, exhibit dominance, while the third gene C has no dominance, the characteristics of the F_1 and F_2 will be still more strongly skewed toward the dominant parent (Fig. 19C).

If dominance is present in all three genes, the expected results in F_1 and F_2 will be as shown in Figure 19D.

ANISOMERIC MODELS Anisomeric genes are multiple genes that have additive but unequal effects, with some individual genes producing stronger effects than others (Sirks, 1956). We examine

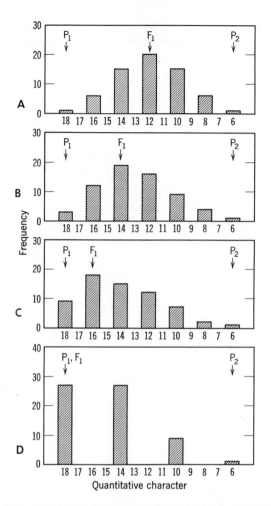

Figure 19. Expected distribution of phenotypes in the F$_2$ progeny of a hybrid segregating for three independent polymeric genes. The characters of the parents and F$_1$ are indicated by arrows. (*A*) No dominance in any gene. (*B*) Dominance in one gene. (*C*) Dominance in two genes. (*D*) Dominance in all three genes. (From Grant, 1964, reproduced by permission of John Wiley and Sons.)

here a series of anisomeric models which are like the preceding polymeric models in every feature except relative strength of the component genes. Let us assume now that the *A* gene has quantitative effects twice as strong as those of *B* or *C*.

The first case is one with no dominance at any gene locus. The parents and F_1 will have the following quantitative values:

$$P_1 \quad AABBCC \quad 9 + 4.5 + 4.5 = 18$$
$$P_2 \quad aabbcc \quad 3 + 1.5 + 1.5 = 6$$
$$F_1 \quad AaBbCc \quad 6 + 3 + 3 = 12$$

The frequency curve in F_2 will be symmetrical with a peak at 12 (Fig. 20A). In this respect it is like the curve for the corresponding case involving polymeric genes. A larger number of phenotypic classes are present in the F_2 generation in this case, however, and the variation is more continuous. Furthermore, the curve is lower and broader here than in the corresponding polymeric case (compare Figures 19A and 20A).

The second case involves dominance in the strong gene A only. The result in F_1 is:

$$P_1 \quad AABBCC \quad 9 + 4.5 + 4.5 = 18$$
$$P_2 \quad aabbcc \quad 3 + 1.5 + 1.5 = 6$$
$$F_1 \quad AaBbCc \quad 9 + 3 + 3 = 15$$

The F_1 is skewed toward the dominant parent, as is the F_2 (Fig. 20B).

If dominance is present in both A and B, the score value for the F_1 hybrid and for the modal type of F_2 individual is 16.5, and is thus still more skewed (Fig. 20C).

With dominance in all three genes, the F_2 distribution is strongly skewed, as in Figure 20D.

OPPOSITIONAL MODELS   In an oppositional multiple gene system, some component genes have additive actions, others subtractive actions, and the quantitative character is the result of a balance between the opposing genes. We assume here, as before, that two parental types differing by three genes (A, B, C) have quantitative characters with values of 18 and 6, respectively. But now the character difference is determined by oppositional gene interactions.

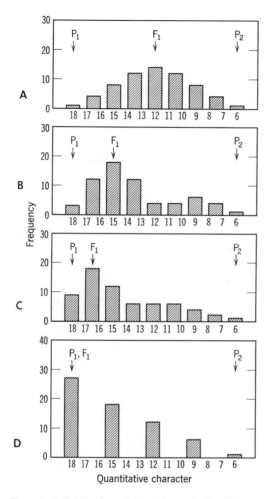

Figure 20. Expected distribution of phenotypes in the F_2 progeny of a hybrid segregating for three anisomeric genes. The characters of the parents and F_1 are indicated by arrows. (A) No dominance. (B) Dominance in the strong gene. (C) Dominance in the strong gene and one of the weak genes. (D) Dominance in all three genes. (From Grant, 1964, reproduced by permission of John Wiley and Sons.)

Let us suppose that the two parental types achieve their respective phenotypic expressions as a result of an interaction between a strongly positive *A* gene and weakly negative *B* and *C* genes. The genes have the following values:

$$\text{P}_1 \quad AABBCC \quad 26 - 4 - 4 = 18$$
$$\text{P}_2 \quad aabbcc \quad 10 - 2 - 2 = \ 6$$

The cross of $\text{P}_1 \times \text{P}_2$ will now yield some F_2 progeny which are more extreme phenotypically than either parent. This phenomenon is known as transgressive segregation. It is due to the formation of new gene combinations in which the balance of forces present in the parental genotypes is broken up by recombination. For instance, the genotype $AAbbcc$ which can arise in the F_2 generation will have the effect $26 - 2 - 2 = 22$, and will thus have a quantitatively greater expression of the character in question than does the greatest parental type. Conversely, the genotype $aaBBCC$, with a value of $10 - 4 - 4 = 2$, will produce a phenotype with a quantitatively smaller score value than that of the smallest parent.

Let us now consider the effects of dominance in an oppositional gene system of this sort. If no dominance is present in any gene, the phenotypic rating of the hybrid as compared with the parents will be:

$$\text{P}_1 \quad AABBCC \quad 26 - 4 - 4 = 18$$
$$\text{P}_2 \quad aabbcc \quad 10 - 2 - 2 = \ 6$$
$$\text{F}_1 \quad AaBbCc \quad 18 - 3 - 3 = 12$$

The F_1 is thus exactly intermediate. The F_2 generation will have the distribution of phenotypes shown in Figure 21A.

If dominance is present in the gene A, the quantitative measure of the F_1 hybrid will be:

$$\text{F}_1 \quad AaBbCc \quad 26 - 3 - 3 = 20$$

This is an interesting situation in which not only some F_2 types but also the F_1s exceed either parent phenotypically (see Figure 21B.).

With dominance in two genes, A and B, the F_1 hybrid is again transgressive, though not as strongly so. The F_2 generation segregates as in Figure 21C.

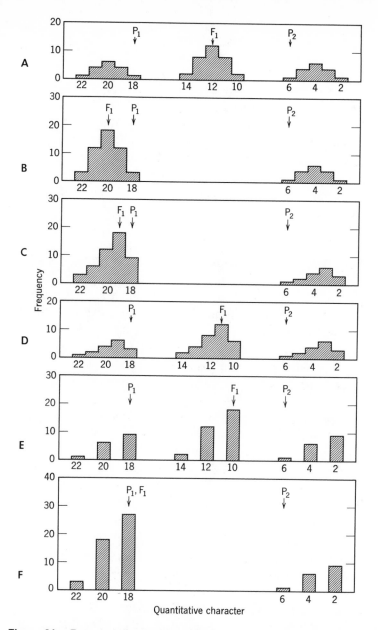

Figure 21. Expected distribution of phenotypes in the F_2 progeny of a hybrid segregating for three independent genes with unequal and oppositional effects. The characters of the parents and F_1 are indicated by arrows. (A) No dominance. (B) Dominance in the one strong gene. (C) Dominance in the strong gene and one weak gene. (D) Dominance in one weak gene. (E) Dominance in two weak genes. (F) Dominance in all three genes. (From Grant, 1964, reproduced by permission of John Wiley and Sons.)

Two other situations involving dominance in genes B or B and C are diagrammed in Figure 21D and E.

With dominance in all three genes, finally, the phenotypic expression of the F_1 hybrid coincides with that of the parent carrying the dominant alleles. The distribution of phenotypes in F_2 is now more strongly skewed toward the dominant parent than in any previous case. Nevertheless, the occurrence of transgressive segregants in the F_2 generation remains as a revealing clue of the type of gene interaction involved (see Figure 21F).

GENERAL CONCLUSIONS DRAWN FROM THE MODELS   A striking difference is found between the oppositional models, on the one hand, and the polymeric and anisomeric models, on the other. In the oppositional multiple gene system, transgressive segregation occurs in the F_2 generation, and sometimes in the F_1. By contrast, the multiple gene systems composed of wholly additive genes give F_2 variations spanning only the range between the parental types.

We examined models of polymeric and anisomeric systems composed of three independent genes. As the number of independently segregating multiple genes with additive effects becomes larger in such systems, the F_2 generation exhibits a lower frequency of the parental types, a larger number of phenotypic classes, and consequently a more continuous series of variations.

For any given number of additive multiple factors, the height and breadth (or kurtosis) of the curve describing the frequency distribution of the F_2 phenotypes is affected by the relative strengths of action of the different multiple genes. Anisomeric gene systems segregate into a larger number of phenotypic classes than do polymeric systems of the same size. Consequently the F_2 variation derived from the segregation of unequal genes is more continuous, more spread out, and exhibits a lower frequency of types with the mean value, than does the F_2 variation resulting from the segregation of the same number of genes with equal effects. In other words, although the frequency distribution curve describing the F_2 variation is bell-shaped for any system of segregating multiple genes with additive effects, this curve has a normal peak when the effects of the different genes are equal, and a lower broader top (i.e., it is more platykurtotic) when the effects of the several genes are unequal.

The difference between polymeric and anisomeric systems is more apparent in theoretical models than in practical experimental work. It would be very difficult to distinguish between polymeric and anisomeric genes on the basis of experimental data.

Skewness of the F_2 variation is related to the presence of dominance. If dominance is present in one of the segregating allele pairs within the multiple gene system, the distribution of phenotypic variations in the F_2 generation will be skewed in the direction of the parental type contributing the dominant allele. In general, the greater the number of genes possessing dominant alleles, the greater is the skewness of the F_2 frequency distribution. If dominance is not present in any of the multiple genes, on the other hand, the bell-shaped curve representing the F_2 variation is symmetrical, the peak and the mean being centered between the two parental values. This is true whether the multiple genes have equal or unequal effects. Statements to the contrary, which relate skewness to unequal gene effects, are present occasionally in the literature but are not well founded.

The bar graphs in Figures 19–21 show the expected distribution of phenotypes when the quantitative character has complete heritability. The assumption of complete heritability is convenient but unrealistic. The bar graphs can also be read as portraying the genotypic component of the observed phenotypic variation in a character with normal, reduced heritability.

Finally, it is important to bear in mind that the segregation of a quantitative character observed in the progeny of a hybrid does not necessarily expose for analysis all of the genes involved in the development of that character. The segregation reveals only those genes that are represented by different alleles in the two parents. The hybrid progeny derived from a cross between two plants differing in stature or growth rate do not segregate for stature or growth rate as such; they segregate with respect to the quantitative differences between the parents in those traits and with respect to the allelic differences responsible for the character differences. The number of segregating genes is probably much smaller in most cases than the total number of multiple genes concerned with the particular character, for the parents can be assumed to be genetically alike and their F_1 hybrids homozygous with respect to some of the genes constituting the multiple factor system. Hybridization experiments thus give us only a glimpse, not a complete picture, of the gene systems controlling various characters; incomplete though it is, it is the only insight we have.

EXPERIMENTAL EXAMPLES OF ADDITIVE GENE EFFECTS   East's (1916) study of the inheritance of corolla tube length in *Nicotiana longiflora* furnishes a good example of a character controlled by multiple genes with additive effects and no dominance. This case has already been described earlier in this chapter (see Tables 18 and 19, and Figure 18). The pattern of segregation in F_2 suggests that several independent genes for corolla size, perhaps four, are involved here, and the intermediacy of the peaks in both the F_1 and F_2 generations indicates absence of dominance.

The contrasting situation of dominance in all multiple genes is illustrated by the duplicate factor inheritance of capsule shape in *Capsella bursa-pastoris* and triplicate factor inheritance of kernel color in wheat, *Triticum aestivum*. These cases were also discussed earlier in this chapter. The strong skewness of the segregation ratios toward one parental type (triangular capsules in Capsella, red kernels in wheat), and the recovery of the opposite parental type in duplicate and triplicate factor ratios (1/16 and 1/64) indicate dominance relations in the two or three multiple genes involved (Nilsson-Ehle, 1909; Shull, 1914).

One would also expect to find cases of additive multiple genes with dominance at some loci only. A probable case is the inheritance of stigma lobe length in interracial hybrids of *Gilia capitata* (Polemoniaceae). One parental race, *G. c. capitata,* has short stigmas, and the other, *G. c. chamissonis,* has long stigmas. Their F_1 hybrid is off-center in the direction of *G. c. capitata* (see Table 21) (Grant, 1950b).

The F_2 generation contains a continuous series of stigma lengths, as seen in Table 21. One parental type is recovered among

TABLE 21

INHERITANCE OF STIGMA LENGTH IN *GILIA CAPITATA* (GRANT, 1950B)

Generation	Frequency distribution of individuals with stigmas of a given length													n
	0.3	0.4	0.5	0.6	0.7	0.8	0.9	1.0	1.1	1.2	1.3	1.4	1.5 mm	
P G.c. capitata	+	+												
P G.c. chamissonis											+	+		
F_1				+	+									
F_2			1	11	43	57	33	24	3	2	1		1	176
B_1 (ca × ch) × ca	1	3	9	1	1	1	1							17
B_1 (ca × ch) × ch				3	1	3			1	1				9

176 F_2 progeny, suggesting that the difference in stigma length between the two races of Gilia is due to differences in three or four genes. The frequency distribution of F_2 types is skewed toward short stigmas. This skewness is manifested again in the backcross progeny derived from crossing the F_1 hybrids with either parental race, as shown by the table. The degree of skewness is not great (about the same as that in Figure 19B), which suggests that dominant alleles are carried by *G. c. capitata* for only a fraction, say one in three, of the additive genes involved (Grant, 1950b).

Two of the races of *Potentilla glandulosa* differ in their seasonal periodicity. A coastal race from Santa Barbara, California, remains active during the winter, while an alpine race from the Sierra Nevada becomes dormant in winter, even when grown in a lowland transplant garden. Clausen and Hiesey (1958) crossed the two races and scored the degree of winter activity of the F_1 hybrids and F_2 progeny. The F_1 hybrids made moderate growth in midwinter and were exactly intermediate between the parents. In the F_2 generation, 975 individuals ranged from fully active to entirely dormant during the winter, with a peak in the intermediate range, as shown in Table 22.

The frequency of one of the parental types recovered in the F_2 generation, 14/975, comes close to the 1/64 ratio expected for three independently segregating genes, and for this reason Clausen and Hiesey attribute the differences in winter activity between the two races to a multiple factor system composed of approximately three genes. None of the F_2 individuals were more extreme in their

TABLE 22
INHERITANCE OF WINTER DORMANCY IN *POTENTILLA GLANDULOSA* (CLAUSEN AND HIESEY, 1958)

Degree of winter activity	Coastal race	Alpine race	F_1	Observed segregation in F_2	Expected segregation with three polymeric genes
Fully active	+			43	15
Fairly active				139 ⎫	⎧
Intermediate		+		601 ⎬ 918 945 ⎨	
Fairly dormant				178 ⎭	⎩
Entirely domant		+		14	15

phenotypic expression than the parental races, indicating that the genes all have additive effects. One cannot decide from the data whether the genes have equal or unequal strengths, but quite possibly, as Clausen and Hiesey suggest, their effects are about equal. Since the modal condition in the F_1 and F_2 generations is unskewed, dominance seems to be absent in each of the several pairs of alleles (Clausen and Hiesey, 1958, pp. 101–2, 109).

Many other cases of multiple factor inheritance with additive gene effects have been described in *Camelina sativa, Mimulus cardinalis* and its relatives, *Mimulus guttatus, Pennisetum glaucum, Pisum sativum, Plantago major, Solidago sempervirens* and *S. rugosa, Zea mays,* and other plants. Additive multiple gene systems are common and widespread. Interesting monographs on multiple factor inheritance in *Potentilla glandulosa* and the *Mimulus cardinalis* group are presented by Clausen and Hiesey (1958) and Hiesey, Nobs and Björkman (1971).

It should be emphasized that experimental results can usually be fitted to hypothetical models only in their broad features. The complexity of actual multiple gene systems and the technical difficulties caused by low heritability of quantitative characters often make it impossible to infer the exact nature of the underlying gene system with certainty. Some of the problems are discussed by Darlington and Mather (1949, ch. 4), Mather (1949), and Comstock (1955).

EXPERIMENTAL EXAMPLES OF TRANSGRESSIVE SEGREGATION

Transgressive segregation in a quantitative character indicates an oppositional multiple gene system consisting of additive and subtractive factors.

A classic case concerns the density of the ear in wheat (Nilsson-Ehle, 1911). The wheat ear varies from compact to open, depending upon the length of the internodes between successive spikelets. A cross between a compact and an open-eared variety produced a compact F_1 hybrid. The F_1 then yielded F_2 and F_3 progenies which included the parental and intermediate types and also some super-compact and some extremely loose-eared forms. The extreme transgressive types appeared in low frequencies in F_2 but gave rise to nonsegregating families in F_3.

Nilsson-Ehle explained the observed results in terms of an oppositional gene system containing two genes (L_1, L_2) with additive effects for internode length in the ear, and a subtractive gene (C) which shortens the internode. The compact parental type has the genotypic constitution $CCL_1L_1L_2L_2$ and the open-eared parent is $ccl_1l_1l_2l_2$. The extreme segregates are recombinations of C and L. The super-compact type is $CCl_1l_1l_2l_2$; the loose-eared type is $ccL_1L_1L_2L_2$ (Nilsson-Ehle, 1911).

The rosette leaves of the perennial herb *Potentilla glandulosa* range in size from small in an alpine race to large in a coastal race. The F_2 progeny of a cross between an alpine race from the Sierra Nevada of California and a race from the California coast shows continuous variation in leaf size, with a low frequency of one parental type. Transgressive segregation is found in the F_2 generation, some individuals having smaller leaves than the alpine parent, and many individuals having larger leaves than the coastal parent (see Figure 22). These facts suggest an oppositional multiple gene system (Clausen and Hiesey, 1958).

An additional interesting feature of leaf-size inheritance in *Potentilla glandulosa* is that the F_1 hybrid is also transgressive, having larger leaves than the coastal parent (Fig. 22) (Clausen and Hiesey, 1958). The observed pattern of transgressive variation in F_1 and F_2 as shown in Figure 22 is similar to the hypothetical model in Figure 21B. This similarity suggests, in turn, that

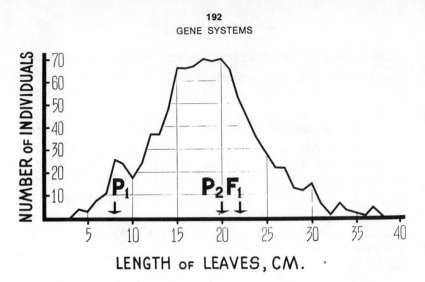

Figure 22. Frequency distribution of the length of the rosette leaf among 996 F_2 progeny of an interracial cross in *Potentilla glandulosa*. Characters of parents and F_1 hybrids are indicated by arrows. (From Clausen and Hiesey, 1958; photograph courtesy of W. M. Hiesey.)

the oppositional system controlling leaf size is complicated by the effects of dominance. Perhaps one or more of the strong genes for leaf size are represented by dominant alleles in the coastal race.

Different cultivated varieties of squash, *Cucurbita pepo*, differ markedly in the shape of the fruit. There are flat disk-shaped forms, spherical ones, elongated ones, and intergrades between these modal types. The various shapes can be resolved into different balances between growth rate in width and growth rate in length. The relative rates along different axes of growth are controlled by a fairly complex oppositional multiple gene system (Sinnott and Hammond, 1930).

Four genes (*A, B, C, D*) act to produce a short longitudinal axis and hence a disk-shaped fruit. A gene *I* acts in the opposite direction to bring about growth in length. The two series of genes work in balance. The equilibrium point between them is affected by dominance at the various loci (Sinnott and Hammond, 1930).

The disk shape is dominant, and elongated shapes are recessive. The dominant alleles of the *A–B* series have a shortening effect whereas the corresponding recessive alleles (*a, b*) produce elongated shapes. Conversely, the dominant *I* allele of the gene *I* brings about elongation, but the recessive *i* allele does not. Spherical fruits result from various allele combinations of the shortening genes and elongating genes, such as *AaCcIi* or *AAccii*. With a greater preponderance of the shortening genes (for instance, in *AABBii*) the fruit becomes disklike. With a stronger dose of the dominant *I* allele, on the other hand (e.g., in *BBII*), the fruit becomes elongated (Sinnott and Hammond, 1930).

Transgressive segregation is observed, as expected, in intervarietal crosses in squash. Thus, two spherical-fruited parents may segregate some elongated types in F_2 (Sinnott and Hammond, 1930).

CHAPTER ELEVEN

POLYGENES

INTRODUCTION
POSTULATED PROPERTIES OF POLYGENES
MULTIPLICATION OF GENE LOCI
UNIDENTIFIED GENETIC MATERIAL IN CHROMOSOMES
HETEROCHROMATIN AND POLYGENES
REPETITIVE DNA
POLYGENES IN DROSOPHILA
BALANCED POLYGENIC SYSTEMS

INTRODUCTION Polygenes are an elusive and ambiguous type of gene. Their properties are certainly very similar to those of multiple genes. If polygenes are merely multiple genes passing under a different name, there is, of course, no problem except the semantic one. If polygenic systems and multiple gene systems simply differ in degree rather than in kind, there is again no particular problem, and a separate designation and treatment are unnecessary. The third possibility is that polygenes, despite resemblances to multiple genes, form a more or less distinct class of genetic elements.

The first two possibilities are in line with most of the available experimental evidence. They obey the law of parsimony, are easy to defend, and probably represent a majority opinion among organismic geneticists. Nevertheless, we have chosen the course of greatest resistance. Following Mather and a few other students (Mather, 1941, 1943; Darlington and Mather, 1949; Grant, 1964), we argue in favor of the third proposition, which is a minority viewpoint.

The case for polygenes, considered as a distinct type of gene, does not stand on an extensive experimental foundation. Nevertheless, the polygene concept should not be dismissed lightly because of other considerations. There are some anomalies in classical genetics which indicate that Mendelian genes, including multiple genes in the strict sense, do not represent the whole story. Some of these anomalies can be accounted for by polygenes. In short, there are grounds for *predicting* the existence of polygenes. The main thrust of our argument in this chapter is to substantiate this prediction; the search for confirmatory evidence comes later in the chapter.

POSTULATED PROPERTIES OF POLYGENES Polygenes in the strict sense are characterized as chromosomal genes which individually have minute phenotypic effects, but which collectively, in polygenic systems, have measurable cumulative effects expressed as quantitative characters. The effects of the polygenes on the phenotypic character are small in comparison with the effects of environmental influences.

Polygenic systems are believed to consist of numerous minute genes of this sort, distributed throughout the genome. The supposedly large number of polygenes makes it inevitable that many of them will be linked together on the same chromosomes. Another consequence of the numbers of loci involved is that character differences controlled by polygenic systems do not segregate into discrete phenotypic classes, but instead show continuous variation in segregating progenies (Mather, 1943; Darlington and Mather, 1949).

Polygenic systems permit the segregation of a finely graded series of phenotypes, and polygenes are therefore referred to as the genes of fine adjustment (Mather, 1943, 1949).

Polygenic systems are presumed to differ from multiple gene systems in having larger numbers of loci with correspondingly smaller individual effects. The segregation of single components of a multiple gene system produces phenotypic variations which can be detected and traced under favorable experimental conditions, whereas the segregation of single polygenes does not. In one example of a multiple gene system, ratios of 63 : 1, 15 : 1, and 3 : 1 were found in different F_3 families of wheat segregating for kernel color (Chapter 10). Polygenic characters display continuous variation only.

I do not pretend to know where to draw the line between multiple gene systems and polygenic systems as regards the number of loci. The polygene hypothesis applies to situations where rather large numbers of genetic factors are apparently involved in the control of a quantitative character. Nor should we attempt to decide at this stage whether multiple gene systems and polygenic systems are completely distinct, with no transitional conditions, or whether they are just very different. I suspect that typical multiple genes and typical polygenes lie at opposite ends of a long spectrum. However, it is sufficient for now to establish, if we can, the existence of a class of polygenic systems which is very different from other known types of gene systems.

The standard methods of factorial analysis fail in attempts to deal with polygenic systems because of the small cumulative effects of the individual polygenes. This poses a logical problem which is well stated by Darlington and Mather (1949, p. 68): "We have been led to postulate systems of genes which cannot be analysed by the Mendelian method. How then can we be sure that these genes are inherited in accordance with Mendelian principles?"

Darlington and Mather (1949) answer this question by laying down the criterion that polygenes should show both segregation and linkage, like other chromosomal genes, and, if they do so, can be concluded to conform to Mendelian principles. As confirmatory evidence they cite experimental results, including East's (1916) study of corolla size in *Nicotiana longiflora,* which I would classify as evidence for multiple gene systems. The problem of a special type of polygenic system thus disappears in the melting pot of multiple gene systems.

I prefer to restate the question in the following fashion: If polygenes cannot be analyzed by the Mendelian method, how then can we know that they exist? The answer is that we do not know for sure, but we predict their existence from some lines of indirect evidence. Direct evidence is also available but is not overwhelming. The indirect evidence is suggestive enough to stimulate a search for new methods which are appropriate for the detection of this type of genetic element.

MULTIPLICATION OF GENE LOCI   Our first approach to the problem of polygenes is to draw inferences from the demands of quantitative inheritance. Quantitative characters require the concurrent action of several or many genic sites to produce growth substances in the required quantity.

The value of having a large number of active gene loci is illustrated by the condition of glandular and nutritive tissues. Such tissues are frequently endopolyploid or polytene.

The salivary-gland chromosomes of Drosophila, which are polytene with about 1000 parallel chromatin strands, provide a well-known example. Higher degrees of polyteny occur in the salivary-gland chromosomes of some other Diptera. Endopolyploidy is found in the silk glands of moths and the gut lining of mosquitoes and water striders (Gerris). Endopolyploid tissues are found again in the intestine of mammals and in the liver of rodents, frogs, and other vertebrates. In flowering plants, endopolyploidy is common in nutritive cells of the anthers and embryo sac, and polyploidy is the usual condition in the endosperm of the seed. For reviews of polyteny and endopolyploidy see Brown (1972, chs. 3 and 12).

An interesting example of the multiplication of gene loci and amplification of gene activity is provided by the salivary glands of a fly, Hybosciara (Sciaridae). This fly has polytene chromosomes in the salivary glands in which each gene is represented by many replicate copies. Certain bands of the salivary chromosomes then go on to produce special micronucleoli in the nucleus. These micronucleoli presumably contain particular genes multiplied to the second order, thus increasing still further the number of metabolically active loci (Da Cunha, Pavan, Morgante and Garrido, 1969).

Glandular and nutritive tissues have an exceptionally high metabolic activity. Their metabolic requirements are met by multiplying the number of active genes through polyploidy or polyteny. Diploid tissues engaged in growth do not, by definition, have the option of increasing the number of genic sites in this way, but they could do so by incorporating a linear series of duplications into the diploid chromosome complement (Grant, 1964).

Suppose that a plant requires some substance, say a vitamin, in the concentration of 12 units per gram of tissue for normal growth in diploid tissues. Multiple gene systems composed of several genes represent one way of accomplishing this result, as noted in Chapter 10. Thus polymeric genes at 3 loci might produce 4 units of the growth substance each, or 6 polymeric gene loci could produce 2 units each. It is important in relation to our present subject to realize that polygenes represent an alternative route to the same goal. The 12 units of growth substance could be produced by 50 or 60 polygenes, each of which yields a fraction of a gram. Or some combination of one or two major genes and numerous polygenes could build up the production of growth substance to the desirable normal level.

Small multiple gene systems therefore do not represent the only possible solution to problems of development of quantitative characters. It is improbable that organisms would "choose" only one of several alternative pathways to reach a given "goal." We would expect sets of polygenes to be assembled in the genotype, in some instances anyway, as a means of achieving certain definite quantitative levels of development.

In fact, the given quantitative norm may be attained more readily in some cases by utilizing polygenes than by utilizing multiple genes. Assume that the plant population is currently ill-adapted in relation to the optimum condition of 12 units of a growth substance per gram of tissue. New mutant alleles of multiple genes with the desirable strength of action might or might not arise by the random process of gene mutation during the history of the plant population. But polygenes with individually minute effects could be duplicated or deleted to the precise quantitative level required for adaptation.

UNIDENTIFIED GENETIC MATERIAL IN CHROMOSOMES The DNA content of individual chromosomes and of haploid nuclei has been measured in a variety of organisms in recent years (see Ris and Kubai, 1970). The total amount of DNA in the genome turns out to be far in excess of the number of genes as estimated by classical genetic methods. The discrepancy could be significant in relation to the polygene problem.

The methods of cytogenetics and mutation genetics lead to estimates of 5,000–15,000 Mendelian genes in *Drosophila melanogaster,* as noted in Chapter 7.

Ris and Kubai (1970) give the length of the DNA in chromosome III of *D. melanogaster* as 51 mm, or 51,000 μ. If we divide this total length by the length of an average cistron, say 0.35 μ, we arrive at an estimate of about 145,000 cistrons in chromosome III. In order to translate this number into numbers of Mendelian genes we should reduce it somewhat to allow for compound loci and for zones of crossing-over. With all due allowances, however, the length of DNA is greatly in excess of the number of genes as estimated by other methods.

Crow (1972) states that the Drosophila nucleus contains enough DNA to make several hundred thousand genes. He considers that about 6000 (Mendelian) genes is a reasonable estimate on the basis of classical genetic evidence. Therefore, the excess of DNA content over probable gene number amounts to an order of magnitude in Drosophila. An even greater discrepancy exists between estimates of DNA quantity and estimates of gene number in mammals (Crow, 1972).

Crow raises the appropriate question, What is all this DNA doing? Various suggestions have been offered by different students —nonfunctional DNA (which, in any considerable amount, seems unlikely to an evolutionary geneticist); regulation (very possible); useful redundancies of DNA sequences (also an attractive possibility to which we will return later).

Among higher plants, Rees and others have discovered substantial variations in the amount of nuclear DNA in Lathyrus, Lolium, Vicia, Allium, and Sorghum. Some species in these genera have more nuclear DNA than related species with the same chromosome number (Rees, Cameron, Hazarika and Jones, 1966; Jones and Rees, 1968; Chooi, 1971a, 1971b; Paroda and Rees, 1971). The role of the excess DNA in the species possessing it is again unknown. We will return to the problem of interspecific differences in nuclear DNA content in Chapter 16.

In summary, there is evidence for an excess of chromosomal DNA which is not incorporated into Mendelian genes. Its genetic role is largely a matter of surmise at present. Some of the extra

DNA could well be engaged in regulatory functions. It is equally possible that some of it forms the material of the polygenes.

HETEROCHROMATIN AND POLYGENES The genetic activity of heterochromatin is a subject of long-standing interest, which has a bearing on the polygene problem. As described earlier (Chapter 3), heterochromatic and euchromatic regions of chromosomes differ in their coiling cycles and in correlated staining properties; heterochromatin remains in a condensed state while euchromatin alternately coils and uncoils.

The primary genetic role of euchromatin is not in doubt. Nearly all of the known Mendelian genes in Drosophila lie in euchromatin. Furthermore, crossing-over seems to take place only in euchromatic regions (Brown, 1966).

The absence or near absence of Mendelian genes in heterochromatin led to the old idea that heterochromatin was genetically inert. This view, taken literally, has long been abandoned. Heterochromatin does have genetic effects. But these effects are general and sometimes vague, rather than definite and specific as in the case of euchromatic genes (Mather, 1944; Darlington and Mather, 1949; Goldschmidt, 1949, 1955).

We have previously noted that heterochromatin tends to depress the activity of Mendelian genes in nearby euchromatic segments. Heterochromatin also induces some types of variegation (see Chapters 3 and 8), and has other effects as well. Fertility in coccid bugs depends upon the presence of a minimum amount of heterochromatin in the chromosomes (Brown and Nur, 1964).

A group of mutants in *Drosophila melanogaster* known as Minutes differ from wild-type flies in having blunt wings, weak legs, rough eyes, and aberrant male genitalia. These Minute types are due to mutations in any one of numerous regions scattered over all the chromosomes and frequently coinciding with heterochromatic regions. It is remarkable that over 50 chromosomal segments, frequently heterochromatic, carry genetic factors which in a mutant form bring about a similar set of characters (Schultz, 1929; Goldschmidt, 1955).

The Y chromosome of *Drosophila melanogaster* is wholly

heterochromatic and carries very few Mendelian genes. Yet it does exert a quantitative effect. Different Y chromosomes from different strains influence the number of sternopleural bristles on the flies. Similar effects are brought about by the heterochromatic regions of the X chromosome. Such quantitative effects are like those expected from polygenes (Mather, 1944; Darlington and Mather, 1949).

The pairing behavior of heterochromatin is also suggestive of a generalized genetic nature. It is well known that euchromatic regions of the chromosomes pair in meiosis, and sometimes also in mitosis, in a highly specific manner reflecting homologies. Pairing occurs between homologous euchromatic segments and presumably between homologous gene alleles, and not normally between nonhomologous segments. Heterochromatin, by contrast, does not show a high degree of specificity in chromosome pairing. Heterochromatic regions of different nonhomologous chromosomes have a tendency to pair, and they may even unite in a single mass, as in the fused chromocenters of the salivary-gland chromosomes of Drosophila. The generalized mode of attraction of heterochromatin is in line with other indications of a generalized genetic constitution (Goldschmidt, 1949, 1955).

These pieces of evidence have led to the suggestion that polygenes are located mainly in heterochromatin (Mather, 1944, 1949; Darlington and Mather, 1949). This hypothesis provides a single explanation for the observed generalized nature of polygenes and of heterochromatin as manifested in both their quantitative phenotypic effects and their pairing affinities.

REPETITIVE DNA   Recent molecular genetic studies reveal the presence of much redundancy in tissues of various animals and plants. Particular nucleotide sequences of DNA are repeated many times, to yield a substantial fraction of redundant or repetitive DNA in the genome.

The repetitive DNA is identified and quantified by a modified form of the technique of DNA hybridization. The original natural DNA of an organism is purified and separated into single-strand molecules, which are mixed in solution. Complementary strands reassociate to form double-strand molecules in the solution. The reassociation thus indicates genic homology at the molecular level between sequences which may or may not have been paired in the original natural DNA. The rate of reassociation then gives a measure of the amount of molecular homology in the sample of DNA. If the DNA of an organism consists largely of unique nucleotide sequences, the reassociation of purified single-strand molecules will be slight and slow. A rapid rate of reassociation, on the other hand, indicates much repetition in the DNA (Britten and Kohne, 1968; Laird and McCarthy, 1968).

On the basis of this technique, Laird and McCarthy (1968) estimate that about 10% of the DNA in the genome of *Drosophila melanogaster* consists of repetitive sequences.

Britten and Kohne (1968) find the same or higher proportions of repetitive DNA in mammals. About 10% of the DNA of mice and 40% of the DNA of cattle consists of repetitive sequences. Repetitive DNA is found as well in other vertebrate and invertebrate animals. Britten and Kohne state that certain genes are repeated between a thousand and a million times each in higher animals.

Britten and Kohne (1968) have also detected repetitive DNA in eight species of higher plants, but they do not specify the amount of repetitiveness. The plant species are *Allium* sp., *Hordeum vulgare, Nicotiana glauca, Phaseolus vulgaris, Pisum sativum, Secale* sp., *Triticum aestivum,* and *Vicia villosa.* Using similar techniques, Chooi (1971b) has found repetitive DNA in *Vicia sativa, V. faba,* and several other species of vetch.

Britten and Kohne (1968) suggest that part of the observed redundancy in DNA could be an expression of cytologically visible duplications, as are well known in Drosophila, but this can be only a minor part of the total. Another part could reflect similarities in different proteins, like different types of hemoglobins, which are produced by different genes with similar nucleotide sequences, but this is also thought to be a minor fraction of the observed redundancy. Britten and Kohne consider the amount of repetitive DNA to be surprisingly great.

It remains to be seen just how minor the second component of redundancies really is. The technique of DNA hybridization permits the reassociation of DNA strands with similar nucleotide sequences as well as the reassociation of identical sequences. The technique, therefore, as Dobzhansky (1970) has pointed out, probably leads to an overestimate of the amount of truly repetitive DNA and the number of identical genes.

The function of the repetitive DNA is unknown. The same possibilities have been suggested for repetitive DNA as for excess DNA not incorporated into Mendelian genes. The repetitive DNA might be inert (but probably is not); it might function in regulation; or it might increase the rate of synthesis of a gene product required in quantity (Britten and Kohne, 1968). Britten, Kohne, and some other recent authors favor a regulatory function.

The location of the repetitive DNA in the cells remains unclear. The repetitive sequences could be partly in the cytoplasm, partly in homologous segments of polytene chromosomes, partly in polyploid nuclei, partly in micronucleoli, or partly in linear series of duplications. It is not clear to me whether it has been possible to distinguish between these alternatives. Some of the materials used by Britten and Kohne (i.e., calf liver and wheat plants) are polyploid. Of greatest significance in relation to the genetic architecture of chromosomes would be estimates of the amount of repetitive DNA distributed in tandem duplications.

Our main interest here is in the possible relationship of repetitive DNA to polygenes. Consequently we are interested in the fraction of repetitive DNA which is arranged in linear series of duplications. The polygene hypothesis calls for linear series of identical or similar genes with individually small but collectively substantial effects. Repetitive DNA arranged as tandem repeats would seem to fulfill the prescribed characteristics of polygenes.

The possible connection between repetitive DNA and polygenes is brought closer by recent findings. Some older indications pointed to a connection between polygenes and heterochromatin, as noted in a previous section. Yunis and Yasmineh (1971) now report that DNA in heterochromatin consists to a large extent of repeated nucleotide sequences.

Molecular geneticists working on redundancy in DNA have not considered the relevance of their findings to the polygene problem. It would seem that they are bringing forth significant new evidence about the genetic material, some of which could be construed as molecular evidence for polygenes.

At least one other plant geneticist has noted the similarities between repetitive DNA and polygenes. H. H. Smith writes (1971, pp. 270–71):

> From classical genetics, compelling similarities are evident between the polygenic systems governing quantitative character inheritance and the preponderance of repeated sequences of DNA. . . . If repetitive DNA were shown to be a physical basis for polygenic systems, then the reason for its prevalence in terms of adaptive significance would become clearer. . . . Both repetitive increases in DNA and polygenic systems of inheritance contribute to broadening the base of genetic variability in plants. The ultimate establishment of a causal link between them would furnish some simplification to what appears at present to be an increasingly complex picture of the genomic composition in higher forms.

POLYGENES IN DROSOPHILA Experimental evidence is available for polygenic control of the density of body hairs, or bristles, in *Drosophila melanogaster*. The Samarkand and Oregon strains of *D. melanogaster* differ in the number of hairs on the underside of two abdominal segments. The average number of hairs per female fly is 59.2 for the Samarkand strain and 43.5 for the Oregon strain. The character difference is accordingly 15.7 hairs. This character difference cannot be traced to any particular major genes, but it segregates as though several or many genes are involved (Darlington and Mather, 1949; Mather and Harrison, 1949).

The haploid set of *D. melanogaster* contains three large chromosomes and a single small one on which the genes controlling abdominal hairs can be located. Each of these chromosomes can be tagged with genetic markers, and its influence on the density of abdominal hairs can then be assayed in controlled crosses (Darlington and Mather, 1949; Mather and Harrison, 1949).

Such assays indicate that the X chromosome of Samarkand determines at least 1.93 more hairs than the Oregon X chromosome. A Samarkand chromosome II determines at least 2.40 more hairs than its Oregon homolog. And a Samarkand chromosome III produces 4.72 more hairs than an Oregon chromosome III. The hair-producing activity of the three Samarkand chromosomes is probably greater than these figures indicate, owing to limitations in the method of assay. In any case, the three large chromosomes combined carry several or many factors determining most or all of the interstrain difference in hairiness (Darlington and Mather, 1949; Mather and Harrison, 1949).

The density of another type of hair in *D. melanogaster,* the sternopleural bristles or chaetae, is controlled by a number of genes with individually small effects. The distribution of these genes on the X chromosome was investigated in crosses between the Oregon, Samarkand, and other strains, which differ phenotypically and genetically with respect to this character. The genes controlling the number of sternopleural bristles are located in five separate regions throughout the X chromosome. The two ends of the X chromosome have a maximum effect on the bristle character, and therefore may harbor a high concentration of the bristle-determining genes. At least five, and probably more, polygenes for the density of sternopleural bristles are thus found on the X chromosome alone (Wigan, 1949). Another set of polygenes for chaeta number occurs on chromosome II (Harrison and Mather, 1950).

By using a larger number of genetic markers and testing progeny on a larger scale, Thoday has been able to locate some of the polygenes for sternopleural chaeta number on the genetic map of *D. melanogaster.* He finds polygenes governing this character at four locations in chromosome III, namely at 28, 38, 49, and 51 centi-Morgans; and in chromosome II at 27, 47, and 64 centi-Morgans (Thoday, 1961).

BALANCED POLYGENIC SYSTEMS   A balanced polygenic system is a special type of oppositional gene system. It consists of plus and minus polygenes balanced in heterozygous condition and linked in repulsion phase, according to the generalized scheme $\frac{+-+-}{-+-+}$ Mather (1941, 1943) has suggested that polygenes are often or usually organized into balanced systems of this sort. The reasons for this supposition are both theoretical and experimental.

The theoretical argument is that a balanced polygenic system is particularly well suited to reconcile the opposing demands for constancy and variability in sexual reproduction.

The ability of an organism to breed true to type, and to multiply the numbers of individuals carrying its genotype, is advantageous insofar as the organism is well adapted to its current environmental conditions. This selective advantage is the basis of the conservative aspects of the hereditary mechanism. But the ability to engender variable progeny is also selectively valuable to organisms living in a world subject to environmental changes. Thus organisms may be confronted, as Darlington, Mather, Stebbins, and others have noted, with diametrically opposed demands for immediate fitness and long-range flexibility (Darlington, 1936; Mather, 1943; Stebbins, 1950).

Mather states the problem as follows (1943, p. 44):

> Heritable variability is necessary for adaptive change, but, in that it implies some individuals departing from the optimum, it lowers present fitness; for . . . departure from the optimum must be correlated with reduction in fitness. In the same way response to fluctuating environmental changes reduces the heritable variability upon which depends adaptation to environmental trends. The success of any organism, in competition with its contemporaries, must depend on the extent to which it reconciles these needs. Failure to achieve an adequate balance spells either its own doom, on the one hand, or that of its descendants, on the other. Existing organisms must, therefore, have descended from those which had most adequately balanced the advantages of fitness and flexibility in the past. The organisms of the future will equally be descended from those which, to-day, best reconcile the needs of fitness and flexibility, the rest dying out sooner or later.

The conflicting demands for fitness and flexibility can be partly reconciled by having variability stored in a potential state. In this connection it is useful to distinguish between potential and free variability. Free variability is that which is manifested by the phenotypes in a population and which is, therefore, exposed to the action of selection. Potential variability, on the other hand, is not manifested in the phenotypes, and hence is not exposed to selection in any given generation, but can become manifest in later generations by passing from the potential to the free state (Mather, 1943).

Potential variability is present in heterozygous genotypes. A homozygote gives rise to genotypically identical progeny; however great its fitness, its potential variability is nil and its flexibility slight. A heterozygous genotype, on the other hand, may possess the potential variability and flexibility which are wanting in homozygotes and may have the same degree of immediate fitness too. Thus individuals of the constitution Aa and AA may be alike phenotypically, assuming dominance of A; but whereas AA breeds true to type, Aa can engender phenotypically dissimilar progeny of the constitution aa. The AA and Aa individuals possess the same degree of immediate fitness resulting from their phenotypic manifestation of A, that is, from their free variability; but the Aa individuals possess in addition future flexibility associated with their potential variability (Mather, 1943).

The next condition to consider is the stability of the system of potential variability. The Aa individuals on intercrossing or selfing produce offspring, one quarter of which are unlike the parents. The potential variability for the trait controlled by the gene A is thus released in a high proportion of the progeny arising in every generation. The system composed of heterozygotes for a single gene is unstable. The stability of a system of potential variability will be increased if the character is determined, not by a single gene, but by a set of polygenes with balanced plus and minus actions (Mather, 1943).

The favorable condition of a character such as size may be a result of the actions of plus polygenes for increased size and minus polygenes opposing them. Assume for the sake of simplicity two polygene loci on one chromosome with a total of two plus polygene alleles and two minus alleles in a diploid heterozygote. These polygenes can be arranged in two types of zygotes which have the same balance of gene effects, and at the same time possess potential variability, namely $\frac{+\,+}{-\,-}$ and $\frac{+\,-}{-\,+}$. The first kind of zygote, like the single-gene heterozygote Aa, is highly unstable, whereas the second kind is stable. Most of the progeny of the latter are

$\dfrac{+\,-}{-\,+}$, $\dfrac{+\,-}{+\,-}$, or $\dfrac{-\,+}{-\,+}$, all of which result in the same net size.

But the $\dfrac{+\,-}{-\,+}$ heterozygote can also release new variability more or less slowly by crossing-over. Crossovers between the two genes on the chromosome can lead to $++$ and $--$ gametes, which can produce $\dfrac{+\,+}{+\,+}$ and $\dfrac{-\,-}{-\,-}$ individuals with a size greater or smaller than normal (Mather, 1943; Thoday, 1958b).

Greater stability is correlated with a larger number of linked heterozygous polygenes, as in the genotype $\dfrac{+\,-\,+\,-}{-\,+\,-\,+}$. Such a genotype permits a more gradual release of stored variability by crossing-over and recombination. New recombination types with a greater than average action in the plus or minus direction can continue to appear over a longer succession of generations.

A gene system controlling a certain character may thus retain potential variability, yet achieve a high degree of stability in its release, by possessing the following combination of properties: The system should be heterozygous for linked polygenes with oppositional interactions and with a spatial arrangement on the chromosome such that neighboring polygenes have opposing directions of action. Such a system of plus and minus polygenes balanced in two ways, first from allele to allele of each gene in heterozygous condition, and secondly from gene to neighboring opposing gene, is a balanced polygenic system on one chromosome (Mather, 1943).

The polygene system may extend to other chromosomes. If each chromosome carrying a part of the polygenic system is internally balanced, i.e., $\dfrac{+\,-\,+\,-}{-\,+\,-\,+}$ and $\dfrac{+\,-\,+\,-}{-\,+\,-\,+}$, then the products of independent assortment of chromosomes will also be balanced (Mather, 1943).

A balanced polygenic system has, as Mather puts it, a statistical stability like the pressure of gas. It contains a great deal of potential variability stored in the heterozygous state. It releases this variability by small increments as a result of crossing-over and recombination between linked polygenes. And the slow and gradual conversion of the variability from the potential to the free state can continue generation after generation (Mather, 1943).

The main experimental evidence for balanced polygenic systems comes from selection experiments. Artificial selection for a quantitative character is carried out over many successive generations. If the population responds to the selection, during a long course of generations, by showing a continual and gradual shift in the quantitative character, one can conclude that the genetic variability underlying the character in question is being released slowly and steadily as expected from a balanced polygenic system.

The predicted results have been obtained in experimental populations of *Drosophila melanogaster* undergoing selection for density of bristles. The bristle characters are those described in the preceding section. The average number of bristles shows a gradual rise in the high selection line, and a gradual decrease in the low line, over many generations (10 or more), until the selection curves eventually level off (Mather, 1941; Mather and Harrison, 1949; Harrison and Mather, 1950). Many later selection experiments with Drosophila and other animals have yielded similar results (see, for example, Scossiroli, 1954). In plants, long-continued response to selection for quantitative characters is exemplified by the well-known Illinois corn selection experiment (Woodworth, Leng and Jugenheimer, 1952; Bonnett, 1954; Leng, 1960).

Gibson and Thoday (1962) have succeeded in locating a small group of balanced oppositional polygenes for chaeta number on chromosome II of *Drosophila melanogaster*. The polygenes occur at two loci in chromosome II at the map poisitions 27.5 and 47.5 centi-Morgans. The flies are heterozygous for the genes at each locus, and the genes are balanced in repulsion phase, $\frac{+\ -}{-\ +}$. The homozygous recombination types are lethal (Gibson and Thoday, 1962).

CHAPTER TWELVE

CYTOPLASMIC INHERITANCE

MATERNAL INFLUENCES
EARLY STUDIES OF PLASTID INHERITANCE
CYTOPLASMIC INHERITANCE IN EPILOBIUM
CYTOPLASMIC MALE STERILITY
HETEROTIC EFFECTS
THE NATURE OF CYTOPLASMIC GENES

MATERNAL INFLUENCES In the vast majority of sexual organisms, fertilization entails the union of two morphologically differentiated gametes, sperm and egg. These dissimilar gametes carry the same amount of nuclear genetic material but different quantities of cytoplasm. Consequently, the amount of chromosomal material contributed to the zygote by the male and female parents is the same or nearly so; but the female parent contributes in addition a large amount of cytoplasm to the zygote, whereas the male parent contributes little or none.

Reciprocal crosses between genetically and phenotypically different parents, A ♀ × B ♂ and B ♀ × A ♂, can be expected to yield similar progeny, insofar as the genetic differences between A and B are carried in the nuclei.

Mendel (1866) found that the products of reciprocal crosses are indeed alike in peas. His statement is quoted here in translation (Mendel, 1965, pp. 13, 15):

> Furthermore, in all the experiments reciprocal crossings were effected in such a way that each of the two varieties which in one set of fertilisations served as seed-bearer in the other set was used as the pollen parent. . . . It was . . . shown by the whole of the experiments that it is perfectly immaterial whether

the dominant character belongs to the seed plant or to the pollen plant; the form of the hybrid remains identical in both cases. This interesting fact was also emphasised by Gärtner, with the remark that even the most practised expert is not in a position to determine in a hybrid which of the two parental species was the seed or the pollen plant.

The similarity of the products of reciprocal crosses has of course been confirmed in many other organisms since then.

Mendel's results, when combined with Hertwig's rule that fertilization involves the union of the nuclei of sperm and egg, point to the nucleus as the chief organ of heredity, a conclusion reached by Hertwig, Weismann, Strassburger, and others in the 1880s.

However, many cases have been found where reciprocal hybrids do differ from one another and resemble their respective maternal parents in some phenotypic character. In such cases the cytoplasm contributed to the zygote by the female parent can be suspected of determining the character in question. Three broad types of maternal influence can be recognized.

The first situation is that where the cytoplasmically determined character persists for only one or two generations. We then have a simple case of maternal effect. The nuclear genotype of the mother determines certain conditions in the cytoplasm of the eggs, and thereby predetermines some particular trait in the hybrid offspring, irrespective of the genes contributed by the father.

A good example of maternal effect, often described in textbooks of genetics, is the direction of coiling in the snail *Limnaea peregra*. The coiling of the shells is either right-handed (R) or left-handed (L). The difference is due to a single chromosomal gene with dominance in the allele for right-handedness. Nevertheless, in reciprocal crosses of R ♀ × L or L ♀ × R, the F_1s are like the mother. F_1s of either the right-handed or the left-handed phenotype give F_2 progeny which all have right-handed coiling. But the F_3s derived from these F_2s segregate in the ratio 3 R : 1 L. The explanation is that the cytoplasm of the egg is determined by the nuclear genotype of the mother long before fertilization takes place. The prepattern for either R or L is then transmitted in the first cleavage divisions of the fertilized egg and expressed later in the phenotype of the shell. The genotype of the sperm has no immediate effect on the phenotype. The male parent does have a delayed effect, however, when its daughters produce eggs of their own.

The second possibility is that the character determined by the cytoplasm persists for a number of generations but eventually disappears. A characteristic having cytoplasmic transmission of limited duration is known as a Dauermodifikation (see Goldschmidt, 1955, pp. 201 ff. for review). The usual explanation of Dauermodifikations is environmental induction of long-lasting but impermanent changes in the cytoplasm. This phenomenon is not well understood.

In the third situation a phenotypic trait which is determined by a particular kind of cytoplasm persists indefinitely through successive generations. The determinants of the phenotypic trait are permanent components of the cytoplasm. We have then true cytoplasmic inheritance, based on the existence of particles in the cytoplasm with the genic properties of self-reproduction and control of specific reactions.

Cytoplasmic inheritance has been established in various groups of animals, plants, fungi, and protistans. The cytoplasmic genes have been identified with particular cell organelles, the plastids and mitochondria, and Sager therefore proposes to call them "organelle genes" (1972, p. 48).

The establishment of plastid inheritance in plants was a relatively easy task and came early in genetics history. It was much more difficult to demonstrate cytoplasmic inheritance when the phenotypic characters involved had no obvious connection with any microscopically visible cell organelles. More elaborate experiments were necessary in such cases, and were designed and carried out by Michaelis and others. Furthermore, a longer lag occurred between completion of this experimental work and general acceptance of the conclusions reached by the experimenters. As recently as 1955 Goldschmidt was able to argue persuasively against the existence of cytoplasmic genes (1955, pp. 194–244).

EARLY STUDIES OF PLASTID INHERITANCE   The progenitors of chloroplasts occur in the cytoplasm and are transmitted mainly through the maternal line. Defective chloroplasts, lacking normal chlorophyll, have an immediate and obvious phenotypic expression in the form of non-green tissue, and pleiotropic effects in photosynthetic ability and viability. These primary and secondary phenotypic characters are also transmitted through the maternal line. The phenotypic effects and their plastid determinants can be correlated in inheritance.

Chloroplasts are not wholly independent of chromosomal genes. Chlorophyll deficiencies may be induced by certain alleles of chromosomal genes. But if the chloroplasts of alternative types—normal or defective—are self-reproducing and maternally inherited, they qualify as cytoplasmic genes. Gene interactions between nucleus and cytoplasm may affect the expression of the chloroplasts, but not their status as genic units, contrary to arguments which were once used by opponents of cytoplasmic inheritance.

Maternal inheritance of chloroplasts and of the resulting green or white leaves was first reported by Correns (1909) in the four-o'clock, *Mirabilis jalapa,* and Baur (1910b) in the snapdragon, *Antirrhinum majus.* Later reports included cases in Oenothera (Renner, 1924, 1937; Schwemmle, 1938), *Zea mays* (Rhoades, 1946), and other groups.

A variegated form of *Mirabilis jalapa* has mottled green and yellowish-white leaves as well as some all-green leaves and some wholly yellowish-white leaves. The variegation is due to the somatic segregation of normal and defective plastids during development of the shoot. In general, the progeny of green flowering branches are all green, while the seedlings derived from white flowering branches are all white, and therefore lethal (Correns, 1909).

Reciprocal crosses between green and white flowering branches yield progeny of the maternal type. Thus the cross G ♀ × W ♂ gives all green sedlings, and the cross W ♀ × G ♂ gives only inviable white seedlings. The F_1 progeny of G ♀ × W ♂ can be grown to maturity and used to produce more advanced backcross generations. The hereditary contribution of the paternal parent can be seen in some characters such as flower color in the G × W hybrids. But the inheritance of plastid type and leaf color is strictly maternal through successive generations (Correns, 1909).

Chloroplast behavior in Oenothera shows interactions between chromosomal genes and plastids. It is possible to make use of the peculiar genetic constitution of many Oenotheras to reveal these nuclear-cytoplasmic interactions. Many microspecies of Oenothera are permanent chromosomal hybrids composed of two genomes or Renner complexes which segregate as intact multichromosome units. The species of Oenothera also differ in their chloroplasts. Reciprocal crosses between the species can therefore put together different combinations of chloroplasts and Renner complexes.

Oenothera lamarckiana has the genomic constitution velans/gaudens. Oenothera hookeri is structurally homozygous with the constitution hookeri/hookeri. The interspecific cross of Oe. lamarckiana × hookeri produces two classes of hybrid offspring in F_1, velans/hookeri and gaudens/hookeri. These two genomic combinations are the same in reciprocal crosses. But reciprocal crosses do differ in the character of the plastids, which are transmitted mainly through the maternal line (Renner, 1924, 1936, 1937).

The results of reciprocal crosses are as follows:

F_1 *hook* ♀ × *lam*   gaudens/hookeri, green leaves
       "   "   velans/hookeri, green leaves

F_1 *lam* ♀ × *hook*   gaudens/hookeri, green leaves
       "   "   velans/hookeri, yellow leaves

Evidently the plastids of Oe. hookeri develop normally in combinations with both gaudens and velans genomes; and the plastids of Oe. lamarckiana also develop normally in combination with either gaudens or hookeri; but the lamarckiana plastids do not develop properly in combination with a velans genome (Renner, 1924, 1936, 1937).

Similar differences are found in reciprocal crosses between Oenothera berteriana and Oe. odorata. Both species are complex heterozygotes, containing different pairs of Renner complexes, and accordingly should produce four classes of hybrids in F_1. The two species also carry different types of chloroplasts. The F_1 progeny of Oe. berteriana ♀ × odorata are normal (with one exception which does not concern us here). In the reciprocal cross, Oe. odorata ♀ × berteriana, one class of F_1 offspring is normal, two classes are pale and weak, and the fourth class is inviable due to chloroplast deficiencies. The odorata chloroplasts do not turn green or divide properly in combination with *berteriana* genomes. In later generations derived from *Oe. odorata* ♀ × *berteriana,* the weak types become greener and stronger, as a result of some adjustment in the genome-plastid interactions (Schwemmle, 1938).

The iojap form of corn (*Zea mays*) has green-and-white-striped shoots. The striped tissue consists of green sectors with normal chloroplasts and white sectors with undeveloped, colorless plastids. The iojap plants are homozygous recessive for a chromosomal gene (*ij/ij*). Nevertheless, the iojap condition shows maternal inheritance since its proximate cause is the plastids, which are transmitted in the ovules. Striped plants arise from ovules containing both normal and undeveloped plastids (Rhoades, 1943, 1946).

Conversely, striped plants can give rise, by somatic segregation in the cell line leading to the ovules, to all-green progeny and all-white progeny as well as striped plants. These three classes of progeny are in fact obtained in crosses of iojap corn ♀ × normal green males. The reciprocal cross of normal ♀ × iojap ♂ gives only green offspring (Rhoades, 1943, 1946).

The nuclear gene *Ij*, in one allelic form or the other (*Ij* or *ij*), exercises ultimate control of the type of plastid. But plastids of a given type, once formed, have a certain autonomy of their own in the cells and a strictly maternal mode of inheritance. It is possible that the action of the *ij/ij* genotype is to induce mutational changes in the plastids (Rhoades, 1943, 1946; Sager, 1972).

In *Pelargonium zonale,* a garden variety of geranium, cytoplasmic inheritance of plastids occurs through the pollen tube as well as the ovule. Consequently the inheritance of green or white leaves is not strictly maternal. Crosses between green-leaved and white-leaved plants give green, white, and variegated F_1s in irregular ratios irrespective of the character of the female parent (Baur, 1909, 1919; see also Sager, 1972, pp. 195 ff., for review of more recent work). Some transmission of plastids in the male line seems to occur also in Oenothera (Renner, 1936; Sager, 1972).

Many other cases of cytoplasmic transmission have now been studied in higher plants. In her recent review of the subject, Sager lists 34 species of angiosperms which show maternal inheritance of plastids (1972, pp. 184–85).

CYTOPLASMIC INHERITANCE IN EPILOBIUM The classic investigations of Michaelis in Epilobium are of great importance since they established the existence of microscopically invisible cytoplasmic genes (or at least microscopically unseen ones) from the results of breeding experiments. Michaelis's experimental method was to make and compare reciprocal backcross progeny through numerous backcross generations.

Striking reciprocal differences are found in the first-generation hybrids derived from crossing *Epilobium hirsutum* and *E. luteum*. The hybrids of *E. luteum* ♀ × *hirsutum* have reclining branches, broad green leaves, large flower petals (10.7 mm long by 9.2 mm wide), and normally developed anthers, which produce 15–20% good pollen. By contrast, the F_1 of *E. hirsutum* ♀ × *luteum* has erect pyramidal branches, narrow yellow-mottled leaves, small flower petals (4.6 mm long by 3.4 mm wide), and abortive and sterile anthers. In general, the products of crossing *hirsutum* ♀ × *luteum* show inhibition of growth in different parts of the plant, as compared with the more normal progeny of the reciprocal cross, *luteum* ♀ × *hirsutum* (Michaelis, 1954). Some of the reciprocal differences are shown in Figure 23.

The egg nucleus in Epilobium is surrounded by cytoplasm, as in other angiosperms, but the sperm nucleus of Epilobium contributes little or no cytoplasm to the fertilized egg (Michaelis, 1954; see also Jensen, 1973). Therefore the reciprocal differences between the F_1 hybrids pointed to a cytoplasmic influence of some kind. In the early stages of the Epilobium experiments it was not possible to tell whether the observed character differences between the reciprocal hybrids were maternal effects or cytoplasmically inherited traits.

In order to distinguish between these possibilities, Michaelis backcrossed the hybrids to the parental types for many generations. In practice it was possible to carry out repeated backcrossing in only one direction. The backcrosses of F_1 (*hirs* ♀ × *lut* ♂) ♀ × *lut* ♂ failed and the lines died out because of lethality and sterility. But the reciprocal backcross was continued through 24 generations. The backcrossing of (*lut* ♀ × *hirs* ♂) ♀ × *hirs* ♂ was continued for over 20 years and ended in the production of the backcross generation B_{23} (*lut* × *hirs*) ♀ × *hirs* (Michaelis, 1954).

Let us consider the changes in hereditary constitution which

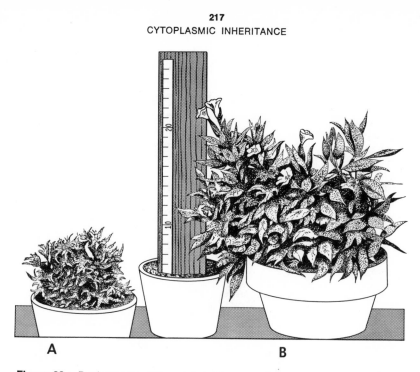

Figure 23. Reciprocally different hybrids of *Epilobium hirsutum* and *E. luteum*. (A) F$_1$ *E. hirsutum* ♀ × *luteum* ♂. (B) F$_1$ *E. luteum* ♀ × *hirsutum* ♂. (Redrawn from Michaelis, 1929; from Grant, 1964, reproduced by permission of John Wiley and Sons.)

take place during repeated backcrossing. The quantity of cytoplasm contributed by the maternal parent remains nearly constant through successive backcross generations. But the proportion of the nuclear genetic material from the recurrent parent rises exponentially in the successive generations, while the proportion of nuclear material of the other parent decreases correspondingly. The relations are as shown in Table 23. It is evident that after a relatively few generations of backcrossing (A × B) as a female to B as a male, the derived individuals consist, for all practical purposes, of the nuclear genotype of B in the cytoplasm of A (Michaelis, 1953).

Now let us compare the genetic constitution of F$_1$ hybrids, straight backcrosses, and certain more complex backcrosses. The crosses to be compared are described and designated by numbers in Table 24. These theoretical comparisons provide the basis for interpreting Michaelis's experimental results.

TABLE 23

AVERAGE PROPORTIONS OF THE CYTOPLASMIC AND NUCLEAR GENES OF TWO PARENTAL TYPES, A AND B, IN SUCCESSIVE BACKCROSS GENERATIONS (MICHAELIS, 1953)

Hybrid generation		Cytoplasm	Nuclear genes
F_1	A ♀ × B ♂	Virtually all A	$\frac{1}{2}B + \frac{1}{2}A$
B_1	(A × B) ♀ × B ♂	"	$\frac{3}{4}B + \frac{1}{4}A$
B_2	(A × B) × B	"	$\frac{7}{8}B + \frac{1}{8}A$
B_3	(A × B) × B	"	$\frac{15}{16}B + \frac{1}{16}A$
B_4	(A × B) × B	"	$\frac{31}{32}B + \frac{1}{32}A$
B_5	(A × B) × B	"	$\frac{63}{64}B + \frac{1}{64}A$
B_9	(A × B) × B	"	$\frac{1,023}{1,024}B + \frac{1}{1,024}A$
B_{19}	(A × B) × B	"	$\frac{1,048,575}{1,048,576}B + \frac{1}{1,048,576}A$

TABLE 24

THE CYTOPLASMIC AND CHROMOSOMAL CONSTITUTION OF THE PROGENY DERIVED FROM VARIOUS RECIPROCALLY DIFFERENT FIRST-GENERATION CROSSES, BACKCROSSES, AND OUTCROSSES BETWEEN TWO PARENTAL TYPES, A AND B

Cross no.	Parents	Progeny	Constitution of progeny
1	A ♀ × B ♂	F_1 A × B	A cytoplasm + A/B nucleus
2	[(A × B) × B] ♀ × B ♂	backcross × B	A cytoplasm + B nucleus
3	(progeny of cross 2) ♀ × A ♂	outcross × A	A cytoplasm + A/B nucleus
4	B ♀ × (progeny of cross 2) ♂	B × backcross	B cytoplasm + B nucleus
5	(progeny of cross 4) ♀ × A ♂	outcross × A	B cytoplasm + A/B nucleus
6	B ♀ × A ♂	F_1 B × A	B cytoplasm + A/B nucleus

The first comparison to consider is that between reciprocal F_1 hybrids and a straight backcross (crosses 1, 2, and 6 in Table 24). The F_1 A ♀ × B ♂ from cross 1 has A cytoplasm and A/B chromosomes. The reciprocal F_1 B ♀ × A ♂ from cross 6 has B cytoplasm and A/B chromosomes. The two F_1 hybrid combinations are thus alike in their chromosomal constitution but different in their cytoplasm. The progeny of cross 2, B_n (A ♀ × B ♂) × B ♂ , consist essentially of A cytoplasm and B chromosomes. Such backcross types resemble the parental type A and the F_1 A × B in their cytoplasm, but are unlike the latter in their chromosomal constitution. If, therefore, the phenotypic characters of the backcross progeny of cross 2 are like those of the F_1 A × B, but different from F_1 B × A, the characters in question are probably determined by hereditary factors in the cytoplasm of A.

The backcross derivatives of *Epilobium luteum × hirsutum,* particularly (*lut × hirs*) ♀ × *hirs* ♂ , did in fact resemble the corresponding F_1 combination, *lut* ♀ × *hirs* ♂ , in most of the characters in which the latter differed from the reciprocally different F_1 hybrid, *hirs* ♀ × *lut* ♂ (Michaelis, 1929, 1953, 1954). Thus the backcross (*lut × hirs*) ♀ × *hirs* ♂ possesses large corollas which are like those of the F_1 *lut* ♀ × *hirs* ♂ and unlike the small corollas of the reciprocal hybrid, F_1 *hirs* ♀ × *lut* ♂ (Michaelis, 1929). These and other characters which behave in a similar fashion are consequently determined by hereditary factors transmitted from generation to generation through the cytoplasm. The cytoplasmic constituents responsible for the characters in question maintain their identity and produce their specific action even though under the influence of a foreign nucleus for 24 generations (Michaelis, 1954).

This conclusion can be tested by crossing the backcross derivatives reciprocally with the parental types. This brings us back to the crosses set forth in Table 24. Crosses 1, 2, and 3 in the table form one consecutive series; crosses 1, 2, 4, and 5 form a reciprocal series. Let us consider these two series in more detail.

The backcross progeny of (A × B) × B can be used as either the female or the male in subsequent backcrosses to the parental type B. If it is used as a female, the progeny will be as shown for cross 2 in Table 24. The progeny of cross 2 can now be outcrossed as a female to parental type A in order to give the outcross type 3.

TABLE 25

PHENOTYPIC CHARACTERS OF RECIPROCALLY DIFFERENT FIRST-GENERATION HYBRIDS AND LATER-GENERATION OUTCROSSES BETWEEN *EPILOBIUM LUTEUM* AND *E. HIRSUTUM.* THE CROSSES ARE NUMBERED AS IN TABLE 24 (MICHAELIS, 1929)

Cross no.	Hybrid generation	Petal size, mm	Leaf width, mm	Anthers	Pollen fertility, %
1	F_1 *lut × hirs*	10.7 × 9.2	23.7	normal	15
3	O_1 [B_4 (*lut × hirs*) × *hirs*] × *lut*	9.6 × 7.8	23.4	normal	22
5	O_1 {*hirs* × B_3 [(*lut × hirs*) × *hirs*]} × *lut*	4.6 × 3.5	20.8	abortive	0
6	F_1 *hirs × lut*	4.6 × 3.4	20.8	abortive	0

The latter possesses A cytoplasm and A/B chromosomes. Accordingly, it is like the F_1 form $A \times B$ (from cross 1) in both its cytoplasmic and chromosomal constitutions.

The reciprocal outcross to A is derived by the following steps. Progeny of cross 2 is crossed as a male to B, yielding backcross type 4. Backcross type 4 is then outcrossed as a female to A, to produce outcross type 5. Outcross type 5 is like the F_1 form $B \times A$ (from cross 6) in possessing B cytoplasm and A/ B chromosomes. By the same token, outcross type 5 differs from outcross type 3 in its cytoplasm but not in its nucleus.

Michaelis (1929) produced the reciprocally different outcrosses from fourth-generation backcross individuals of Epilobium. One of these had the derivation indicated by the formula O_1 [B_4 (*lut × hirs*) × *hirs*] ♀ × *lut* ♂. It corresponds to outcross type 3 in Table 24. The other outcross was O_1 {*hirs* ♀ × B_3 [(*lut × hirs*) × *hirs*]} ♀ × *lut* ♂. It corresponds to outcross type 5.

The phenotypic characters of the two outcross types and of the two F_1 hybrids are summarized in Table 25 and illustrated in Figure 24. It is evident that the two outcross progeny differ phenotypically in the same ways that the reciprocal F_1 hybrids differ. Outcross types 3 and 5 differ from one another not in their nuclei but in their cytoplasm and in several phenotypic traits. And they resemble phenotypically the type of F_1 hybrid which has the same kind of cytoplasm (Michaelis, 1929).

Figure 24. Flowers of reciprocally different F_1 hybrids and later-generation backcrosses between *Epilobium luteum* and *E. hirsutum*. (A) F_1 *E. luteum* ♀ × *hirsutum* (cross type 1). (B) Backcross-outcross with *luteum* cytoplasm (cross type 3). (C) Backcross-outcross with *hirsutum* cytoplasm (cross type 5). (D) F_1 *E. hirsutum* ♀ × *luteum* (cross type 6). See Tables 24 and 25 for further explanation of crosses. (Redrawn from Michaelis, 1929; from Grant, 1964, reproduced by permission of John Wiley and Sons.)

The nucleus is the same in all hybrid combinations that were compared and can therefore be eliminated as a variable factor. The correlation is between the inheritance of certain phenotypic characters and the inheritance of the cytoplasm. This observed correlation shows that the two kinds of cytoplasm in *Epilobium luteum* and *E. hirsutum* carry genic units with different specific effects (Michaelis, 1929).

The cytoplasmic-genetic differences between *Epilobium hirsutum* and *E. luteum,* and between different races of *E. hirsutum,* affect not only external morphology and pollen fertility of the plants, but also various physiological features. Among the latter are temperature tolerance, sensitivity to poisons, susceptibility to parasites, permeability of protoplasm, viscosity of protoplasm, isoelectric point, redox potential, and oxidative enzyme activity. Differences in the activity of oxidative enzymes between different cytoplasms were considered to be a possible cause of the dwarfness of some Epilobium hybrids and the heterosis of others (Michaelis, 1953, 1954).

Reciprocal differences between hybrids are found in combinations of other species pairs in Epilobium, e.g., *E. parviflorum* × *roseum* and *E. hirsutum* × *roseum;* interspecific differentiation in the cytoplasm thus appears to be widespread in this genus (Lehmann, 1932).

Grafts between strains of Epilobium with different cytoplasms do not result in the infiltration of the cytoplasmic factors from one strain into the other. The cytoplasmic genes are not transmitted as infectious agents in sap infiltrations. Instead they are regular, integrated components of the cells (Michaelis, 1953).

Michaelis recognized that cytoplasmic genes can be affected by chromosomal genes, and vice versa. Cytoplasm and nucleus form an integrated system within any given race. The normal balance between the two components is upset in hybrids and may lead to some altered course of development (Michaelis, 1950, 1954).

CYTOPLASMIC MALE STERILITY Differences between reciprocal F_1 hybrids in pollen fertility are widespread in higher plants. A given pair of species often produces a male-sterile hybrid when crossed in one direction and a hybrid which is fertile or semifertile as to pollen when crossed in the opposite direction. Examples have been reported in such genera as Aquilegia, Capsicum, Diplacus, Epilobium, Gilia, Linum, Nicotiana, Secale, Solanum, Sorghum, Streptocarpus, Tragopogon, Triticum-Aegilops, and Zea.

Specific references are: Skalinska (1929) for Aquilegia; Peterson (1958) for Capsicum; McMinn (1951) for Diplacus; Lehmann (1932) and Michaelis (1933) on Epilobium; Grant (1950b, 1956a) on Gilia; Gairdner (1929) on Linum; Clayton (1950) and Smith (1962, 1968) on Nicotiana; Putt (1954) on Secale; Lamm (1941), Grun, Aubertin and Radlow (1962), and Grun (1970) on Solanum; Stephens and Holland (1954) on Sorghum; Oehlkers (1940) on Streptocarpus; Ownbey and McCollum (1953) on Tragopogon; Kihara (1958) on Triticum-Aegilops; Rhoades (1933, 1954), Jones (1956) and Duvnick (1965) on Zea. General reviews are given by Edwardson (1956, 1970) and Jain (1959).

The development of anthers and pollen in hybrids appears to be sensitive to interactions between nucleus and cytoplasm. The role of the cytoplasm in this interaction could be that of maternal effect or of cytoplasmic genes. These possibilities are probably mixed in the series of examples above. Hybridization experiments have not been carried far enough to distinguish between the two in many cases. Experimental evidence for the involvement of cytoplasmic genes in male sterility has been obtained in Epilobium, Nicotiana, Secale, Streptocarpus, Triticum-Aegilops, and Zea.

In Zea and Nicotiana it has been possible to narrow the nucleus-cytoplasm interaction down to particular Mendelian genes and specific kinds of cytoplasm. A good example is provided by Smith's analysis of male sterility in hybrids between *Nicotiana langsdorffii* and *N. sanderae* (1962, 1968). These species have two chromosomal genes (Rf_1 and Rf_2) for male fertility, and these genes occur in two allelic forms (Rf and rf). The genotype of *N. langsdorffii* is Rf_1/Rf_1 rf_2/rf_2; that of *N. sanderae* is rf_1/rf_1 Rf_2/Rf_2. The species also have different kinds of cytoplasm. Hybrid plants carrying the dominant alleles Rf_1 or Rf_2 in combination with either specific type of cytoplasm are male fertile. Hybrids with *sanderae* cytoplasm in combination with any genotype for Rf_1 and Rf_2 are also male fertile. But the combination of *langsdorffii* cytoplasm with the double recessive homozygous genotype rf_1/rf_1 rf_2/rf_2 gives male-sterile hybrids (Smith, 1962, 1968).

A similar situation is found in maize. The cytoplasm in certain varieties, known as Texas cytoplasm or *cms-T,* induces pollen sterility. The expression of this condition is also affected by different alleles of a nuclear gene, *Rf.*

The Texas cytoplasm has come to have considerable economic importance. Strains carrying *cms-T* cytoplasm have been used widely to produce hybrid corn. More recently these same strains have proven to be particularly susceptible to a new race of southern corn blight (*Helminthosporium maydis,* race T), and have been hard hit by epidemics in the United States corn belt. The susceptibility to the fungus blight resides in the corn cytoplasm. Recently it has been possible to identify the cytoplasmic sites of blight susceptibility as being the mitochondria, which are probably also the carriers of the cytoplasmic factors for male sterility (Miller and Koeppe, 1971).

HETEROTIC EFFECTS Heterosis is found in some hybrids of Epilobium with certain combinations of nuclear and cytoplasmic genes. Michaelis (1951) compared the growth of backcross derivatives containing a *hirsutum* nucleus in *luteum* cytoplasm with that of straight *E. hirsutum* under several controlled environmental conditions. The backcross hybrid exceeded *E. hirsutum* in most environments. It produced nearly twice as much herbage as *E. hirsutum* when grown in warm long days. It also exceeded its *E. luteum* parent in growth.

Heterosis appears again in some nucleus-cytoplasm combinations assembled from different races of *E. hirsutum*. The nucleus of a Jena race was transplanted by repeated backcrossing to the cytoplasm of numerous other races of *E. hirsutum*. The combination consisting of a Jena nucleus in South African cytoplasm was heterotic. Several other interracial nucleus-cytoplasm combinations in *E. hirsutum* did not show heterosis (Michaelis, 1951).

Michaelis and his co-workers found that hybrids with different cytoplasms exhibit different physiological features, as we have noted. Among these physiological features are the activity of oxidative enzymes and rate of respiration. Michaelis (1954) suggested that the cytoplasmically induced heterosis of some hybrids and the inhibited growth of others could well be due to these physiological characteristics.

Recent investigations conducted independently and without reference to the earlier Epilobium work confirm this idea. Some combinations of varieties in maize produce heterotic hybrids while other combinations do not. Mitochondria can be extracted from the parental varieties and from the hybrids and measured for rate of respiration. The mitochondria of different varieties differ in respiration rate and related characteristics. Hybrids have mixtures of the parental mitochondria. And heterotic hybrids have mixtures of mitochondria which produce favorable complementary effects on respiration (Sarkissian and Srivastava, 1967).

Artificial mixtures of isolated mitochondria extracted from the parents of heterotic hybrids in maize show a complementary enhancement in respiration as compared with the isolated mitochondria of either parent alone. Some parental varieties do not yield heterotic hybrids, and their mitochondria do not show complementation when mixed artificially. Thus heterosis in maize hybrids can be correlated with mixtures of parental types of mitochondria which have enhancing complementary effects on respiration (Sarkissian and Srivastava, 1967). Mitochondrial complementation has also been found in *Neurospora crassa* (Leiter, LaBrie, Bergquist and Wagner, 1971).

THE NATURE OF CYTOPLASMIC GENES Cytoplasmic inheritance has been established in representatives of all kingdoms of eukaryotic organisms. The phenomenon is known in Paramecium among protistans, in Drosophila among animals, and in several groups of fungi, namely Saccharomyces, Neurospora, Podospora, and Aspergillus (see reviews by Caspari, 1948; L'Heritier, 1948; Sonneborn, 1950a, 1950b; Wagner and Mitchell, 1964; Sager, 1972). In the plant kingdom, cytoplasmic inheritance has been found not only in the angiosperms, as emphasized in this chapter, but also in lower plants. One of the earliest cases of cytoplasmic inheritance was described in the moss Funaria (Wettstein, 1925) and one of the most thoroughly analyzed is in the green alga Chlamydomonas (Sager, 1972, ch. 3).

The nature of cytoplasmic genes has been a subject of interest and speculation since the earliest days of cytoplasmic genetics. One class of cytoplasmic genes, the plastids, could be identified as such in the early period, since they were visible both microscopically and exo-phenotypically in their normal or mutant form. Other types of cytoplasmic genes could not be connected so readily with any known cell organelles. Michaelis did put forward the speculation in 1954 that some cytoplasmic genes might be chondriosomes (or mitochondria, in modern terminology).

The approaches of molecular and cell biology have made the picture somewhat clearer in recent years and hold the promise of further clarifications. Of particular significance is the finding that two types of organelles in the cytoplasm, chloroplasts and mitochondria, contain DNA. This indicates that these organelles are the sites of most, or perhaps all, cytoplasmic genes (Gibor and Granick, 1964; Sager, 1972; Wagner, 1972). Genic units were associated more definitely with mitochondria by studies of cytoplasmically transmitted respiratory mutants in Neurospora and yeast in the 1950s (see Wagner, 1969).

There are some microscopically visible particles in the cytoplasm of Paramecium and Drosophila which fulfill most of the criteria of cytoplasmic genes—they are transmitted through the female line, contain DNA, and in most cases produce known, specific phenotypic effects—but nevertheless do not fall into the categories of plastids and mitochondria. Among these anomalous cytoplasmic particles are the kappa particles in *Paramecium aurelia* and small inclusions associated with altered sex ratio and male sterility in Drosophila (Preer, 1950; Hanson, 1954; Wolstenholme, 1965; Kernaghan and Ehrman, 1970). The suggestion has been made by these and other workers that the particles in question are microbial symbionts or parasites living in the cytoplasm of the eukaryotic host organism. In such cases we may not be dealing with cytoplasmic inheritance in the strict sense, but with hereditary infections, or Dauermodifikations, or special symbiotic relationships.

The situation in *Drosophila paulistorum* is very interesting in this regard. Cytoplasmic male sterility appears in certain interracial hybrids in this species (Ehrman, 1963). The determinants of the male sterility are transmitted from generation to generation through the egg cytoplasm. These determinants can also be transmitted artifically by injection. Female flies of a strain which normally produces fertile male offspring will, when injected with cytoplasm from a strain carrying the sterility factor, yield sterile sons (Ehrman and Williamson, 1965; Williamson and Ehrman, 1967). Electron-microscopic studies reveal bodies which look like Mycoplasma bacteria in Drosophila cytoplasm carrying the sterility factor (Kernaghan and Ehrman, 1970).

Kernaghan and Ehrman suggested that the cytoplasmic male sterility factor in *Drosophila paulistorum* is a Mycoplasma-like symbiont. Some races are adjusted to this symbiont and produce normal fertile males despite its presence; other races are not adjusted to it, and, when exposed to it in interracial hybrids, respond by expressing the male sterility (Kernaghan and Ehrman, 1970).

This suggestion brings us to a more general idea which was considered in the early period of cytoplasmic genetics and is under active discussion again today. The hypothesis proposed is that chloroplasts and mitochondria are themselves of symbiotic origin in eukaryotic cells (see Margulis, 1970; Wagner, 1972; opposing view by Raff and Mahler, 1972). The distinction between regular cytoplasmic genes and impermanent or unintegrated cytoplasmic particles tends to break down somewhat if this is the case.

The symbiont hypothesis of the origin of cell organelles, being essentially a phylogenetic and historical hypothesis, probably cannot be verified rigorously by evidence available to us in the living world, but it can be made more or less probable by reference to comparative evidence from the fields of cell and molecular biology. That evidence has been well reviewed in the recent books of Margulis (1970) and Sager (1972), to which the reader is referred for further information regarding these aspects of the problem. A comparison of much interest is that the DNA of mitochondria in animals and fungi is organized into ringlike threads, as in bacteria (Sager, 1972).

Some of the DNA in mitochondria synthesizes protein directly. Other proteins in mitochondria are, in the final analysis, coded for by chromosomal DNA. Malic dehydrogenase in maize occurs in the mitochondria but shows Mendelian inheritance; apparently a chromosomal gene controls the mitochondrial DNA concerned with the synthesis of this protein. In cases such as this the genetic function of the mitochondria is under nuclear control (Longo and Scandalios, 1969; Wagner, 1969). Autonomy of cytoplasmic genes is probably the exception, and interaction of these entities with chromosomal genes is probably the general rule.

The transmission features of cytoplasmic genes are irregular. Heterozygous mixtures of cytoplasmic genes arise when some cytoplasmic genes are contributed by the sperm to eggs of a different cytoplasmic constitution. Segregation in respect to the resulting cytoplasmic differences can then occur in irregular, non-Mendelian ratios as a result of various processes within the cells. The two types of cytoplasmic genes may or may not be distributed equally to daughter cells during mitosis or meiosis. Unequal distribution and somatic segregation are common. Furthermore, the two types of cytoplasmic genes may not reproduce themselves at equal rates in a given cytoplasm, and this difference leads to further irregularities in segregation (Michaelis, 1954). As a result of these processes, cytoplasmic heterozygotes usually do not remain heterozygous for long, but tend to segregate one or both homozygous types in rapid succession (Sager, 1972, pp. 219–21).

CHAPTER THIRTEEN

ALTERED SEGREGATION

INTRODUCTION A simple genotypic ratio of $1:2:1$ depends on various conditions in the genetic system which are often, but not always, fulfilled. Among these conditions are the location of the segregating genes in chromosomes, a diploid chromosome constitution, regular chromosome disjunction at anaphase of meiosis, equal viability of different types of gametophytes or gametes, random fertilization, and equal viability of different classes of zygotes. Deviations in these conditions lead to deviations from a simple Mendelian ratio.

Such deviations are not rare. We noted in the preceding chapter that cytoplasmic genes give irregular segregation ratios. In this chapter we will discuss briefly some of the causes of altered segregation ratios for chromosomal genes.

PREFERENTIAL SEGREGATION The $1:1$ gametic ratio for a segregating gene pair depends, in the first place, on the circumstance that the chromosome pair carrying this gene is oriented and distributed at random to the poles in meiosis. The contrasting situation is preferential segregation. Here one type of chromosome in a homologous pair passes to one particular pole in a nonrandom fashion, that is to say, in significantly more than 50% of the meiotic products.

The detection of preferential segregation usually requires prominent cytological differences between homologous chromosomes which can be followed through successive cell and individual generations. Genetic markers associated with the morphologically differentiated chromosomes are useful in any case and are necessary to demonstrate an altered segregation. A further consideration is that the preferential segregation has genetic effects mainly on the female side, where only one of the four potential products of meiosis functions as a gamete.

Preferential segregation was first discovered and well documented in *Zea mays* (Rhoades, 1942, 1952; Longley, 1945). The short chromosome 10 of maize occurs in two forms. A form of chromosome 10 from Latin America has a large heterochromatic segment or knob near one end; the form from temperate North America lacks the knob. Plants homozygous for either type of chromosome 10 give normal Mendelian ratios for genes located on this chromosome. But heterozygotes containing a standard and a knobbed chromosome 10 have a nonrandom segregation of the two chromosomes at meiosis on the female side (Rhoades, 1952).

The functioning megaspores resulting from female meiosis in chromosome 10 heterozygotes possess the knobbed chromosome in about 70% of the cases, instead of the expected 50%. Some possible factors such as lethal genes or megaspore competition have been ruled out. The disjunction and distribution of the two members of the chromosome 10 pair can be traced through megasporogenesis in a sample of dividing cells. In 70% or more of the dividing megasporocytes, the knobbed chromosome 10 passes to the basal cell in the linear tetrad, which is the cell destined to develop into the functioning embryo sac with the included egg (Rhoades, 1952).

Gene segregation ratios are biased in a parallel way. The gene *R* is located close to the knobbed region on chromosome 10. Segregation is studied in plants heterozygous for the allele pair *Rr* and for chromosome type. Either *R* allele can be linked with the chromosome knob. The proportions of R and r female gametes produced by the Rr heterozygote are determined from the results of the testcross $Rr\ ♀ \times rr\ ♂$ (Rhoades, 1952).

The heterozygote with the r allele on the knobbed chromosome (knob–r/std–R) produces female gametes in the observed proportions 2717 r : 1118 R. The complementary heterozygote, knob–R/std–r, produces female gametes in proportions of 252 r : 841 R. The allele on the knobbed chromosome enters 71% of the functioning female gametes in the first heterozygote and 77% in the second. The distortion in gene segregation ratios parallels that observed cytologically in the chromosomes themselves (Rhoades, 1952).

Other chromosomes of the complement in maize may also have knobs. Heterozygotes containing a knobbed and a knobless member of these other chromosomes also show preferential segregation of the knobbed member. The preferential segregation is especially well marked for chromosomes 4 and 9 (Longley, 1945). The knobbed chromosome 10 in heterozygous or homozygous condition promotes preferential segregation in the other chromosomes of the complement (Rhoades and Dempsey, 1966).

The mechanism responsible for the preferential segregation of the knobbed chromosomes is believed to reside in the heterochromatic segment itself. The knob is thought to enhance the mobility of the chromosome on the spindle at anaphase I. This mobility presumably allows the chromosome to move to the pole that eventually becomes the basal functioning cell of the linear tetrad in a high proportion of the meiotic divisions (Rhoades, 1952).

A number of cases of preferential segregation are known in Drosophila. These cases, which present features different from those in maize, are reviewed by Grant (1963, pp. 231 ff.) and John and Lewis (1965, pp. 243 ff.), among others. In plants, univalent chromosomes and supernumerary B chromosomes often show preferential segregation (See Brown, 1972).

RENNER EFFECT Meiosis on the female side in angiosperms usually results in a linear tetrad of megaspores, only one of which develops into the embryo sac. The megaspore destined to become the source of the embryo sac lies in a particular position in the linear row of daughter spores. The functioning megaspore is commonly the basal or chalazal cell; in some genera it is the outer or micropylar cell. Many deviant modes of embryo-sac development occur in the angiosperms, and are described in textbooks of plant morphology and embryology. An interesting exceptional mode of development in Oenothera is known as the Renner effect.

In species of Oenothera with a normal genetic system, the megaspore in the micropylar position is the one which produces the embryo sac. The deviation from this mode occurs in Oenothera species or microspecies with a permanently hybrid constitution for their genomes or Renner complexes.

A microspecies with the heterozygous genomic constitution A/B segregates two classes of meiotic products at megasporogenesis, A and B, and either one of these types may come to lie in the micropylar position. Nevertheless, only one type of megaspore, say B, is genetically able to develop into a functioning embryo sac. The B megaspore develops irrespective of its position in the tetrad. If it lies in an unfavorable position in the tetrad it develops anyway, gaining predominance over the micropylar type A megaspore, and it goes on to produce the embryo sac and egg. This competition between megaspores in a tetrad, and preferential development of one genotypic class irrespective of position in the tetrad, is the Renner effect (Darlington, 1937).

The microspecies *Oenothera muricata,* for example, has the hybrid constitution rigens/curvans. Rigens and curvans are two haploid genomes differentiated with respect to successive translocations. These genomes also carry different sets of gene alleles. The two genomes segregate as distinct entities at meiosis on both the male and female sides. On the male side the only functioning pollen is the curvans type.

On the female side the functioning embryo sacs are all or nearly all rigens. In embryogeny certain megaspores not in the micropylar position, but presumably carrying the rigens genome, often develop at the expense of the micropylar cell. Consequently *Oe. muricata* transmits only its rigens genome and the genes borne on it through the eggs. When *Oe. muricata* is outcrossed as a female parent to *Oe. hookeri,* the F_1 hybrids all have the genomic constitution rigens/hookeri (Darlington, 1937).

POLLEN-TUBE COMPETITION Pollen grains are normally delivered to the stigma of a flower in numbers greatly exceeding the numbers of ovules. In *Phlox pilosa,* for example, an average of 320 pollen grains is delivered per pollinating visit by Colias butterflies; the average number of pollen tubes per style, however, is 15.8. The ratio of pollen grains to pollen tubes is thus about 20 : 1. A similar ratio is found in *Phlox glaberrima.* A further sharp reduction in numbers of male gametophytes takes place between pollen-tube growth and fertilization (Levin and Berube, 1972).

These conditions lead to competition between pollen tubes. If the pollen tubes differ genetically in ways affecting their growth rate in the style, the competition will have a selective component, some genetic types of pollen being more likely than others to reach the ovules and effect fertilization. The differential growth rate of different genetic classes of pollen or pollen tubes, which affects their chances of accomplishing fertilization, is called certation. Certation will clearly bias the segregation ratios of any genes linked with the factors determining the relative competitive ability of the different pollen types.

Certation is a common process in flowering plants. It is readily demonstrated in cases of artificial interspecific pollination. When flowers are pollinated with mixtures of conspecific and alien pollen, the latter is usually slower growing and less effective in fertilization than the former. Thus the pollen of *Iris tenuis* requires 50 hours to reach the ovules in flowers of *I. tenax,* whereas the domestic *I. tenax* pollen reaches the same ovules in 30 hours (Smith and Clarkson, 1956). Similar differences between domestic and species-foreign pollen have been found in Petunia, Streptocarpus, Haplopappus, and other genera (see Smith, 1970).

Certation is also clearly apparent in the transmission of the monosomic condition through the pollen as compared with the ovules in intraspecific crosses. Monosomic plants lack a single chromosome and thus have the chromosome number $2n - 1$. The transmission features of the monosomic condition have been extensively studied in wheat and Nicotiana.

The monosomic plant produces two types of gametes with respect to the monosomic chromosome: normal haploid (n) gametes and deficient $(n - 1)$ gametes. Theoretically the two types could be formed in equal frequencies. Actually the monosomic plant produces more than 50% of $n-1$ gametes, often between 60 and 80% of such gametes, on both the female and male sides. The excess of $n-1$ gametes is due to the fact that the monosomic chromosome frequently lags in meiosis and consequently does not become incorporated in half of the daughter nuclei (Nishiyama, 1928; Lammerts, 1932; Clausen and Cameron, 1944; Sears, 1944).

The cross between a monosomic parent and a diploid parent could be expected to yield an F_1 generation consisting of monosomics and diploids in the proportion of $> 50\%$ monosomics to $< 50\%$ diploids. This expectation is realized if, but only if, the cross is made in the direction of monosomic ♀ × diploid ♂. Two monosomic types in *Nicotiana rustica,* used as female parents and pollinated by normal diploid males, produced 63 and 71% monosomic progeny (Lammerts, 1932).

The reciprocal cross, in which the monosomic type is used as the pollen parent, yields, by contrast, a greatly reduced proportion of monosomic progeny. Table 26 shows the contrasting rates of transmission of the monosomic condition through the pollen and the ovules for two monosomic types in *Nicotiana rustica.* The monosomic condition, though readily transmitted through the female line, is passed on to the next generation through the pollen to only a very limited extent (Lammerts, 1932). A similar reduction in the transmission rate of monosomics through the pollen is found in *Nicotiana tabacum* and wheat (Nishiyama, 1928; Clausen and Cameron, 1944; Sears, 1944).

Thus a monosomic female pollinated by a diploid male produces monosomic and diploid progeny approximately in proportion to the frequency of $n-1$ and n ovules. But a monosomic male parent, which forms over 50% $n-1$ pollen grains, yields only a very low proportion of monosomic progeny. The pollen tubes produced by the monosomic plant, unlike the ovules, enter into a strong competition with one another. This pollen-tube competition results in a selective elimination of the male gametophytes deficient for one chromosome (Nishiyama, 1928; Lammerts, 1932; Clausen and Cameron, 1944; Sears, 1944).

It might be predicted that where the pollen competition is relieved somewhat, in a sparsely pollinated flower, a greater variety

TABLE 26
RELATIVE NUMBERS OF MONOSOMIC PROGENY PRODUCED IN RECIPROCALLY
DIFFERENT CROSSES BETWEEN MONOSOMIC AND DIPLOID PARENTS IN
NICOTIANA RUSTICA (LAMMERTS, 1932)

Cross	Progeny		
	monosomic	diploid	total
Monosomic D ♀ × diploid ♂	65	38	103
Diploid ♀ × monosomic D ♂	0	94	94
Monosomic G ♀ × diploid ♂	77	31	108
Diploid ♀ × monosomic G ♂	1	104	105

of pollen types could function in fertilization than under conditions
of dense pollination. Some preliminary evidence suggests that this may
be the case. The pollen of an F_1 hybrid of tomato was applied in two
different concentrations on the stigmas to make F_2 and B_1 seeds.
Some flowers were pollinated with fewer pollen grains than ovules,
thus eliminating pollen competition, while other flowers were pollinated heavily, creating a competitive condition among the pollen
grains. Various quantitative characters, such as leaf length, plant
height, flower number, and fruit weight, were measured in the F_2
and B_1 progeny (Lewis, 1954b).

The results differed according to the method of pollination.
Although the mean for each character was the same in progenies
derived from dense or sparse pollinations, the variation around the
mean as expressed in the coefficient of variation was lower in all
characters under conditions of dense pollination. The extreme large
or small classes were more poorly represented in the progeny produced by pollination with superabundant pollen grains. Where the
pollen competition is severe, therefore, the pollen grains carrying
gene alleles determining extreme phenotypes tend to be eliminated
(Lewis, 1954b).

Similar effects of pollen density on variability of the progeny
have been reported in cotton (*Gossypium hirsutum*) and wheat
(*Triticum aestivum*) (Ter-Avanesjan, 1949; Ter-Avanesjan and
Nigmatullin, 1959). Emasculated flowers of cotton were pollinated
with known numbers of pollen grains, ranging from 5 to 1000.
The progeny of flowers pollinated with less than 300 grains showed
marked variation from plant to plant. This individual variation be-

came progressively greater as the number of pollen grains became smaller (Ter-Avanesjan, 1949).

A GENE FOR POLLEN-TUBE GROWTH IN *ZEA MAYS* The gene *Ga* in *Zea mays* controls the rate of pollen-tube growth, the *Ga* allele bringing about rapid growth and the *ga* allele poor growth. This gene is linked with the gene *Su* which determines the type of endosperm in the corn kernels. The dominant allele *Su* gives rise to starchy endosperm and the recessive allele *su* to sugary endosperm. The F_2 and testcross progeny of *Su/su* heterozygotes normally segregate in simple Mendelian ratios of 3 : 1 and 1 : 1 where *Ga* is not a complicating factor. The double heterozygote for both *Su* and *Ga,* however, gives altered segregation ratios for *Su* under certain conditions (Mangelsdorf and Jones, 1926).

The original cross was between rice popcorn with the constitution *Su Ga/Su Ga* and sugary corn with the constitution *su ga/ su ga,* to give the double heterozygote *Su Ga/su ga.* The F_1 was self-pollinated to produce an F_2 and was backcrossed in various directions to give three different backcross progenies. The backcross *Su/su* ♀ × *su/su* ♂ yielded 1374 starchy and 1397 sugary kernels. The backcross of *Su/*su ♀ × *Su/Su* ♂ yielded 207 heterozygotes and 213 homozygotes for *Su* in B_1. In both of these crosses the theoretical 1 : 1 backcross ratio was actually obtained (Mangelsdorf and Jones, 1926).

In the F_2 generation derived by selfing the F_1, however, the observed numbers of starchy and sugary kernels deviated markedly from the expected 3 : 1 ratio, as follows:

expected   2761 starchy : 920 sugary
observed   3085 starchy : 596 sugary

The F_2 generation thus has an excess of starchy kernels and a deficiency of sugary kernels. The backcross progeny from the cross *Su/Su* ♀ × *Su/su* ♂ also shows a deviation from a 1 : 1 genotypic ratio in the direction of a deficiency of *Su/su* types (Mangelsdorf and Jones, 1926).

These comparisons show that Mendelian segregation ratios are produced when the F_1 is pollinated with one type of pollen alone. Both pollen types, *Su–Ga* and *su–ga,* are functional in themselves. But a mixture of the two types of pollen on the same style gives distorted ratios in the next generation. Although both classes of pollen are able to accomplish fertilization, they are not equally effective at this task, and the differences between them come to expression under conditions of pollen competition (Mangelsdorf and Jones, 1926).

The gene *Ga,* which is subject to gametophyte selection, is also linked with the gene De_1 for defective seeds, which is subject to zygote selection. The deleterious alleles of the two genes may be linked in either coupling or repulsion phase. The amount of distortion in the frequency of the De_1 alleles in segregating populations is determined, first, by the phase of linkage, and second, by interaction between gamete and zygote selection (Jain, 1967).

POLLEN LETHALS Genes which act as lethal factors in the male gametophyte are known in a number of plant groups and are suspected in others. Morphological marker genes linked with such pollen lethals show distorted ratios, including, in the extreme case, the absence of one class of segregates.

The gene *Ge* is a gametophyte factor located on chromosome 4 in the tomato, *Lycopersicon esculentum.* Three alleles are known: the normal allele (Ge_n) and the alleles for gametophyte elimination $(Ge_c$ and $Ge_p)$. These alleles occur in different cultivated varieties of the tomato, in wild populations of *L. esculentum* in northwestern South America, and in the closely related *L. pimpinellifolium* (Rick, 1966, 1971a). Gametophyte elimination occurs only in Ge_c/Ge_p plants. In such plants the pollen and also the ovules abort. Penetrance is incomplete and the amount of aborted pollen is about 50% (Rick, 1966).

Various marker genes on chromosome 4 show distorted segregation ratios in F_2 and F_3. The genes in question are: $Afl, Di_1, E, Ful_1,$ and W_4; and the recessive alleles used as markers are: afl (for white cotyledon), di_1 (whitish stems), e (entire leaves), ful (bright foliage), and w_4 (wiry-4). The last two genes are located quite close to *Ge* and exhibit severe distortion (Rick, 1966).

Single-flowered stocks (*Matthiola incana*) heterozygous for a gene S produce a double-flowered type which is sterile. There are two types of heterozygotes. Single-flowered plants with the heterozygous constitution Ss yield, on selfing, 3 singles : 1 double. Another type of heterozygote known as an eversporting single, when selfed, segregates in an altered ratio of 1 single : 1 or slightly more doubles. The eversporting single stocks are of interest to us here (Johnson, 1953).

Reciprocal crosses between eversporting and true-breeding single-flowered stocks indicate that the S allele is transmitted through the eggs but not through the pollen. The altered ratio in eversporting stocks is then due to the nonfunctioning of the pollen that carries the S allele. The S allele in eversporting singles is apparently linked with a pollen lethal (Johnson, 1953).

The linkage is not complete but is broken on rare occasion by crossing-over, to give exceptional normal heterozygous singles and exceptional true-breeeding singles. The latter have the genotype SS in some cases. In one case, where unequal crossing-over in the Ss heterozygote apparently produced an abnormal Ss gamete, the genotype of the derived true-breeding single form is probably SSs (Johnson, 1953).

In certain varieties of wheat a gene for resistance to rust is linked with a pollen lethal, and this linkage causes an altered segregation ratio for the rust-resistance gene (Loegering and Sears, 1963).

Altered ratios for leaf color in *Gossypium hirsutum* and for growth form in *Phaseolus lunatus* result from linkage of the segregating major genes involved with pollen growth factors (Finkner, 1954; Bemis, 1959).

Pollen lethals have also been reported in *Nicotiana tabacum, Datura stramonium,* and *Plantago insularis* (Cameron and Moav, 1957; Avery, Satina and Rietsema, 1959; Murr and Stebbins, 1971).

In segregating progenies derived from interspecific crosses, a correlation is sometimes evident between pollen sterility and certain morphological characters. Such correlations have been observed in the crosses of *Phaseolus vulgaris* × *multiflorus, Chrysanthemum carinatum* × *coronarium,* and *Nicotiana langsdorffii* × *sanderae* (Lamprecht, 1941, 1944, 1956; Smith, 1952a; Smith and Daly, 1959). This correlation suggests a linkage between separate gene systems governing morphological characters and pollen development (Smith, 1952a; Grant, 1967).

PREFERENTIAL FERTILIZATION   The term selective or preferential fertilization has been widely and loosely used to include phenomena discussed here under pollen-tube competition and other headings. Preferential fertilization should be defined more narrowly as being a direct reaction between certain genetic types of male gametes or gametophytes and certain types of female gametes or gametophytes which leads to nonrandom fertilization or special avoidance of fertilization (Arnold, 1958; Schwemmle, 1968). Some reported cases of selective fertilization do not conform to this more restricted and therefore more useful definition. However, a few examples of true preferential fertilization remain after the others are excluded.

Preferential fertilization was first established by Schwemmle (1938, 1968) in the cross *Oenothera berteriana* ♀ × *Oe. odorata* ♂. The parental species are permanent heterozygotes for Renner complexes with the constitutions *B/l* and *v/I*, respectively. Each parent produces two classes of gametes, and their F_1 generation is expected to include four classes of zygotes. Actually only three classes appear (*lv, lI, BI*), and the expected *Bv* genotype is absent. Various possible causes of the absence of *Bv*, such as pollen lethals or zygote lethals, can be ruled out. The actual cause is that the *v* pollen tubes do not grow to the *B* ovules. The *Bv* zygote can be obtained when the cross is made in the opposite direction so that *B* pollen has a chance to combine with *v* eggs (Schwemmle, 1938, 1968; Schwemmle and Koepchen, 1953).

It has been shown experimentally in Oenothera that pollen-tube growth toward the micropyle and embryo sac is guided by a chemical stimulus. Where preferential fertilization occurs, differences probably exist between ovules in the strength or type of the chemical stimulus, and between pollen tubes in their chemotropic response to this stimulus, so that certain combinations of ovules and pollen are exceptionally frequent or infrequent (Arnold, 1958; Schwemmle, 1968).

In the garden pea, *Pisum sativum,* the allele pair *Uni/uni,* which usually gives a 3 : 1 ratio in F_2, has a pronounced deficit of the homozygous recessive type in one line, with ratios of about 7 : 1. The gene *Lac* also gives a deficit of the recessive type in segregating F_2s in one line of garden peas, and the gene *Obo* segregates to yield a great surplus of recessives. No gametic sterility could be found in the lines with abnormal segregation. The seed fertility and seed germination are normal. Pollen-tube competition has not been excluded but is deemed unlikely. The most probable cause of the altered ratios is considered to be preferential fertilization (Lamprecht, 1954).

Small supernumerary B chromosomes commonly occur in the nucleus along with the normal chromosome complement in *Zea mays.* Such plants produce some B-containing sperm nuclei and some no-B sperm. Roman (1948) found that sperm nuclei with two B chromosomes were more successful than no-B sperms in fertilizing the egg nuclei.

SELECTIVE ELIMINATION OF ZYGOTES *Oenothera lamarckiana* (or *Oe. erythrosepala*) is the classic example of a genomically heterozygous plant with a balanced lethal system which operates at the zygote stage. The plants are semisterile as to seeds, and the abortive seeds represent the classes of genomic homozygotes. Consequently, *Oe. lamarckiana,* though highly heterozygous and predominantly self-pollinating, does not normally segregate but breeds true for its heterozygous genomic constitution.

Mechanisms which dampen the segregation exist in a number of other plant species. In *Trifolium subterraneum* two ovules are formed in each ovary; normally one of these develops and suppresses the other. In the event of sterility in one ovule, the other is able to continue growth and develop into a seed (Brock, 1956). In *Paeonia californica,* which is highly heterozygous and usually semisterile as to seeds, the elimination of ovules and young seeds apparently has a strong selective component (J. Walters and V. Grant, unpublished data).

Selective elimination of zygotes as a factor distorting segregation ratios leads into the wide subject of natural selection, which is beyond the scope of this book. An important aspect which does lie within our purview, namely linkage between morphological markers and viability factors, is discussed in Chapter 19.

TETRASOMIC INHERITANCE Segregation ratios in autotetraploids are vastly different from those in diploids and are more complex. The complexities have several causes. There are three possible heterozygous types ($AAAa$, $AAaa$, $Aaaa$) instead of one as in diploids. For any heterozygous type the phenotypic ratios will be affected by dosage effects and the completeness or incompleteness of dominance.

Further variations in ratios result from different possible modes of chromosome pairing in the duplex heterozygote $AAaa$. If the four homologous chromosomes pair and assort at random, the expected gametic ratio will be 1 AA : 4 Aa : 1 aa. Preferential pairing of A/A and a/a gives all Aa gametes. Pairing of A/a and A/a gives a gametic ratio of 1 AA : 2 Aa : 1 aa. Each type of pairing of course leads to a different zygotic ratio in F_2.

Still another complicating factor is the possibility of either chromosome or chromatid segregation in an autotetraploid. The tetraploid has four homologous chromosomes but eight chromatids. The duplex heterozygote $AAaa$ has the genic constitution $AAaa$ with respect to the chromosomes but the constitution $AAAAaaaa$ with respect to chromatids. The random combining of the four alleles in sets of two in the diploid gametes is chromosome segregation; chromatid segregation is the random combining of the eight alleles in sets of two. Chromosome and chromatid segregation yield different gametic ratios. In general, chromatid segregation yields a higher frequency of aa gametes and a higher frequency of the recessive type in F_2 than does chromosome segregation.

Genes located close to the centromere in an autotetraploid show chromosomal segregation. Chromatid segregation is made possible by regular crossing-over between a gene and its centromere. Therefore genes located on the distal segments of chromosomes are expected to show chromatid segregation.

The various types of expected segregation ratios in autotetraploids have been reviewed by Crane and Lawrence (1934, 1938), Lindstrom (1936), and Little (1945, 1958). Following these authors, we list below some of the theoretical ratios for a duplex heterozygote, *AAaa*.

1. *AAaa* with chromosome segregation
   Gametic ratio, 1 *AA* : 4 *Aa* : 1 *aa*
   Zygotic ratio in F_2, 1 *AAAA* : 8 *AAAa* : 18 *AAaa* : 8 *Aaaa* : 1 *aaaa*
   Phenotypic ratio in F_2 with complete dominance of *A*, 35 *A* : 1 *a*
   Phenotypic ratio in backcross to *aaaa* with complete dominance of *A*, 5 *A* : 1 *a*
2. *AAaa* with chromatid segregation
   Gametic ratio, 3 *AA* : 8 *Aa* : 3 *aa*
   Phenotypic ratio in F_2 with complete dominance of *A*, 21 *A* : 1 *a*
   Phenotypic ratio in backcross to *aaaa* with complete dominance of *A*, 11 *A* : 3 *a*

Note that the recessive type segregates in a much lower frequency in autotetraploids than in diploids.

Empirical data on tetrasomic inheritance were obtained in early studies of tetraploid tomatoes by Sansome (1933) and Lindstrom (1936). Some of Sansome's data on autotetraploid *Lycopersicon esculentum* are presented in Table 27. The observed numbers of recessive types for the genes *S, D,* and *P* approach the expected numbers on chromatid segregation. Gene *Y* approaches chromosome segregation, and *R* appears to be intermediate. Lindstrom's additional data point to the same conclusion for most of these genes but indicate chromatid segregation for *R*.

Little's summary of the problem of chromosome versus chromatid segregation is worth quoting here (Little, 1945, pp. 79–80):

> It is rather common in the literature for geneticists to refer to 'chromosome segregation' and 'chromatid segregation' as though these were two fixed ratios between which one must choose in fitting a set of data. As a matter of fact, these two types of segregation are, in reality, only ideal limiting ratios which are seldom attained, and probability favors the occurrence of ratios which are intermediate between these two limits. The important statistical problem is, therefore, not to determine which ratio the data fit more closely, but to find out to what extent the two opposing forces of reductional and equational separation have affected the data.

TABLE 27

PHENOTYPIC SEGREGATION RATIOS IN F_2 IN AUTOTETRAPLOID TOMATO (*LYCOPERSICON ESCULENTUM*). THE SEGREGATING HETEROZYGOTE HAS THE DUPLEX CONSTITUTION *AAaa* (SANSOME, 1933)

Gene	Character	Observed no.		Expected no. of recessives with:	
		dominant	recessive	chromosome segregation	chromatid segregation
S	inflorescence branching	787	39	23	38
D	normal or dwarf stature	995	43	28	48
P	hairiness of fruit	988	50	28	48
R	color of fruit flesh	647	26	18	31
Y	color of fruit epidermis	554	17	15	26
A_1*	anthocyanin in stems and leaves	172	4	5	8
A_2*	anthocyanin in stems	207	8	6	10

* Gene A_1 is now referred to as *A,* and A_2 *as Al* (Barton, Butler, Jenkins, Rick and Young, 1955).

More recently Moens (1964) has repeated the work on segregation in autotetraploid tomato on a more extensive scale. He used a series of marker genes on chromosome 2; these were in the duplex condition in the F_1 heterozygote. The observed segregation ratios in F_2 are given in Table 28. It will be noted that all ratios are approximately 30 : 1, which is an intermediate condition between the theoretical chromosome and chromatid ratios. The observed ratios agree with expectation, however, when cytological observations of pairing and crossing-over in chromosome 2 are taken into account (Moens, 1964).

TABLE 28

SEGREGATION RATIOS IN F_2 FOR FIVE GENES ON CHROMOSOME 2 IN TETRAPLOID TOMATO (*LYCOPERSICON ESCULENTUM*). THE F_1 WAS A DUPLEX HETEROZYGOTE OF THE TYPE *AAaa* (MOENS, 1964)

Gene	Observed no.		Ratio
	dominant	recessive	
D	37,200	1,236	30.1 : 1
Dv	9,189	300	30.6 : 1
M	13,263	444	29.9 : 1
Aw	9,180	309	29.8 : 1
Wv	37,220	1,216	30.6 : 1

Tetrasomic inheritance has been observed in artificial tetraploids of *Antirrhinum majus, Datura stramonium, Oryza sativa, Primula sinensis,* and *Zea mays;* in natural tetraploid species, *Lotus corniculatus, Medicago sativa,* and *Solanum tuberosum;* in the octoploid *Dahlia variabilis;* and in other plants. The available evidence up to 1958 is reviewed by Little (1945, 1958).

SEGREGATION IN ALLOPOLYPLOIDS An allotetraploid is derived by hybridization and summation of chromosome sets from two diploid species. The parental diploids must be fairly closely related in order to hybridize successfully, and, being related, will have many genes in common. Consequently, the allotetraploid also carries a number of homologous genes in the different genomes. The mode of segregation of these homologous genes in the allotetraploid depends upon the type of chromosome pairing, which in turn depends upon various structural and genic factors.

Let us say that the allotetraploid carries the allele pair AA in one genome and the homologous allele pair aa in the other. If the tetraploid plant is a strict genomic allopolyploid, with chromosome pairing taking place exclusively between members of the same genome (A/A and a/a), it will form all Aa gametes and all $AAaa$ zygotes. In short, under these conditions the amphiploid will not segregate at all for its internally heterozygous constitution in A.

A contrasting possibility is that the chromosomes or chromosome segments carrying the A and a alleles in the tetraploid are homologous to such an extent that they undergo random pairing and assortment at meiosis. In this case the tetraploid segregates for A in tetrasomic ratios.

Another possibility is that chromosome pairing in the A segment is predominantly intragenomic (A/A and a/a), but rarely intergenomic (A/a). The tetraploid then shows limited segregation for A. The observed ratios are then intermediate between tetrasomic segregation and no segregation.

The three modes of segregation described are characteristic of different types of tetraploids: genomic allotetraploids and autotetraploids as the two extremes, and segmental allotetraploids as the intermediate condition (see Grant, 1971, ch. 15, for further discussion). Alternatively, the three modes may be exhibited by genes located in different regions of the genomes of a single tetraploid.

Restricted segregation has been recorded in experimental allopolyploids in Elymus-Sitanion, Fragaria, Geum, Gilia, Madia, Rubus, Saxifraga, and other groups. The progenies were too small in most of the earlier studies to permit identification of the mode of segregation. More recently Gerstel and co-workers have carried out a series of studies of synthetic allopolyploids in Gossypium and Nicotiana which yield more precise data on this question.

One study concerns segregation at the R_1 locus in the synthetic allohexaploids *Gossypium hirsutum-raimondii* and *G. hirsutum-thurberi*. *Gossypium hirsutum* is tetraploid with the genomic constitution AADD, and *G. raimondii* and *G. thurberi* are related diploid species with the constitution DD. The hexaploids are produced by doubling the chromosomes in the triploid interspecific hybrids, and thus have the genomic constitution AADDDD. The R_1 gene is in the D genomes of the three parental species (Gerstel, 1956).

The R_1 gene controls anthocyanin formation in various parts of the plants, including, in some cases, the petal spots. It is represented by three alleles: R_1, r_1, R_{rai}. The synthetic allopolyploids had various heterozygous combinations of these alleles, such as $R_1R_1R_{rai}R_{rai}$ and $r_1r_1R_{rai}R_{rai}$. Gametic ratios were obtained by backcrossing them to recessive types of *G. hirsutum* (r_1r_1) (Gerstel, 1956).

The observed gametic ratios for R_1 were intermediate in all cases between genomic allopolyploid ratios ($\infty : 0$) and tetrasomic ratios (5 : 1 to 3.7 : 1). The observed ratio in *G. hirsutum-thurberi* was 604 dominant : 9 recessive (= 67.1 : 1); and that in *G. hirsutum-raimondii* was 1163 : 134 (= 8.7 : 1). The ratios thus differ significantly between the two allopolyploids; the ratio in *G. hirsutum-raimondii* is closer to tetrasomic segregation. This result is in line with independent taxonomic evidence indicating a close relationship between *G. raimondii* and *G. hirsutum* and a more distant relationship of *G. thurberi* to *G. hirsutum* (Gerstel, 1956).

Much more extensive data on segregation in allopolyploids in Gossypium are given in subsequent papers (Gerstel and Phillips, 1957, 1958; Phillips and Gerstel, 1959; Phillips, 1962). The data are partly summarized in Tables 29 and 30.

TABLE 29

TESTCROSS RATIOS IN SYNTHETIC ALLOTETRAPLOIDS IN GOSSYPIUM (REAR-
RANGED FROM GERSTEL AND PHILLIPS, 1958)

Hybrid combination	Genomic constitution	Gene	Segregation	Ratio of dominant to recessive
G. arboreum × herbaceum	$A_1A_1A_2A_2$	R_1	309 : 78	4.0 : 1
		L	49 : 5	9.8 : 1
		Y	36 : 7	5.1 : 1
		P_a	51 : 13	3.9 : 1
		R_2 (leaf)	16 : 104 : 34	3.5 : 1
G. thurberi × raimondii	$D_1D_1D_5D_5$	R_1 (R_{rai})	363 : 27	13.4 : 1
G. arboreum or herbaceum × anomalum	AABB	R_1	317 : 7	45.3 : 1
		L	1 : 404 : 3	135 : 1
		Y	446 : 5	89.2 : 1
		P_a	405 : 6	67.5 : 1
		R_2 (leaf)	5 : 167 : 1	172 : 1
G. arboreum × thurberi	AADD	L	117 : 1	117 : 1
		P_a	118 : 0	118 : 0
		R_2 (leaf)	158 : 0	158 : 0

Table 29 presents the backcross ratios to the recessive type, and by inference the gametic ratios, in four synthetic allotetraploids derived from different diploid interspecific hybrids. The allotetraploid combinations are listed according to the degree of relationship and pairing between their component genomes. The *Gossypium arboreum-herbaceum* tetraploid, composed of very similar genomes, shows tetrasomic segregation or an approach to it in several genes. At the opposite extreme is the genomic allotetraploid *G. arboreum-thurberi* with little or no segregation in some of the same genes. Intermediate ratios are exhibited by two other allotetraploid combinations with intermediate degrees of genomic affinity (Gerstel and Phillips, 1958).

Table 30 gives ratios in a series of synthetic allohexaploids involving *G. hirsutum* and various diploid species. Here again we see tetrasomic ratios for genes in the closely related A genomes in *G. hirsutum-arboreum* and *G. hirsutum-herbaceum*. Approaches to tetrasomic ratios occur in *G. hirsutum-raimondii* for genes located in the somewhat differentiated D genomes of these two parental species. Steeper ratios occur for segregating genes distributed in the still more

TABLE 30

TESTCROSS RATIOS IN HEXAPLOID DERIVATIVES OF *GOSSYPIUM HIRSUTUM* ×
DIPLOID SPECIES (REARRANGED FROM GERSTEL AND PHILLIPS, 1958)

Hybrid combination	Genomic constitution	Gene	Character	Ratio of dominant to recessive
G. hirsutum × *arboreum* or *herbaceum*	AAAADD	H_2	pilose	4.9 : 1
		R_2	leaf color	5.0 : 1
		L	leaf shape	5.1 : 1
		Y	petal color	5.2 : 1
G. hirsutum × *raimondii*	AADDDD (genes in D genomes)	Cl	clustered inflorescence	7.5 : 1
		Cn	crenate	8.0 : 1
		Rd	red dwarf	8.3 : 1
		Yg	yellow green	8.3 : 1
		Cr	crinkle	8.9 : 1
		R_1	plant color	9.5 : 1
		V	virescent	10.1 : 1
		L	leaf shape	11.5 : 1
		Gl_1	glandless	16.2 : 1
G. hirsutum × *thurberi*	AADDDD	L	leaf shape	22.4 : 1
		R_1	plant color	42.7 : 1
G. hirsutum × *anomalum*	AADDBB	N	seed fuzz	146.5 : 1
		Lc_1	lint color	148.5 : 1
		R_1	plant color	346 : 1
		R_2	petal spot	346 : 1
		L	leaf shape	347 : 0
G. hirsutum × *raimondii*	AADDDD (genes in A genome)	N	seed fuzz	102 : 0
		Lc_1	lint color	102 : 0
		Y	petal color	429 : 0
		Pa	pollen color	431 : 0

differentiated D genomes of *G. hirsutum* and *G. thurberi* in their allohexaploid. Approaches to genomic allopolyploid segregation are
found in *G. hirsutum-anomalum*. And genes in the A genome in
G. hirsutum-raimondii (AADDDD), unlike those in the D genomes
in the same plants, do not segregate at all (Gerstel and Phillips,
1958).

In some instances a single gene, such as *Y*, R_1, or *L*, segregates
in very different ratios in different allopolyploid combinations
(Table 30). A comparison not given in Table 30 concerns the gene

H_2 in the A genome. It segregates in a tetrasomic ratio in *G. hirsutum-arboreum* or *G. hirsutum-herbaceum* (AAAADD); and in a ratio of 443 : 1 in *G. hirsutum-raimondii* (AADDDD) (Gerstel and Phillips, 1958).

The segregation ratios are correlated with the average degree of chromosome pairing between genomes carrying the genes in the different hybrid combinations (Gerstel and Phillips, 1958; Gerstel, 1966).

Further studies reveal a similar range of segregation ratios in synthetic allopolyploids in Nicotiana (Gerstel and Phillips, 1958; Gerstel, 1960, 1963). Gerstel produced a series of synthetic allohexaploids consisting of *N. tabacum* (4*x*) and a diploid species. The diploid species represent varying degrees of systematic and genomic relationship with *N. tabacum*. Allohexaploids containing an *N. tabacum* and a closely related diploid, like *N. tabacum-tomentosiformis,* give tetrasomic ratios or approaches to such ratios for genes on the common genomes. The same genes show little or no segregation in allohexaploids like *N. tabacum-glauca,* composed of three distinct genome groups. Intermediate ratios are found in allohexaploids containing *N. tabacum* and a moderately closely related diploid, *N. tabacum-otophora* (Gerstel and Phillips, 1958; Gerstel, 1960, 1963).

Nicotiana sylvestris is the closest living relative to one of the diploid ancestors of *N. tabacum*. The allohexaploid *N. tabacumsylvestris* has the genomic constitution SSSSTT and segregates for various genes on different chromosomes in the S genome. The different genes segregate in modes ranging from tetrasomic ratios to steep ratios. The probable explanation of these differences is that some chromosomes in the S genomes of *N. tabacum* and *N. sylvestris* have remained homologous during the evolutionary history of the two species, while other chromosomes in the S genomes have become differentiated (Gerstel, 1963).

DUPLICATE FACTORS IN ALLOPOLYPLOIDS Since allotetraploids contain the genomes of two related diploid species, they can be expected to have some independent but homologous genes which show duplicate factor inheritance. Allohexaploids, by the same reasoning, might exhibit triplicate factor inheritance for some genes and duplicate factor inheritance for others.

The principles of multiple factor inheritance were outlined in Chapter 10. Two independent duplicate genes, D_1 and D_2, in a double heterozygote $D_1/d_1 \, D_2/d_2$, with dominance of D in each allele pair, will segregate to give an F_2 phenotypic ratio of 15 dominant : 1 recessive. For triplicate factors with dominance in each allele pair the corresponding ratio is 63 : 1 in F_2.

In an allotetraploid, where the duplicate genes occur in different genomes, and where heterozygosity and dominance exist in each allele pair, the duplicate factor ratio of 15 : 1 is expected under two conditions. The first and probably most common condition is chromosome pairing and independent assortment in the mode D_1/d_1 and D_2/d_2. An alternative possibility is regular intergenomic chromosome pairing (or allosyndesis) in the mode D_1/d_2 and D_2/d_1, followed by independent assortment.

Duplicate and triplicate factor inheritance has been found in a number of allopolyploid species. In fact, one of the first cases of multiple factors to be discovered was that of the triplicate factors (R_1, R_2, R_3) for kernel color in allohexaploid *Triticum aestivum* (see Chapter 10). Other early cases were found in the allohexaploid *Avena sativa*.

Sansome and Philp (1939) compared a number of genes governing homologous characters in *Triticum aestivum, Avena sativa,* and *Hordeum vulgare.* The first two species are allohexaploids and the third is diploid on the same common basic chromosome number of $x = 7$. Duplicate or triplicate factors are rather common in wheat and oats, but uncommon in diploid barley. A list of 24 genes in barley includes only two or possibly a few more cases of duplicate or triplicate factors. In wheat, by contrast, a corresponding gene list includes nine cases of duplicate or triplicate factors (Sansome and Philp, 1939, pp. 207–10).

Allotetraploid tobacco (*Nicotiana tabacum*) contains duplicate factors for chlorophyll formation, as indicated by the segregation of green seedlings and white seedlings in a 15 : 1 ratio in an intervarietal cross (Clausen and Cameron, 1950).

In an allopolyploid species which has been in existence for a long time, the members of an original duplicate factor system may diverge in gene function. Then the two genes will have closely related phenotypic effects but will not be allelic. Instead of exhibiting duplicate factor inheritance, they will segregate in a simple dihybrid ratio (9 : 3 : 3 : 1) or a modified dihybrid ratio such as complementary factor inheritance. Sansome and Philps' (1939) list of genes and gene systems in oats includes several cases of epistasis and one of complementary inheritance (see also Chapter 9).

Interesting evidence for divergence from duplicate factors during the evolutionary history of a species is found in Gossypium. The American group of cultivated cottons (*G. hirsutum* and *G. barbadense*), as is well known, are allotetraploids with the genomic constitution AADD. The Old World group of diploid cultivated cottons (*G. arboreum* and *G. herbaceum*) have the genomic constitution AA. The D genome is carried by a group of American wild diploid species including *G. raimondii* and others.

A linkage group including the gene R for anthocyanin pigment and the gene Cl for inflorescence congestion occurs in both the A and D genomes and again in the AADD tetraploids. The R gene is present as a single locus in the AA diploids and the DD diploids, but as two loci (R_1 and R_2) in the AADD tetraploids. Similarly the gene Cl is present as a single locus in the AA diploids, and probably also in the DD diploids, but occurs in two loci (Cl_1 and Cl_2) in the AADD tetraploids (Stephens, 1951a, 1951b).

The genomic location of these genes in the tetraploids is R_2 and Cl_2 in the A genome, R_1 and Cl_1 in the D genome. The mode of gene interaction between R_1 and R_2 and between Cl_1 and Cl_2 in the tetraploids is not, contrary to what might be expected, that of simple duplicate factors (Stephens, 1951a, 1951b).

The genes R_1 and R_2 were probably duplicate factors when tetraploid cotton first originated. If they were still duplicate factors in modern tetraploid cotton, the double heterozygote $R_1/r_1\ R_2/r_2$ should segregate in a phenotypic ratio of 15 : 1 in F_2, but it does not. Instead R_1 and R_2 have complementary effects. The petals of R_1 plants are spotless, and R_2 plants have a ghost spot, but the combination R_1R_2 produces red spots (Stephens, 1951a, 1951b).

Similarly, the Cl_1 and Cl_2 genes, which were probably duplicate factors when tetraploid cotton originated, now give complementary effects. The inflorescence is clustered with Cl_1, and also clustered with Cl_2; but the combination Cl_1Cl_2 gives normal inflorescence branching. The genes Cl_1 and Cl_2, like the genes R_1 and R_2, are no longer strictly allelic but behave as complementary factors. Therefore they must have diverged in gene function from their putative original condition of duplicate factors (Stephens, 1951a, 1951b).

CHAPTER FOURTEEN

SEX DETERMINATION

INTRODUCTION Hermaphroditism and self-compatibility are common and probably original in the angiosperms (East, 1940; Lewis, 1942). Hermaphroditism and self-compatibility in combination with diverse types of floral mechanisms lead to various mixtures of biparental and uniparental reproduction. Breeding systems in the angiosperms have evolved from this flexible original condition toward strict uniparental reproduction and toward obligate biparental reproduction (see Chapter 24). We are concerned with the latter development in this chapter.

The arrangements for securing obligate biparental reproduction in angiosperms are diverse. They have in common that the plant population consists of different sex types, or mating types, or flower forms; and that the only fertile crosses are those between different types. But the details vary greatly.

In fact several different breeding systems are involved in achieving strict biparental reproduction and outcrossing:

1. *Dioecism*. Some plants bear pistillate flowers and others bear staminate flowers. In gymnosperms and bryophytes the appropriate corresponding terms for the morphological female and male

organs can be substituted for "pistillate" and "staminate" flowers. In any case, the plant population consists of female individuals and male individuals.

2a. *Self-incompatibility with hermaphroditism.* The flowers contain both stamens and pistil, and the floral mechanism is the same on different individual plants. But self-fertilization is blocked by an incompatible reaction between pollen and pistil of the same individual. This incompatibility can be reduced to antagonistic reactions between like alleles of the S gene in pollen and pistil of the same plant. Cross-fertilization between different individuals with the same genotype for the S gene is likewise blocked. The plant population thus consists of different mating types determined by the different multiple alleles of the S gene. Successful crosses are possible only between individuals belonging to different mating types.

2b. *Self-incompatibility with monoecism.* The flowers are unisexual, but male and female flowers occur on the same individual. Self-fertilization is again blocked by a self-incompatibility reaction.

3. *Heterostyly.* The flowers are hermaphroditic but occur in two or three forms differing in the length of the style and height of the anthers. An individual plant bears flowers of one type. In the simplest case the population consists of two kinds of individuals, those with long-styled flowers and those with short-styled flowers, or pins and thrums respectively, and successful cross-pollination is possible only between pin and thrum plant. Or there may be individuals with three different style lengths in the population. Self-incompatibility is usually associated with heterostyly.

An enormous body of literature has accumulated on each of the foregoing breeding systems and on other related breeding systems, such as subdioecism and incomplete self-incompatibility, which permit some selfing. This body of information cannot be covered comprehensively here. A number of good reviews are available which summarize the details and give extensive bibliographies.

These review papers are classified by topic as follows. Systematic distribution of dioecism in angiosperms (Yampolsky and Yampolsky, 1922; Allen, 1940) and bryophytes (Allen, 1945). Systematic distribution of self-incompatibility in angiosperms (East, 1940; Lewis, 1949). Systematic distribution of heterostyly (Vuilleumier, 1967). General surveys of breeding systems in angiosperms (Yampolsky and Yampolsky, 1922), seed plants (Fryxell, 1957), and green plants (Whitehouse, 1959). Genetics of sex determination in dioecious angiosperms (Allen, 1940; Lewis, 1942; Westergaard, 1958). Genetics of heterostyly and self-incompatibility (Lewis, 1949).

SYSTEMATIC DISTRIBUTION OF DIOECISM Dioecism is a rare condition in the angiosperms. A survey of the British flora by Lewis (1942) revealed only 54 species of dioecious angiosperms, or 2% of the angiosperm species in this flora. In the world flora 4% of the species of dicotyledons and 3% of the species of monocotyledons are dioecious (Yampolsky and Yampolsky, 1922).

Dioecism rises to higher frequencies in a few island floras in the Pacific region. Carlquist (1966) finds 405 dioecious species in the Hawaiian Islands, representing 27% of the Hawaiian flora, and a similar high frequency of dioecious species occurs in New Zealand. For some unknown reason dioecism must have a particularly high selective value in these insular biotas.

Dioecism is not only rare in most floras but also sporadic in systematic distribution, being scattered throughout numerous groups. The 54 species of dioecious angiosperms in Great Britain are distributed in 26 genera belonging to 18 families (Lewis, 1942). The same pattern of wide systematic distribution is found in the world flora (Yampolsky and Yampolsky, 1922).

The widespread systematic distribution of the dioecious condition is shown to some extent by Table 31. This table gives a partial list of dioecious angiosperms grouped into orders and subclasses according to the system of classification of Cronquist (1968). A perusal of the list reveals a large representation of families, many orders, and quite a few subclasses from the Magnoliidae to the Liliidae. A much longer list of families and genera containing dioecious species is given by Yampolsky and Yampolsky (1922, pp. 39–48).

The dioecious condition is usually characteristic of individual species or species groups in the angiosperms. Occasionally it characterizes a whole subgenus, as in Rumex, or a genus, as in Humulus. Only rarely in the angiosperms does dioecism characterize a whole family, like the Salicaceae (Fig. 25) (Westergaard, 1958).

TABLE 31

SOME DIOECIOUS ANGIOSPERMS [COMPILED FROM LISTS OF ALLEN (1940)
AND WESTERGAARD (1958)]

DICOTYLEDONS

1. Subclass Magnoliidae
   Clematis virginiana (Ranunculaceae, Ranunculales)
   Thalictrum fendleri (Ranunculaceae, Ranunculales)
   Cocculus trilobus (Menispermaceae, Ranunculales)

2. Subclass Hamamelidae
   Cercidiphyllum japonicum (Cercidiphyllaceae, Hamamelidales)
   Cannabis sativa (Moraceae, Urticales)
   Humulus japonicus, lupulus (Moraceae, Urticales)
   Morus bombycis (Moraceae, Urticales)
   Cudrania triloba (Moraceae, Urticales)
   Urtica dioica (Urticaceae, Urticales)
   Myrica cerifera (Myricaceae, Myricales)

3. Subclass Caryophyllidae
   Melandrium rubrum, album, divaricatum, glutinosum
     (= *Lychnis* spp.) (Caryophyllaceae, Caryophyllales)
   Silene otites, roemeri (Caryophyllaceae, Caryophyllales)
   Spinacia oleracea, tetrandra (Chenopodiaceae, Caryophyllales)
   Atriplex hymenelytra (Chenopodiaceae, Caryophyllales)
   Acnida tuberculata (Amaranthaceae, Caryophyllales)

4. Subclass Dilleniidae
   Actinidia polygama (Dilleniaceae, Dilleniales)
   Eurreya japonica (Theaceae, Theales)
   Napaea dioica (Malvaceae, Malvales)
   Carica papaya (Caricaceae, Violales)
   Datisca cannabina (Datiscaceae, Violales)
   Bryonia dioica (Cucurbitaceae, Violales)
   Trichosanthes cucumeroides, and other species (Cucurbitaceae, Violales)
   Ecballium elaterium, in part (Cucurbitaceae, Violales)
   Coccinea indica (Cucurbitaceae, Violales)
   Populus balsamifera, and other species (Salicaceae, Salicales)
   Salix caprea, and other species (Salicaceae, Salicales)
   Empetrum nigrum (Empetraceae, Ericales)

5. Subclass Rosidae
   Ribes saxatile (Grossulariaceae, Rosales)
   Sedum rhodiola (= *S. roseum*) (Crassulaceae, Rosales)
   Fragaria elatior (Rosaceae, Rosales)
   Shepherdia canadensis (Elaeagnaceae, Proteales)
   Aucuba chinensis (Cornaceae, Cornales)
   Garrya elliptica (Garryaceae, Cornales)
   Buckleya joan (Santalaceae, Santalales)
   Viscum album (Loranthaceae, Santalales)
   Phoradendron villosum (Loranthaceae, Santalales)

5. Subclass Rosidae, continued
   Ilex serrata, in part (Aquifoliaceae, Celastrales)
   Mercurialis annua (Euphorbiaceae, Euphorbiales)
   Daphniphyllum macropodum (Daphniphyllaceae, Euphorbiales)
   Vitis vinifera (Vitaceae, Rhamnales)
   Acer negundo (Aceraceae, Sapindales)
   Picrassima quassioides (Simarubaceae, Sapindales)
   Zanthoxylum piperitum (Rutaceae, Sapindales)
   Trinia hispida (Umbelliferae, Umbellales)
6. Subclass Asteridae
   Valeriana dioica (Valerianaceae, Dipsacales)
   Cirsium arvense (Compositae, Asterales)

MONOCOTYLEDONS

7. Subclass Alismatidae
   Sagittaria montevidensis (Alismaceae, Alismatales)
   Vallisneria spiralis (Hydrocharitaceae, Hydrocharitales)
   Elodea canadensis, gigantea (Hydrocharitaceae, Hydrocharitales)
   Hydrilla verticillata (Hydrocharitaceae, Hydrocharitales)
   Najas marina (Najadaceae, Najadales)
8. Subclass Commelinidae
   Carex grallatoria (Cyperaceae, Cyperales)
9. Subclass Arecidae
   Phoenix dactylifera, canariensis (Palmaceae, Arecales)
   Trachycarpus excelsus, fortunei (Palmaceae, Arecales)
10. Subclass Liliidae
    Asparagus officinalis (Liliaceae, Liliales)
    Smilax herbacea, china, and other species (Smilacaceae, Liliales)
    Dioscorea gracillima (Dioscoreaceae, Liliales)

In the gymnosperms, by contrast, dioecism is fairly common and often occurs at high taxonomic levels. The cycads (Cycadales) are dioecious. Ginkgo, representing the Ginkgoales, is dioecious. In the Gnetales, Gnetum, Welwitschia, and most species of Ephedra are dioecious. Many of the conifers are monoecious. But in this large group dioecism is characteristic of Taxus and Torreya (Taxaceae), Podocarpus (Podocarpaceae), most Araucariaceae, and some Cupressaceae such as Juniperus.

In the ferns, dioecism is characteristic of the heterosporous ferns (Marsileales and Salviniales), which have unisexual gametophytes. The homosporous Filicales, on the other hand, are nearly all monoecious, with bisexual gametophytes, but one exceptional species of Polypodiaceae is dioecious in the gametophyte stage. Selaginella and Isoetes in the Lycopsida are also dioecious (see Bold, 1967).

Figure 25. Flowering shoots of female and male individuals of a dioecious willow, *Salix fragilis*. Pistillate plant (*left*) and staminate plant (*right*). (From Kerner, 1894–1895.)

Bryophytes, like ferns and fern allies, have either bisexual or unisexual gametophytes, and are thus either monoecious or dioecious in the gametophyte stage. Dioecism is fairly common and widespread in different main groups of bryophytes. Some examples are given in Table 32 (Allen, 1945).

SEX CHROMOSOMES Morphologically differentiated sex chromosomes are associated with dioecism in many but not all cases. The reported occurrence of sex chromosomes in bryophytes is indicated in Table 32. The observations are old and no doubt need reinvestigation.

TABLE 32
SOME DIOECIOUS BRYOPHYTES. SPECIES WITH DISTINGUISHABLE SEX CHRO-
MOSOMES ARE ANNOTATED XY, THOSE WITHOUT ARE ANNOTATED 0
(REARRANGED FROM ALLEN, 1945)

1. Sphaerocarpales
   Riella helicophylla (XY)
   Sphaerocarpus cristatus, and other species (XY)

2. Marchantiales
   Marchantia diptera, polymorpha (XY)
   Marchantia cuneiloba, radiata (0)
   Riccia bischoffii (XY)
   Lunularia cruciata (XY)

3. Jungermanniales
   Pellia fabbroniana, radiculosa (XY)
   Frullania dilatata, japonica (XY)
   Frullania densiloba, moniliata (0)

4. Anthocerotales
   Aspiromitus sampalocensis (XY)

5. Bryales
   Mnium punctatum (XY)
   Mnium maximowiczii (0)
   Pogonatum grandifolium (XY)
   Polytrichum formosum (XY)
   Polytrichum commune, juniperinum (0)
   Bryum caespiticium (0)

As regards the dioecious angiosperms, Westergaard (1958) has critically reexamined the older reports of sex chromosomes, with the results summarized in Table 33. Sex chromosomes do occur in several genera of angiosperms. But in many other groups, like Salix and Urtica, the earlier reports of sex chromosomes are considered to be dubious. The list of dioecious angiosperms which lack cytologically distinguishable sex chromosomes is a long one, including Asparagus, Bryonia, Empetrum, Mercurialis, Spinacia, and Thalictrum (Allen, 1940; Lewis, 1942; Westergaard, 1958).

Several sex chromosome mechanisms are represented. Male heterogamety is the rule; that is, the sex which is heterozygous for the sex chromosomes is usually the male. *Fragaria elatior* is the only known dioecious angiosperm with possible or probable sex chromosomes, in which the female is heterogametic (Lewis, 1942; Westergaard, 1958).

TABLE 33
SEX CHROMOSOMES IN ANGIOSPERMS
(REARRANGED FROM WESTERGAARD, 1958)

I. Well established cases of morphologically differentiated sex chromosomes

Cannabis sativa,   XX ♀ + XY ♂
Humulus lupulus,   XX ♀ + XY ♂, or $X_1X_1X_2X_2$ ♀ + $X_1Y_1X_2Y_2$ ♂
Humulus japonicus,   XX ♀ + XYY ♂
Rumex angiocarpus,   XX ♀ + XY ♂
Rumex tenuifolius, tetraploid,   (XX)XX ♀ + (XX)XY ♂
Rumex acetosella, sens. str., hexaploid,   (XXXX)XX ♀ + (XXXX)XY ♂
Rumex graminifolius, octoploid,   (XXXXXX)XX ♀ + (XXXXXX)XY ♂
Rumex hastatulus,   XX ♀ + XYY ♂
Rumex acetosa, and related species,   XX ♀ + XYY ♂
Rumex paucifolius, tetraploid,   XXXX ♀ + XXXY ♂
Melandrium album,   XX ♀ + XY ♂
Melandrium rubrum,   XX ♀ + XY ♂

II. Reported cases of sex chromosomes which are uncertain and require further study

Cocculus trilobus (Menispermaceae)
Cercidiphyllum japonicum (Cercidiphyllaceae)
Morus bombycis (Moraceae)
Cudrania triloba (Moraceae)
Urtica dioica (Urticaceae)
Silene otites, densiflora (Caryophyllaceae)
Spinacia oleracea, tetrandra (Chenopodiaceae)
Atriplex hymenelytra (Chenopodiaceae)
Actinidia polygama, kolomicta (Dilleniaceae)
Eurreya japonica (Theaceae)
Datisca cannabina (Datiscaceae)
Trichosanthes cucumeroides, and other species (Cucurbitaceae)
Coccinea indica (Cucurbitaceae)
Populus (Salicaceae)
Salix (Salicaceae)
Fragaria elatior (Rosaceae)
Buckleya joan (Santalaceae)
Phoradendron villosum, flavescens (Loranthaceae)
Ilex serrata (Aquifoliaceae)
Daphniphyllum macropodum (Daphniphyllaceae)
Acer negundo (Aceraceae)
Picrassima quassioides (Simarubaceae)
Zanthoxylum piperitum (Rutaceae)
Valeriana dioica (Valerianaceae)
Elodea canadensis, gigantea (Hydrocharitaceae)
Hydrilla verticillata (Hydrocharitaceae)
Trachycarpus excelsus, fortunei (Palmaceae)
Smilax spp. (Smilacaceae)
Dioscorea spp. (Dioscoreaceae)

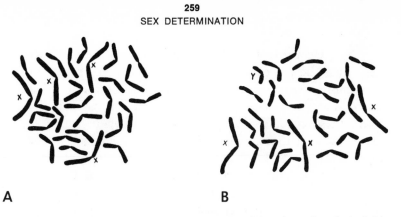

A   **B**

Figure 26. Somatic chromosome complements of female and male individuals of *Rumex paucifolius* (2*n* = 4*x* = 28). (*A*) Female, XXXX. (*B*) Male, XXXY. (From Löve and Sarkar, 1956.)

The most common system of sex chromosomes is that of XX ♀ and XY ♂ as in Melandrium (= Lychnis). A modification involving two XX pairs and two XY pairs occurs in a strain of *Humulus lupulus* (hops). Another modification is the addition of X chromosomes in both sexes in polyploid species of Rumex (Löve, 1943; Löve and Sarkar, 1956) (see Table 33 and Fig. 26). A system of XX ♀ and XYY ♂ occurs in *Humulus japonicus, Rumex acetosa,* and other species. Here the sex chromosome trivalent in the male pairs in the order Y_1-X-Y_2 and disjoins in the pattern X versus Y_1Y_2.

In dioecious bryophytes the XY pair segregates at meiosis in the sporophyte to produce female gametophytes with the constitution X and male gametophytes with the constitution Y in nearly all known cases (Allen, 1945).

The Y chromosome is often smaller than the X. This is true, for example, in the mosses and liverworts (Allen, 1945) and in *Rumex paucifolius* (Löve and Sarkar, 1956). In Melandrium the Y is longer than the X. In *Humulus japonicus* the Y chromosome is heterochromatic. The genetic role of the Y chromosome also varies from group to group. In *Rumex acetosa* and *Humulus japonicus* the Y chromosome is inert as regards sex, which is determined by the balance between X chromosomes and autosomes. By contrast, the Y chromosome is active and carries positive genes for maleness in Melandrium and Vitis (Lewis, 1942; Westergaard, 1958).

The segments of the sex chromosomes which carry the sex genes must be prevented from crossing over if the sex-determining mechanism is to work with precision. Other segments of the sex chromosomes not involved in sex determination can pair and cross over. These conditions are achieved by a subdivision of the sex chromosomes into nonpairing differential segments and homologous segments, respectively (Darlington, 1932).

In Melandrium the homologous segments of the X and Y chromosomes are short and occur at the distal ends. Most of the length of the X and Y is made up of the differential segments. Two genes or gene clusters—M_1 for maleness and Su^F for female-suppression—are located in separate regions of the differential segments. In the Y chromosome, M_1 occurs in the middle region of the differential segment and Su^F in the distal part. M_1 and Su^F are completely linked because they are located in the differential segments (Westergaard, 1958).

SEX GENES   Hermaphroditic and monoecious plants have both female and male organs on the same individual. The development of the two sex forms is controlled directly by auxins and indirectly by various external environmental factors (Heslop-Harrison, 1957). In a number of bisexual plants, genes are also known which suppress selectively the development of one type of sex organ or the other (Lewis, 1942). Perhaps these sex genes regulate the auxin level.

Lewis (1942) lists the following hermaphroditic or monoecious species as containing known genes which suppress the formation of stamens or of staminate flowers:

Antirrhinum sp.	*Prunus persica*
Lathyrus odoratus	*Rubus idaeus*
Lycopersicon esculentum	*Sorghum* sp.
Oenothera organensis	*Zea mays*
Oryza sativa	

The following species, according to Lewis (1942), have genes which suppress the pistils or pistillate flowers:

Oryza sativa   *Pisum sativum*
Pharbitis nil   *Rubus idaeus*
Phleum pratensis   *Zea mays*

By recombination of the two sex genes in the same species, a normally bisexual strain can be converted into a dioecious strain. In this way Emerson (1932) and Jones (1932) produced dioecious strains of *Zea mays* from mutant types of a normal monoecious stock.

The ts_2 allele of the Ts gene in maize changes the male tassels into female organs, thus producing female plants. The gene Sk (silkless) suppresses the styles in the female ear, and the gene Ba (barren stalk) has a similar effect. Combinations of Ts and Sk or Ba give female plants and male plants. The genotypes are as follows:

male	Ts_2/ts_2	sk/sk
female	ts_2/ts_2	sk/sk

or

male	Ts_2/ts_2	ba/ba
female	ts_2/ts_2	ba/ba

The segregation of one allele pair, Ts_2/ts_2, in the heterozygous male in these arrangements determines male or female sex in the progeny (Emerson, 1932; Jones, 1932; Lewis, 1942).

In the normally hermaphroditic *Rubus idaeus,* the raspberry, the sex gene F suppresses the carpels, and M suppresses the stamens. Recombinations of F and M are all female or all male. Female plants have the genetic constitution *Fmm.* Males are *ffM* (Crane and Lawrence, 1931; Lewis, 1942).

So-called subdioecious species consist of females and males along with some regularly occurring bisexual plants, either hermaphroditic or monoecious. *Vitis vinifera,* the European wine grape, is subdioecious; the wild populations are dioecious but some cultivated varieties are hermaphroditic. *Cannabis sativa* is dioecious in the wild form but either monoecious or hermaphroditic in cultivation. *Ecballium elaterium* contains dioecious and monoecious races. Subdioecism may represent a transitional stage between an ancestral bisexual condition and the strictly dioecious condition (Westergaard, 1958).

The evidence concerning sex genes in bisexual plants, taken together with the phenomenon of subdioecism, lends support to the idea that dioecious plants carry the potential for femaleness and for maleness in every individual. One of these potentials, however, is suppressed by sex genes in each individual, which thereupon gives expression to the opposite sex form (Westergaard, 1958).

In *Vitis vinifera* the sex genes are designated M and F. The dominant allele F suppresses the pistil and recessive m suppresses the anthers. There are two possible hypotheses as to the constitution of females, males, and hermaphrodites, depending on whether M and F are linked or independent. If linked, the genotypes would be:

female	mf/mf
male	Mf/mf
hermaphrodite	Mf/Mf and Mf/mf

If independent and epistatic, the genotypes would be (Lewis, 1942):

female	$mmff$
male	$Mmff$ (and some other genotypes)
hermaphrodite	$mmFF$ and $mmFf$

Modifying genes are also involved in sex determination in Vitis (Negi and Olmo, 1972).

The sex genes in Melandrium are M_1 for maleness and Su^F for suppression of female organs. These genes or possibly gene clusters are linked on the differential segments of the sex chromosomes. The genotypes of females and males are, according to Westergaard (1958),

female	XX	m_1su^F/m_1su^F
male	XY	m_1su^F/M_1Su^F

The alleles determining maleness are on the Y chromosome and the alleles for femaleness on the X (Westergaard, 1958).

Westergaard (1958) suggests the same basic genotypes for females and males in *Ecballium elaterium*, and, with modifications, for *Carica papaya*.

EVOLUTION OF SEX-DETERMINING MECHANISMS   As we have seen, a wide range of sex-determining mechanisms is found in the angiosperms. Lewis (1942) has arranged the various mechanisms in a series of probable evolutionary steps. Lewis' suggested evolutionary sequence is developed further by Westergaard (1958).

Dioecism frequently occurs together with monoecism in the same genus or family (e.g., Caricaceae), and in such cases the evolutionary pathway has probably been from monoecism to dioecism. But in some cases (e.g., Rumex) the pathway has probably been from hermaphroditism to dioecism (Lewis, 1942).

The simplest mode of sex determination in dioecious plants is by the segregation of a single allele pair. This system could be only two mutational steps removed from the ancestral bisexual condition: one mutation in the F gene to suppress the pistil and another in the M gene to suppress the stamens. Then one sex, usually the male, remains heterozygous for one of these genes (i.e., Mm), which thereupon segregates to yield male and female progeny. Or both sex genes could remain permanently heterozygous in the heterogametic sex (i.e., MF/mf) and segregate together (Lewis, 1942; Westergaard, 1958).

Dioecism is sometimes controlled genetically in these ways. Dioecious strains can be synthesized from bisexual stocks in the experimental garden, indicating that the same change could also occur in nature. Naturally occurring subdioecious species with segregating sex genes but no sex chromosomes (e.g., Ecballium) may represent actual transitional stages to simple gene-controlled dioecism (Lewis, 1942; Westergaard, 1958).

Further differentiation of the two sexes requires more gene differences. And then these genes must be linked in opposing blocks to prevent the formation of undesirable recombination types. By means of chromosomal rearrangements all the sex genes can be grouped on a single segregating chromosome pair. This chromosome pair, which is cytologically homologous and homomorphic at this stage, represents the beginnings of sex chromosomes (Lewis, 1942). The primitive sex chromosomes do not completely prohibit recombination between the alternative sets of sex genes. In order to prevent these genes from recombining it is necessary to group them in special nonpairing differential segments in the sex chromosomes. This change is manifested in the development of cytologically differentiated X and Y chromosomes (Lewis, 1942).

The Y chromosome at first plays a positive role in determining maleness. But a new disadvantage appears in the early stages of the XY sex-determining mechanism. Although the genes on the X chromosomes can recombine, in XX females, those on the Y chromosomes cannot. The Y chromosome then gradually becomes inert, and as this happens, the autosomes come to replace the Y chromosome as the carriers of the male-determining genes. Sex is now determined by the X–autosome balance, as in Rumex and Humulus (Lewis, 1942; Westergaard, 1958).

SEX RATIO The cross between a heterogametic male and a homogametic female is essentially a backcross and is expected to segregate in a 1 : 1 ratio. The expected proportion of 50% females and 50% males is approximated in Bryonia, Melandrium, Mercurialis, and Spinacia (Correns, 1928; Allen, 1940; Lewis, 1942).

Deviations from the 1 : 1 sex ratio also occur in populations of a number of angiosperm species. Usually there is an excess of females, more rarely an excess of males. Lewis (1942) cites the following observed proportions of males in several dioecious species:

	% males
Silene otites	62
Melandrium album	44
Cannabis sativa	42
Humulus japonicus	29
Humulus lupulosus	10
Rumex acetosa	28
Rumex thyrsiflorus	10

A deficiency of females occurs in Silene otites. The other species show varying excesses of females.

More recently Godley (1964) has measured the frequencies of the two sexes in 17 dioecious species native in New Zealand. In 4 species the sex ratio is close to 1 : 1. These species are *Plagianthus betulinus* and *P. divaricatus* (Malvaceae), *Coprosma australis* (Rubiaceae), and *Dacrydium cupressinum* (Podocarpaceae). The 1 : 1 ratio also occurs in particular populations of some other species. A preponderance of females is found in 2 species of Aciphylla (Umbelliferae). The remaining 11 species show a deficiency of females in natural populations. For example, *Macropiper excelsum* (Piperaceae) has between 12 and 24% females in three populations, and *Griselinia littoralis* (Cornaceae) 31% females.

Zarzycki and Rychlewski (1972) determined the sex ratio in 22 natural populations of *Rumex acetosa* and *R. thyrsiflorus* throughout Poland. An excess of females was found in all natural populations. The sex ratios varied from population to population. The proportion of females ranged from 52 to 93% in 16 populations of *R. acetosa,* and from 62 to 93% in 6 populations of *R. thyrsiflorus.*

The selective elimination of the minority sex could occur at various stages: at meiosis, as in the sex-ratio type in Drosophila; during pollen-tube growth; in the early zygote or early embryo stage; or during the growth of the young plants. In most cases of altered sex ratio in plants the stage and the causes of the distortion are unknown.

Zarzycki and Rychlewski (1972) also determined the sex ratios in seedling progeny grown from seeds harvested in several natural populations of *Rumex acetosa* and *R. thyrsiflorus.* The sex ratios in the seedling samples were close to 1 : 1. The altered sex ratios found in adult samples thus arise during the establishment and growth of the seedlings. Some conditions in nature appear to be operating selectively against the males (Zarzycki and Rychlewski, 1972).

In a few cases the selective elimination is known to take place in the stage of pollen-tube growth: *Melandrium rubrum, M. album,* and *Humulus japonicus.* Here X and Y pollen grains are produced in equal numbers, but they differ in growth rate of the pollen tubes. The Y pollen tubes grow more slowly down the style than do X pollen tubes. Sparse pollination, involving little or no pollen-tube competition, yields female and male progeny in equal numbers. Dense pollination, however, sets up a condition of competition which is unfavorable for the Y types, and this condition is reflected in a reduced frequency of males in the next generation (Correns, 1928; Lewis, 1942; Darlington, 1971a). Two of these species were listed earlier as examples in which excesses of female plants normally occur.

One can approach the problem of altered sex ratios in dioecious plants from a different standpoint, namely that of the selective value of a given sex, ratio rather than the stage at which selective elimination occurs. Dioecism is an inefficient system of sexual reproduction in a sedentary organism, as Lewis (1942) and others have pointed out. It entails many unfruitful cross-pollinations or visitations, in combinations of $\male \times \male$ and $\female \times \female$, and consequently results in a reduced seed output of the population. The inherent disadvantages of the dioecious condition can be mitigated and optimized, however, by alterations in the sex ratio (Lewis, 1942, with modifications).

It may be that a surplus of female plants in a dioecious population is advantageous to ensure the maximum participation of a fixed number of plants in seed production. Alternatively, it may be that a surplus of male plants is advantageous to ensure maximum pollination of the female flowers available. Probably a compromise must be reached between the opposing advantages and disadvantages. We know very little about the ecological situations which would affect the compromise point and shift the sex ratio in one direction or the other in dioecious plant populations.

In any case, the differences in Melandrium and some other plants in the growth rate of X and Y pollen tubes provide a mechanism for adjusting the sex ratio in the population. When the balance shifts toward a preponderance of male plants, the female flowers will be densely pollinated; under these conditions the female-determining pollen is favored in the pollen-tube competition. The result is more female plants in the next generation. On the other hand, when the proportion of male plants becomes low, the female flowers are sparsely pollinated, and more male progeny are produced (Correns, 1928; Lewis, 1942).

GYNODIOECISM   Gynodioecious species consist of females and hermaphrodites. This breeding system occurs in various species of Mentha, Origanum, Satureia, and Thymus in the Labiatae, and in a few other groups (e.g., Araceae). It does not enforce strict bi-parental reproduction, as dioecism does, but it certainly promotes outcrossing. Gynodioecious and dioecious plants have some similar problems as regards mechanisms of sex determination and optimum sex ratios.

The flowers on the female individuals in a gynodioecious population often contain abortive stamens and are obviously derivatives of hermaphroditic flowers. The suppression of the stamens could be brought about either by nuclear sex genes, as in dioecious plants, or by cytoplasmic genes. In fact, as Lewis (1941) has pointed out, the suppression of the stamens in most naturally occurring gynodioecious species is due to cytoplasmic factors. The female plants in gynodioecious species mostly have cytoplasmic male sterility. (The general features of cytoplasmic male sterility are discussed in Chapter 12.)

Lewis (1941) explored the theoretical consequences, for population fecundity, of nuclear-gene versus cytoplasmic control of male sterility in a gynodioecious system. He found that when male sterility is due to a nuclear gene the females cannot persist in the population unless they are more than twice as seed-fertile as the hermaphrodites. This restrictive condition does not apply when the male sterility is determined by the cytoplasm. In this case the females need to be only slightly more seed-fertile than the hermaphrodites in order to persist. Therefore it is easier to establish male sterility in a population on the basis of cytoplasmic genes rather than nuclear sterility factors (Lewis, 1941; Lewis and Crowe, 1956; see also Jain, 1968).

This difference depends on the fact that females and hermaphrodites in a gynodioecious population contribute equal amounts of cytoplasmic genetic material to the next generation but unequal numbers of nuclear genes. A hermaphroditic individual contributes nuclear genes through its own line and through a female line. It can contribute three times as many nuclear genes to the next generation as a female. The genes for sex type in the population gene pool will become strongly biased in favor of the hermaphroditic condition if the sex genes are transmitted with the chromosomes but not if they are transmitted through the cytoplasm (Lewis and Crowe, 1956).

The sex ratio in a gynodioecious population is correlated with the amount of outcrossing. The higher the frequency of females, the greater the amount of outcrossing. But if the females are at a selective disadvantage in reproduction compared with the hermaphrodites, as will occur with nuclear-gene control of male sterility, then the females will fall to a numerically minor status in the population. The population will not attain the maximum rate of outcrossing. This selective disadvantage of the females is avoided when the male sterility is controlled by the cytoplasm. With cytoplasmic control, a high proportion of females can be maintained in the population and the amount of outcrossing can be maximized (Lewis and Crowe, 1956).

Gynodioecious populations of *Origanum vulgare* (Labiatae) occur widely throughout England and Western Europe. The proportion of female plants in natural populations is high, lying between 30 and 50%. Artificial crosses of females × hermaphrodites yield progenies segregating into females and hermaphrodites. One set of crosses yielded 227 females and 339 hermaphrodites (Lewis and Crowe, 1956).

The analysis of additional breeding results suggests that sex type in *Origanum vulgare* is controlled by two independent nuclear genes, *F* and *H*. The dominant allele *F* causes anther abortion, and dominant *H* is a suppressor of *F*. Female plants are *FFhh* and *Ffhh*. Hermaphrodites are *FH* or *fH* (Lewis and Crowe, 1956).

Insofar as sex determination in *O. vulgare* is under nuclear gene control, the females tend to decrease in frequency in the populations. But this tendency is counteracted by other factors. Female plants have a higher average seed output than hermaphrodites. Furthermore, some of the possible hermaphrodite genotypes arising in the course of segregation of *F* and *H*, particularly *ffhh* and *ffHH,* are in part inviable or infertile. These two factors bias the ratio in populations in favor of females (Lewis and Crowe, 1956).

Godley (1955b) reports gynodioecism in *Fuchsia excorticata* and *F. perscandens* (Onagraceae) in New Zealand. He gives numbers of the two sex types in 12 natural populations of *F. excorticata* in widely separated districts. The frequency of female plants ranges from 19 to 40% in 11 of these populations, and is 4% in 1 population. The sex ratio in the total sample of *F. excorticata* throughout New Zealand is 375 females and 1019 hermaphrodites, or 27% females.

SELF-INCOMPATIBILITY   Self-incompatibility in hermaphroditic angiosperms involves the inhibition of pollen growth in some part of the pistil of the same individual plant or of a different plant belonging to the same mating type. Pollen growth is blocked on the stigma in *Brassica oleracea, Raphanus sativus,* and *Secale cereale,* for example; in the style in *Nicotiana sanderae* and *Petunia violacea;* and in the ovary in Lilium and other plants with hollow styles (Brewbaker, 1957).

The inhibition of pollen growth is due to an immunological reaction between proteins in the pollen and pistil. A pistil with a given constitution for the S gene, say S_1S_2, produces certain antibodies. Pollen derived from the same plant has the corresponding antigens. The reaction between antigens and antibodies is specific. Hence pollen growth is blocked when an S_1S_2 plant is selfed or outcrossed with another S_1S_2 plant. But cross-pollinations with plants carrying other S alleles, e.g., S_3S_4, are successful (see reviews of Lewis, 1949, and Brewbaker, 1957, 1964).

The antigen-antibody hypothesis of self-incompatibility has been confirmed by biochemical evidence. In Petunia, after incompatible pollinations a protein was found in the style which did not occur in styles of compatibly pollinated flowers. Pollens of different S genotypes were injected into rabbits. The antisera were then extracted from the rabbits and combined with pollen extracts. Nonhomologous combinations of antisera and pollen extracts gave weak precipitin reactions, whereas homologous combinations gave strong precipitin reactions (Lewis, 1954a).

Mutations induced by x-rays in the S gene in *Oenothera organensis* and *Prunus avium* indicate that this gene is compound. It appears to consist of two subgenes governing different functions.

One subgene controls the pollen reaction and the other the style reaction (Lewis, 1949). Further studies of the compound structure of the S gene in Nicotiana have been made by Pandey (1968, 1969, 1970 and earlier papers cited therein).

The antigen-antibody theory suggests an explanation for the frequent breakdown of the self-incompatibility system in polyploids. It has been observed in a number of plant groups that self-incompatible diploids, on being converted to tetraploids, become self-compatible. Examples include Pyrus, Trifolium, Solanum, Petunia, Antirrhinum, and Campanula; but Raphanus, Brassica, and Oenothera in part do not respond to chromosome-number doubling by losing their self-incompatibility (Lewis, 1949, p. 481).

A self-incompatible diploid, S_1S_2, when doubled, gives a tetraploid with the constitution $S_1S_1S_2S_2$, and the tetraploid produces diploid pollen. These changes open up the possibility of interactions between the S alleles in the diploid pollen and in the tetraploid styles. These allele interactions could block the normal incompatibility reaction between pollen and style, allowing the pollen tube to grow down the style (Lewis, 1949).

It is obviously advantageous for a plant population to contain many alleles of the S gene, so as to reduce the chances of incompatible cross-pollinations. The S gene is in fact represented by multiple alleles in several known cases. In *Oenothera organensis*, 45 S allelles were found in 500 plants (Lewis, 1949). Whitehouse (1950) has estimated the number of S alleles in populations of Trifolium as 75–175 in *T. repens* and 150–170 in *T. pratense*.

Two types of self-incompatibility systems are found in hermaphroditic, homomorphic (non-heterostylous) angiosperms. They are known as the gametophytic and sporophytic systems. In the gametophytic system the pollen phenotype for the incompatibility reaction is determined by its own S allele. In the sporophytic system the pollen phenotype is determined by the maternal genotype rather than by its own haploid constitution.

Gametophytic self-incompatibility is the commonest and longest-known system. It is found in Nicotiana, Oenothera, and many other groups. Here a plant of genotype S_1S_2 develops an S_1S_2 pistil and produces two classes of pollen, S_1 and S_2. The S_1 allele in the pistil produces antibodies which block S_1 pollen tubes, usually in the style; and the S_2 allele in the pistil blocks the S_2 pollen similarly. The cross-pollination of S_1S_2 ♀ $\times S_1S_3$ ♂ results in inhibition of the S_1 pollen but uninhibited growth of the S_3 pollen.

In the sporophytic system an S_1S_2 plant produces pollen with a phenotype implanted by the S_1S_2 genotype. The pollen does not segregate for antigenic reaction into two classes, S_1 and S_2. Instead, all the pollen produced by an S_1S_2 parental plant has the same antigens and the same incompatible reaction to an S_1S_2 pistil. The block in this case usually occurs at the stage of pollen germination on the stigma.

One consequence of sporophytic control of pollen reaction is that it is subject to dominance relations between S alleles. Thus an S_1S_2 plant with dominance in the S_1 allele will produce pollen that all exhibits the S_1 mode of reaction. The sporophytic system occurs in Parthenium, Crepis, and a few other groups (Babcock and Huges, 1950; Gerstel, 1950; Brewbaker, 1957).

The gametophytic and sporophytic systems also differ in the time of action of the S gene. In the sporophytic system, the S gene acts at an early stage, perhaps early in or prior to microsporogenesis, so that all four microspores and pollen grains derived from the same meiosing pollen mother-cell carry the same antigenic substances in their walls or cytoplasm. The four grains then react in the same manner on contact with a genotypically incompatible (or compatible) stigma. The S alleles come into action later in the gametophytic system, perhaps in the second meiotic division in microsporogenesis. Consequently the spore walls are free of the antigenic substance, and the pollen grains derived from the same pollen mother-cell can all germinate on the stigma of an incompatible pistil. The incompatible reaction occurs when a pollen tube comes into contact with the style. Furthermore, the mode of reaction, whether incompatible or compatible, is a characteristic of the individual pollen tube rather than the whole tetrad of pollen grains, as in the sporophytic system. Thus in the cross S_1S_2 ♀ $\times S_1S_3$ ♂, the S_1 pollen tubes are blocked but the S_3 pollen tubes are not (Brewbaker, 1957; Pandey, 1960).

The time of action of the S gene is related to the population of ribosomes in the cells. Both the S gene and the ribosomes synthesize protein products. The population of ribosomes undergoes a drastic reduction in numbers during the early stages of meiosis in microsporogenesis (Heslop-Harrison, 1968).

Gametophytic self-incompatibility requires pollen-tube growth before inhibition sets in. Hence it also requires much protein synthesis. The sporophytic system is much more economical of protein products and of ribosome activity. The occurrence of ribosome elimination during meiosis would therefore seem to favor the sporophytic system as a biochemically more efficient method of bringing about the incompatibility reaction (Heslop-Harrison, 1968).

PHYLOGENETIC POSITION OF SELF-INCOMPATIBILITY East (1940) surveyed the systematic distribution of self-incompatibility in a large sample of species belonging to many orders of angiosperms. He found that self-incompatibility is widespread; moreover, it is relatively rare in primitive woody dicotyledons but becomes more prevalent in the advanced groups, especially in herbaceous forms. He concluded that self-incompatibility is probably a derived condition in the angiosperms.

Several later authors have implied or stated that self-incompatibility is an original condition in the angiosperms (Whitehouse, 1950, 1959; Crowe, 1964; Heslop-Harrison, 1968; and others). This conclusion is inconsistent with East's evidence, which is not brought into the recent discussions, and it is not supported by any convincing new evidence. In my opinion, it is probably more correct on the basis of several considerations, ecological as well as genetical, to regard self-incompatibility as one of the derived breeding systems for securing obligate outcrossing and maximum recombination in the angiosperms (see Chapter 24).

Brewbaker (1957) made the remarkable observation that gametophytic self-incompatibility is closely correlated with the binucleate type of pollen grain in the angiosperms, and sporophytic self-incompatibility with trinucleate pollen. The combination of binucleate pollen and gametophytic control is found in the Commelinaceae, Leguminosae, Onagraceae, Rosaceae, Scrophulariaceae, and Solanaceae. The combination of trinucleate pollen and sporophytic control occurs in the Compositae, Cruciferae, and Sterculiaceae (Theobroma). The Gramineae are exceptional in having trinuclear pollen with gametophytic control (Brewbaker, 1957).

A phylogenetic sequence is suggested by these correlations. Binucleate pollen is primitive and trinucleate pollen derived. Probably, therefore, sporophytic self-incompatibility is also derived in the angiosperms (Pandey, 1960). As noted in the preceding section, the sporophytic system seems to be more efficient than the gametophytic system in ribosome utilization for antigen production (Heslop-Harrison, 1968).

HETEROSTYLY Among the classic examples of heterostyly are *Fagopyrum esculentum* (Polygonaceae), *Linum perenne* (Linaceae), *Lythrum salicaria* (Lythraceae), *Oxalis valdiviensis* and other species (Oxalidaceae), *Primula vulgaris* and other species (Primulaceae), and *Pulmonaria officinalis* (Boraginaceae). These examples were studied by Darwin and his followers in the nineteenth century and have been reinvestigated by various workers in recent times.

The list of heterostylous species has been greatly extended since the early period of investigation. Among more recent studies of heterostyly we call attention to those of Baker on Armeria, Limonium, and related genera (Plumbaginaceae), Byrsocarpus (Connaraceae), and Psychotria, Mussaenda, and other genera (Rubiaceae) (Baker, 1948, 1953, 1958, 1962, 1966). Attention is likewise called to the detailed and interesting studies of Ornduff in Jepsonia (Saxifragaceae), Nymphoides (Menyanthaceae), Oxalis (Oxalidaceae), and Pontederia (Pontederiaceae) (Ornduff, 1964, 1966a, 1966b, 1970a, 1970b, 1971, 1972). A comprehensive list of genera containing heterostylous species is given by Vuilleumier (1967).

Most heterostylous plants are dimorphic (e.g., Primula, Pulmonaria, Fagopyrum, Linum). A few are trimorphic (e.g., Lythrum, Oxalis, Pontederia).

The flowers of different individuals differ with respect to a whole complex of characters involved in the pollination mechanism. A dimorphic population consists of pins and thrums. Pins have a long style, anthers in a low position, small pollen grains, and large stigmatic papillae. Thrums have the complementary characters: short style, high anthers, large pollen grains, and small stigmatic papillae (see Figs. 27, 28). These character combinations reduce the chances of cross-pollinations of pin × pin or thrum × thrum, and, by the same token, promote cross-pollinations of pin × thrum.

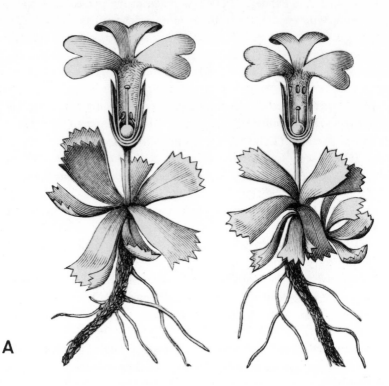

A

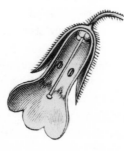

B

Figure 27. Two classic examples of heterostyly. (*A*) *Primula minima*. (*B*) *Pulmonaria officinalis*. (From Kerner, 1894–1895.)

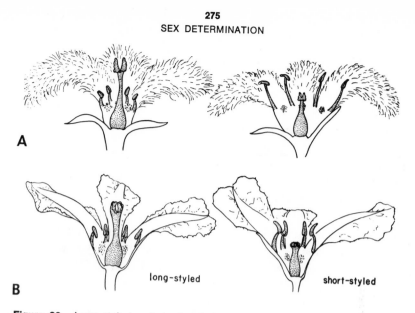

A

B   long-styled   short-styled

Figure 28. Long-styled and short-styled forms in two species of Nymphoides (Menyanthaceae). (A) N. humboldtiana. (B) N. peltata. (From Ornduff, 1966a.)

In the vast majority of cases self-incompatibility is associated with heterostyly. In at least some of these cases the self-incompatibility appears to be of the sporophytic type (Lewis, 1949; Pandey, 1960). A few instances are known of heterostyly without self-incompatibility, for example, *Amsinckia* spp. and *Oxalis suksdorfii* (Ray and Chisaki, 1957a; Ornduff, 1964).

The character combination in most heterostylous plants thus includes the physiological character of compatibility reaction as well as several morphological features. The morphological and physiological characters are correlated in inheritance. Pins and thrums have different genotypes for the self-incompatibility gene, S. Crosses of pin × pin and thrum × thrum are incompatible. The only compatible combinations are pin × thrum.

In Primula, for example, thrum plants are permanently heterozygous Ss, and pin plants are homozygous recessive ss. Crosses of $Ss × ss$ yield Ss and ss progeny in a 1 : 1 ratio. The self-incompatibility gene is closely linked with the genes governing the various components of the pollination mechanism in Primula. Therefore S symbolizes a whole character complex in this case. And pin × thrum crosses yield progeny segregating into pins and thrums in a 1 : 1 ratio. The whole character complex, including both the morphological and the physiological features, segregates as a block, indicating that the genes controlling the different features must be closely linked. In fact, the genetic determinants of pin and thrum types furnish a prime example of a supergene, or segregating gene block, as will be described in Chapter 21. This supergene segregates like the sex chromosomes in dioecious species.

Ornduff found that the proportions of pin plants and thrum plants in eight populations of *Jepsonia parryi* in southern California closely approximate 1 : 1 ratios in all cases. Typical counts for three of these populations are 86 pins : 81 thrums, 74 : 62, and 78 : 77 (Ornduff, 1970a).

MONOECISM Monoecism is another relatively uncommon, but widely distributed, derived condition in the angiosperms. In monoecism, unlike the other systems discussed in this chapter, the alternative sex forms (in this case unisexual female and male flowers) occur on the same individual plant.

As a breeding system, monoecism does not enforce outcrossing obligatorily unless it is combined with self-incompatibility. Most monoecious angiosperms appear to be self-compatible, according to East's survey (1940). Some are self-incompatible, namely, *Castanea crinita, C. mollissima* (Fagaceae), *Euphorbia cyparissias* and *Hevea brasiliensis* (both Euphorbiaceae) (Godley, 1955a).

The female and male sex forms are expressions of the same genotype in a monoecious plant group. Monoecism is thus of interest to us in exemplifying a mode of sex determination which is not based on genetic segregation.

Flower formation is under the direct control of chemical growth regulators such as auxin and ethylene. In a monoecious plant the gradient in auxin or other growth hormone controls the linear sequence of female and male flowers on the shoot. The growth hormones are themselves subject to influence by various external factors, and these can modify sex expression by acting through the hormones (Heslop-Harrison, 1957).

Cucumis melo, the muskmelon, is mostly monoecious. Some strains also have hermaphroditic flowers. The main biochemical basis of sex determination in this cucurbit is an auxin–gibberellin balance. It has recently been found that natural ethylene also regulates sex expression, promoting femaleness. Chemical inhibitors of ethylene promote maleness. A large amount of ethylene, relative to the inhibitors, produces pistillate flowers, while a small amount leads to staminate flowers. With intermediate levels of ethylene, hermaphroditic flowers appear (Byers, Baker, Sell, Herner and Dilley, 1972).

The influence of external environmental factors on sex expression is illustrated by *Cucurbita pepo* (acorn squash). High temperatures and long days promote the formation of male flowers in this monoecious plant. Low temperatures and short days promote female flowers (Westergaard, 1958).

MOSAICISM

INTRODUCTION A multicellular body may consist of cells, cell sectors, or tissues with different genetic constitutions. The genetic differences between the cells involved can reside in their nuclear genes, in their chromosome complements, or in the hereditary particles in their cytoplasm, including the plastids. An individual consisting of genetically different cells is known as a mosaic.

The genetic differences between cells in a mosaic may or may not be expressed in a readily visible differentiation of tissues. In respect to the phenotypic expression of the intercellular genetic differences, it is convenient to distinguish between the exo-phenotype and the endo-phenotype. Some intercellular genetic or chromosomal differences are not manifested at the exo-phenotypic level; in other words, the mosaicism is not visible in the external morphology of the organism. We are concerned with externally visible mosaicism in the present chapter.

Mosaicism comprises a very heterogeneous array of phenomena. This is true whether we consider the modes of phenotypic expression, as seen in chimeras, variegation, bud sports, tissue differentiation and the like, or whether we consider the underlying genetic causes, which also vary widely.

CHIMERAS   In Greek mythology, a chimera was an animal with the head of a lion, the body of a goat, and the tail of a serpent or dragon. The botanical counterpart is similar in principle, but less bizarre. A graft chimera in plants is an individual containing tissues belonging to two or more species or varieties. More generally, a chimera in plants is an individual composed of two or more genotypically different tissues. Cramer (1954, p. 223) adds the important qualification that the genotypically different tissues are represented by their respective cell lines in the growing point in a plant chimera.

The subject of plant chimeras has been reviewed by Baur (1919), Jones (1934), Cramer (1954), Sirks (1956), Avery, Satina and Rietsema (1959), Tilney-Basset (1963), and Howard (1970).

Depending on the spatial relations between the component tissues, chimeras are classified into two main types, periclinal and sectorial, and a third type, mericlinal. In a periclinal chimera, one tissue occurs as a central core and the other grows around it as an enveloping layer. In a sectorial chimera the two kinds of tissue occur side by side in adjacent sectors. A mericlinal chimera consists mainly of one kind of tissue, forming the central core and much or most of the enveloping layer, with the second kind of tissue occurring as a partial enveloping layer of limited extent. A mericlinal chimera is thus a modified periclinal chimera (see Tilney-Basset, 1963; Rieger, Michaelis and Green, 1968).

The terms chimera and mosaic are partially synonymous. For our present purposes we are following a common usage in regarding mosaicism as the more general set of phenomena. Chimeras then become a special class of mosaics in which different kinds of tissues occur intermixed in a periclinal or sectorial relationship.

The differences between the tissues in a chimera may arise by grafting, in a graft chimera or graft hybrid. Or the tissue differentiation may arise by somatic mutations, or by a sorting out of different cell lines in a variegated region.

Many examples of chimeras have been described in Pelargonium, Datura, Solanum, Musa, Antirrhinum, and other plants (Baur, 1919; Cramer, 1954; Avery, Satina and Rietsema, 1959; Tilney-Basset, 1963; Simmonds, 1966; Stubbe, 1966; Howard, 1970). Three historical examples of graft chimeras are reviewed in the next section.

In Solanum, Lycopersicon (tomato), and Datura (Solanaceae), the growing point consists of a tunica of two cell layers (L1 and L2) covering a central corpus. Several studies of periclinal leaf chimeras in these plants indicate that leaf shape is determined mainly by the constitution of the L2 cell layer. Experimental periclinal chimeras of *Lycopersicon esculentum* and *Solanum luteum* were produced in which the constitution of each cell layer—the L1, L2, and corpus—was controlled and varied (Jørgensen and Crane, 1927; Howard, 1970).

The tomato and the Solanum species have very different leaf shapes, the former being large and pinnately compound, and the latter small and entire-margined. Their chimeras have a leaf shape similar to that of the parent species which contributed the L2 layer. By the same token, striking differences in leaf form occur between chimeras with the same combination of tissues in different positions. A chimera with a tomato L1 layer and a Solanum L2 layer has a leaf very similar to the Solanum species. Conversely, a leaf chimera with Solanum L1 and tomato L2 is pinnate like the tomato although somewhat smaller in size (Jørgensen and Crane, 1927; Howard, 1970).

GRAFT CHIMERAS   One of the classic graft chimeras is *Laburnum adami*. This plant was produced by grafting of *Cytisus purpureus* onto *Laburnum anagyroides* in the early nineteenth century. It exhibited intermediate characteristics between the stock and scion, and was therefore called a graft hybrid. The term graft hybrid was used and construed in a literal sense in the nineteenth century. It was believed that a fusion of vegetative cells of the parental species had given rise to the intermediate *Laburnum adami* and to other known graft hybrids with similar intermediate features. The true nature of *Laburnum adami* was established in the early twentieth century by Baur (1909, 1919) and others. The plant was shown to be a periclinal graft chimera.

Laburnum anagyroides (= *L. vulgare, Cytisus laburnum,* Leguminosae) is a large shrub or small tree with spreading branches, large trifoliate leaves, and drooping racemes bearing large golden flowers (Fig. 29A). *Cytisus purpureus* is a low shrub with small trifoliate leaves, erect terminal racemes, and small purple flowers (Fig. 29C). It was a common practice to graft *Cytisus purpureus* onto *Laburnum anagyroides* in order to place the purple flowers of the former species on the tall stems of the latter. Such grafts usually have the normal result; pure *purpureus* shoots develop out of the grafted stem or bud.

A French nurseryman by the name of Adam budded plants of *Laburnum anagyroides* over to *Cytisus purpureus* on an extensive scale in the 1820s, with the normal result in all but one case. One exceptional grafted bud developed into a branch with a mixture of the characters of the two species, as in a hybrid. This exceptional branch originated in 1829 and was later propogated vegetatively and widely distributed in the horticultural trade under the name *Laburnum adami* (or *Cytisus adami*).

Laburnum adami is like *L. anagyroides* in habit, foliage, raceme, and flower size (Fig. 29B). The flower color is dull purplish, marking the influence of *Cytisus purpureus*. The flowers are nearly always sterile. The rare seeds which are produced on *Laburnum adami* give *L. anagyroides* seedlings. Furthermore, *L. adami* has the peculiar tendency to segregate vegetatively; that is, occasional branches appear which are like either pure parental type; and the flowers on these parental-type branches are mostly fertile (Darwin, 1875; Baur, 1919).

The nineteenth-century students of hybridization and heredity, including Gaertner, Darwin, and Focke, concluded that *Laburnum adami* was a vegetative hybrid. Its hybrid nature was indicated by its morphological intermediacy, its seed sterility, the tendency to segregate or "revert" to one parental type in its rare seedling progeny, and its ability to revert to both parental types in vegetative branches. The Adam laburnum was known to have originated from a graft.

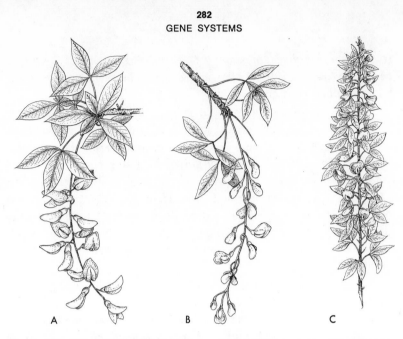

Figure 29. The graft hybrid *Laburnum adami* and its parents. (*A*) *Laburnum anagyroides*. (*B*) *Laburnum adami*. (*C*) *Cytisus purpureus*. (Redrawn from Baur, 1919.)

Hence, the hybrid constitution must have arisen from a fusion of vegetative cells.

Darwin summed it up in *The Variation of Animals and Plants under Domestication* (1875, ch. 11):

> We must admit the extraordinary fact that two distinct species can unite by their cellular tissue, and subsequently produce a plant bearing leaves and sterile flowers intermediate in character between the scion and stock, and producing buds liable to reversion. . . . The formation of hybrids between distinct species or varieties, without the intervention of the sexual organs, . . . is possible.

This case was reexamined by Baur (1910a, 1919) and others in the early twentieth century. The then new science of Mendelian genetics cast considerable doubt on the hypothesis of vegetative hybridization and literally true graft hybrids. Baur had been investigating chimeral tissue differences in variegated Pelargonium leaves. Winkler was performing graft experiments with different species of

Solanum, and had obtained occasional graft products with mixtures of characteristics comparable to those observed in *Laburnum adami*. Baur (1910a) therefore proposed the alternative interpretation that *Laburnum adami* was a periclinal chimera.

Baur's interpretation agreed well with the evidence of external morphology. *Laburnum adami* is basically *Laburnum anagyroides* with an epidermis of *Cytisus purpureus*. Anatomical sections confirmed this. The epidermis of *L. adami* is histologically identical with that of *C. purpureus,* and its central tissue is the same as that of *L. anagyroides*. One tissue layer or the other might occasionally break out by itself to produce a new branch. This event explains the observed somatic segregation toward the two original species. The fact that the ovules originate in a deep tissue layer accounts for the observation that the few seeds produced yield *Laburnum anagyroides* seedlings (Baur, 1910a, 1919).

A parallel example is the Bizarria orange which originated from a grafted seedling in Florence, Italy, in 1644. The Bizarria orange had fruits with odd mixtures of the characters of the sour orange (*Citrus aurantium*) and the citron (*C. medica*). It also produced some segregate branches like the sour orange in leaves, flowers, and fruits, and other segregate branches like the citron. The Bizarria orange was long supposed to be a graft hybrid in the literal sense. But later anatomical studies established that it is a periclinal or mericlinal chimera of the two Citrus species (Cramer, 1954).

Another famous case is the cultivated tree known as *Crataegomespilus asnieresii* (Rosaceae). This handsome flowering tree is intermediate between *Crataegus monogyna* and *Mespilus germanica* in its leaf and fruit characters. Its fruits resemble Crataegus in their small size, and Mespilus in their brown color. Its leaves have the general size and shape of Crataegus but have margins approaching the condition in Mespilus. Anatomical sections reveal that Crataegomespilus has a leaf and fruit epidermis of Mespilus covering an inner layer of Crataegus tissue. Baur (1919) concluded that Crataegomespilus is a periclinal chimera with the composition indicated above. After some debate and the consideration of alternative possibilities, Baur's interpretation was finally confirmed by additional anatomical evidence (Cramer, 1954).

HEREDITARY TRANSMISSION IN PERICLINAL CHIMERAS   A

somatic mutation may appear in a cell which divides so as to form part or all of one cell layer in a multilayered growing point. A layer of mutant cells then exists side by side with one or more layers of normal cells in the growing point of a bud. The growth of such a bud yields a periclinal chimera.

Many mutant types in the potato (*Solanum tuberosum*) are periclinal chimeras. The mutant condition in these chimeras may be expressed in characters of the leaves, flowers, tubers, or tuber sprouts. The periclinal and mutant condition for various vegetative and floral characters is often propagated through the tuber. In vegetative reproduction in potatoes, different layers of tissue enter into the formation of a new shoot arising from a tuber, so that a mutant condition may be perpetuated in the form of a chimera (Simmonds, 1965a; Howard, 1970).

The chimeral nature of potato mutants can be tested by excising the eyes from the tubers of a mutant strain and letting them regenerate from adventitious buds forming in the excision. Such adventitious buds may develop in either deep or surface layers of tissue. Consequently, if the mutant type is a periclinal chimera, the plants developing from a sample of adventitious buds on excised tubers should comprise a mixture of mutant and nonmutant types. This result was obtained for a number of mutant types in the potato, and they are therefore inferred to be chimeras (Simmonds, 1965a).

The foregoing method of analyzing mutant chimeras in the potato can be checked by breeding experiments. A mutant type of potato, suspected of being a chimera, is outcrossed to a normal type. It is known that pollen mother cells and embryo-sac mother cells in the potato are derived ultimately from the second or L2 cell layer in the tunica of the growing point (Howard, 1970).

The potato variety Rode Star has pink tubers. It has given rise to two chimeral mutants, Groene Rode Star and Bonte Rode Star, which have tubers with mixed white and colored areas. The mutant condition in Groene and Bonte is perpetuated by vegetative reproduction.

To determine the breeding behavior of these two mutant types, they were outcrossed to a tester variety, Triumf, which has white tubers. The difference between pink and white tubers is monofacto-

rial. A control cross of normal Rode Star × Triumf yielded 19 seedlings with white tubers and 17 with colored tubers (Dorst, 1952;
Howard, 1970).

The cross of Bonte Rode Star × Triumf gave 13 seedlings with
white tubers and 11 with colored tubers, as in the control cross of
normal Rode Star × Triumf. Here the somatic mutation from pigmented to white must be carried in the L1 layer, where it is not
transmitted sexually. The mutant white condition in Bonte Rode
Star is thus transmitted through the tubers but not through the seeds
(Dorst, 1952; Howard, 1970).

By contrast, the cross of Groene Rode Star × Triumf yielded 64
seedlings with white tubers and no seedlings with colored tubers. The
somatic mutation from pigmented to white tubers is transmitted
through both the tubers and the seeds in this chimeral mutant. Therefore, the somatic mutation must be carried in the L2 layer, which is
included in both modes of reproduction (Dorst, 1952; Howard,
1970).

The leaves of *Pelargonium zonale* are variable in color pattern,
as is well known. One common type of leaf has a green center and
broad white border. Solid green leaves are also common, and pure
white leaves develop frequently on green plants. The difference between green and white leaf areas is due, as expected, to the presence
or absence of chloroplasts. Baur (1919) showed that the variegated
type of Pelargonium leaf is a periclinal chimera of green and white
tissue components.

The color pattern of the leaf can be traced back to the tissue
structure of the growing point in the bud. A solid green leaf develops from a bud with one outer layer of colorless cells. The mature
green leaf has one epidermal layer covering the green mesophyll
tissue (Fig. 30A, B). Leaves with a green center and broad white
margin develop from buds containing two outer colorless layers of
cells. The additional cell layer persists as a white subepidermis in
the mature leaf. It lies over the green mesophyll in the green central
part of the leaf, and replaces the green mesophyll in the leaf margin,
which is consequently white (Fig. 30C, D) (Baur, 1919).

The variegated leaf condition in Pelargonium can be propagated vegetatively by cuttings. But the chimeral condition is not
transmitted through the seeds. The pollen and eggs are derived from

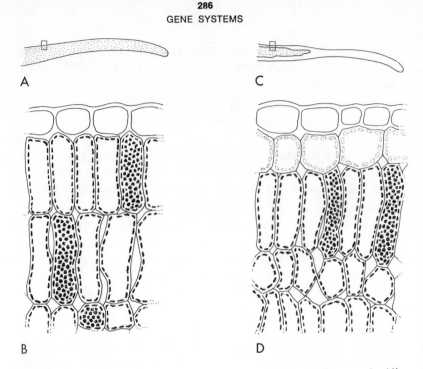

Figure 30. Cross-sections of two types of leaves in *Pelargonium zonale*. (*A*) Section through margin of solid green leaf. (*B*) Enlarged view of box in A, showing chloroplasts in subepidermal layer. (*C*) Section through margin of white-bordered leaf; green tissue stippled. (*D*) Enlarged view of box in C, showing subepidermal layer without chloroplasts. (Redrawn from Baur, 1919.)

the second (L2) cell layer of the growing point in Pelargonium, as in Solanum and many other angiosperms. Consequently, the sexual progeny of a chimeral plant all have the genetic constitution of the L2 layer. A chimera with variegated leaves and a genetically white L2 layer gives all white seedlings, which die in the early seedling stage. Plants with a potentially green L2 layer yield uniform seedling progeny with solid green leaves (Baur, 1919).

A chimera with green-and-white leaves occasionally segregates somatically to all-green or all-white shoots. The white outer layer is sometimes replaced by the potentially green third layer in a growing point. Then the bud produces an all-green shoot, which is constant for the altered condition. Conversely, the two outer cell layers sometimes produce all the leaf tissue. The result is a pure white leaf or shoot. Such an albino leaf or branch shoot can survive so long as it is attached to the variegated green mother plant (Baur, 1919).

Later work revealed some complications and led to controversies which we need not go into here. For reviews of these aspects of the Pelargonium story, see Cramer (1954) and Tilney-Basset (1963).

SOMATIC MUTATIONS IN CLONES   Plants which reproduce by means of vegetative propagation—by stolons, rhizomes, tubers, bulbs, root crowns, stem joints, leaf buds, inflorescence bulbils, and so on—form clones consisting of genotypically identical individuals in the majority of cases. Sometimes, however, exceptional subclones derived from the same original clone differ phenotypically and genetically. Such exceptional cases are explained as a result of bud sports occurring in the ancestry of the deviating subclone.

A somatic mutation may occur in a cell line which becomes incorporated in the growing point of a bud. The branch shoot, flower, tuber, or other organ developing from that bud then exhibits the mutant characteristics. Such bud sports are fairly common and widespread in higher plants. Perennial plants which have been extensively grown and observed in agriculture or horticulture afford numerous examples. Bud sports for novel fruit characters have been observed in virtually all the commonly cultivated fruit plants.

A red-flowered strain of *Dahlia variabilis* was found with one aberrant branch bearing a flowering head with white-and-red ray florets (Sirks, 1956). A potato plant of the Eigenheimer variety had one rhizome bearing a blue tuber and another rhizome terminating in a white tuber (Sirks, 1956). In Citrus, bud sports are fairly common, and have given rise to some cultivated varieties including the Washington Naval orange (Reuther, Batchelor and Webber, 1968).

In cultivated bananas (cultivars of *Musa acuminata* and *M. balbisiana*), bud sports are likewise rather common and sometimes important agriculturally. The Dwarf Cavendish banana, which is one of the most widespread and abundant varieties, arose as a somatic mutant of a less widespread clone (Simmonds, 1966). Simmonds lists the known somatic mutants in cultivated bananas. The 50 mutant types included in the list affect all plant parts. There are somatic mutants for dwarf stature, drooping leaves, black stems, waxiness, pink fruits, and various inflorescence characters (Simmonds, 1966, pp. 58–60). Some varieties like Gros Michel which have been grown on a large scale for a long time have given rise to several different mutant types.

Some figures are available on the frequency of bud sports in the potato. We will examine the figures for seven mutant types of leaves, such as spinach-leaf and blistered leaf. The frequency of origin of aberrant plants from bud sports for these leaf mutations lies in the range 1.2–8.4×10^{-5} per mutant type. This means that one or several plants with mutant leaves of a given type may be expected to appear in a clonal population of approximately 100,000 normal plants (Heiken, 1958; Howard, 1970). Howard notes that the somatic mutation rates in the potato are of the same order of magnitude as germinal mutation rates for various genes in maize (Howard, 1970, p. 70).

A bud sport in a vegetatively reproducing plant species has a chance of being replicated and multiplied in all the members of a new subclone. In this way an original transitory mosaicism between different buds on a single plant can be perpetuated in their respective clonal derivatives.

We have already seen that cultivated bananas, which are propagated vegetatively, include many varieties which have arisen as bud sports in an old standard variety and have multiplied as new clones. The Cavendish group of cultivars arose in this fashion from a more primitive native clone; and the Gros Michel variety has given rise by bud sports to the Weeping Jamaica, Pink Jamaica, Plantain Jamaica, and other clones. In these and other cases the newer clone resembles an older standard variety in all characters except the mutant one (Simmonds, 1966).

As in potatoes, banana subclones sometimes differ significantly in yield and disease resistance. Thus, two subclones of the Majestic variety differ in yield by as much as one-half to one ton per acre. They also differ in their reactions to virus Y and leaf-roll diseases. These differences are necessarily due to bud mutations becoming fixed in the deviant subclone (Simmonds, 1966).

VARIEGATION   Variegation is the occurrence of patches or streaks of tissue with different pigmentations and colors on the surface of an organ or organism. It is convenient to distinguish between zonal and streaky variegation. In zonal variegation there are relatively large areas of each color type of tissue, as in the green-and-white leaves of Pelargonium and other plants. Numerous small streaks, stripes, or flecks of one color on the background of another color are called streaky variegation, exemplified by the flecked flower petals in many plant species.

Variegation in the form of many small streaks reflects genetic changes which occur or are expressed at a relatively late state of development when they can affect only a few cells in each site. Zonal variegation results from changes which have their onset at an early stage of development. We have examined a case of zonal variegation in Pelargonium leaves. Many cases of variegation which we encounter in plants are of the streaky type.

Variegation is a heterogeneous phenomenon in terms of both appearances and causes. Similar variegated phenotypes may arise from different causes. There is no easy way of distinguishing the cause of variegation from the outward appearance.

The manifold causes of variegation in plants have been listed by several authors (Ar-Rushdi, 1957; Smith and Sand, 1957; Darlington, 1971b). The main causes given by these authors are regrouped in the list that follows.

1. Segregation, loss, or defects of plastids (Pelargonium, Mirabilis, Oenothera)
2. Somatic mutations in mutable nuclear genes (Delphinium, Antirrhinum, Nicotiana)
3. Position effects when a gene is transposed to a site near heterochromatin (Oenothera, Zea)
4. Chromosomal aberrations of various types: (a) aberrations involving rings or bridges (Zea); (b) sticky chromosomes (Zea); (c) deletion of chromosomes or fragments (Nicotiana)
5. Somatic crossing-over in heterozygotes (Drosophila, fungi)
6. Sorting out of tissues in chimeras (Pelargonium, Hedera, Ilex)
7. Virus infection (tulip petals, honeysuckle leaves)

Green-white variegation in leaves and stems can arise from

mutations or irregular segregations in plastids. Plastids sometimes mutate to a chlorophyll-lacking condition and yield white cell lines and sectors. Or mixtures of green and white plastids may sort out during a series of cell divisions to give pure green and pure white stripes. Variegated green-white leaves due to plastid mutation or segregation are found in *Acer pseudoplatanus, Antirrhinum majus, Hedera helix, Humulus japonicus, Ligustrum vulgare,* and *Mirabilis jalapa* (Correns, 1909; Baur, 1919; Darlington, 1971b).

Flower petals become striped or flecked when a gene governing pigment formation mutates frequently from one allelic form to another or from recessive to dominant condition in a heterozygote. *Delphinium ajacis* sometimes has variegated petals with purple streaks on a rose background. This variegation is attributed to mutability of the rose-alpha gene for flower color (Demerec, 1931). The normal allele of this gene produces rose-colored flowers. It mutates frequently to an allele for purple color. The mutation may occur early or late in petal development, giving purple areas that vary in size from large stripes to small flecks (Demerec, 1931).

An allele of the pallida (*Pal*) gene for flower color in *Antirrhinum majus* is mutable. The normal allele of this gene (*pal* +) produces red flowers. The mutable allele is *pal-rec*. Plants of the constitution *pal-rec/pal-rec* have ivory flowers as the basic condition. But *pal-rec* mutates back to *pal* + at a high rate. This gives streaks or sectors of red tissue on an ivory background. The mutability of the *pal-rec* allele increases at higher temperatures (Stubbe, 1966).

Position-effect variegation is not a form of mosaicism according to the definition adopted at the beginning of this chapter. In position-effect variegation the cells of different phenotype do not differ in the sum total of their genic and chromosomal components. The essential feature here is a difference in gene expression associated with gene location, as described in more detail in Chapter 8. We should note, however, that various reported cases of gene mutability in variegated plants could really be instances of variegated position effect.

CYTOLOGICAL CAUSES OF VARIEGATION IN *ZEA MAYS* Three types of chromosomal aberrations which lead to variegation are well

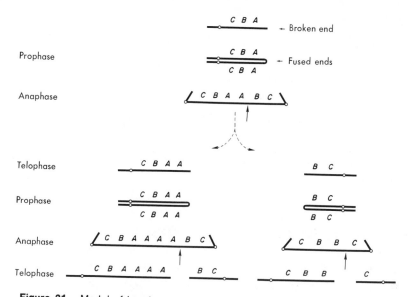

Figure 31. Model of breakage-fusion-bridge cycle in a chromosome carrying three genes, A, B, and C. Centromere is indicated by small circle, and breakage point by short solid arrow. The cell line on left accumulates duplications for A, while the right-hand cell line carries a deficiency for A. (Redrawn from McClintock, 1941.)

illustrated by examples in maize. The cytogenetic abnormalities in question are breakage-fusion-bridge cycles, ring chromosomes, and sticky chromosomes. (A fourth types, position-effect variegation, is discussed in Chapter 5.)

Breakage-fusion-bridge cycles are illustrated diagrammatically in Figure 31, following McClintock (1941). A break occurs at a given point distal to a gene A. The two sister chromatids with the homologous broken ends then refuse. The refusion gives a dicentric chromosome at prophase which is stretched out into a dicentric bridge at anaphase. The product of refusion now carries A in duplicate condition. Breakage occurs again at the same point, yielding two chromosomes with single centromeres which pass to daughter cells. One daughter cell line has a duplication for A, and the other daughter cell line a deficiency for A. Such cycles have been shown to occur in gametophytic and endosperm tissues in maize, though not in the sporophyte (McClintock, 1941).

Suppose that the gene A controls pigment formation in corn endosperm. Suppose further that the maize plant is heterozygous for

A, and that the dominant allele is on the chromosome involved in breakage-fusion-bridge cycles. Then some sectors will be pigmented, while other sectors carrying deficiencies for dominant *A* will have the recessive unpigmented phenotype. Some cases of variegation in corn kernels can be attributed to breakage-fusion-bridge cycles in a chromosome carrying dominant alleles for endosperm pigmentation (McClintock, 1941).

Ring chromosomes occur in maize. They may or may not undergo exact reduplication in somatic cell divisions. The sister strands in a small ring chromosome sometimes become connected in such a way that they produce a pair of interlocked small rings or one large dicentric ring. Neither product separates cleanly at anaphase. The small ring chromosomes may become lost. Or breaks may occur at some level in a large dicentric ring, followed by reunion of the broken ends, so as to restore the condition of small single-centromere rings in the daughter cells (McClintock, 1932, 1938).

Most of the small daughter rings derived from a large dicentric ring will carry duplications and deficiences for various genes, depending on the location of the breaks in the large ring. The loss of small rings in an interlocked pair of rings likewise brings about deficiencies in the daughter cell lines. The deficiencies may uncover recessive alleles in the normal homologs of the ring chromosomes. The cell lines with the deficiencies will then appear phenotypically as sectors of mutant or recessive tissue (McClintock, 1932, 1938).

If the ring chromosome carries a gene for pigment formation in some plant part, and if the plant is heterozygous for this gene, the irregular reproduction of the ring chromosome will result in variegated tissue. Several cases of variegation of leaves, stalks, and kernels are associated with ring chromosomes in maize (McClintock, 1932, 1938).

The condition of sticky chromosomes, where chromosomes tend to stick together extensively in cell divisions, leads to elimination of whole chromosomes or fragments in some cell lines. Genes for pigment formation on the sticky chromosomes are subject to elimination at the same time. Beadle has shown that the behavior of sticky chromosomes leads to color variegation in the leaves and endosperm of maize (Beadle, 1937).

FLOWER-COLOR VARIEGATION IN NICOTIANA An early study by Clausen (1930) was concerned with the cause of carmine-coral variegation in the flowers of *Nicotiana tabacum*. Some lines of tobacco have solid carmine corollas and others have solid coral corollas. The two color types represent differences in a gene *Co,* the dominant type being carmine (*Co*) and the recessive type coral (*co*). There are also occasional plants with variegated corollas marked by light coral stripes on a darker carmine background.

Cytogenetic analysis reveals that *Co* is on the F chromosome in *N. tabacum*. The chromosome complement of variegated carmine-coral plants contains a fragment of this F chromosome in addition to normal F chromosomes. The fragment happens to carry the *Co* gene, and in the plants studied by Clausen it carried the dominant *Co* allele. The fragment frequently lags and becomes lost in cell divisions (Clausen, 1930).

Variegated corolla tissue results from the dominance relations between the *Co* allele in the fragment and the *co* alleles in the normal F chromosomes in heterozygous plants containing both types of alleles. The presence or absence of the fragment in different cell lines permits the expression of one allele or the other, and this in turn gives sectors with different pigments (Clausen, 1930).

Gerstel and Burns (1966) have studied a parallel form of flower color variegation in interspecific hybrids of *Nicotiana tabacum* and *N. otophora*. The hybrids display variegated corollas with light coral streaks on a darker carmine background (Fig. 32). The carmine color is due to the allele Co^v and the coral color to *co,* with Co^v dominant over *co*.

Nicotiana otophora is a diploid species close to one ancestor of tetraploid *N. tabacum*. The hybrid generations studied were the F_1 triploid hybrid *tabacum* ♀ × *otophora,* its artificial hexaploid *tabacum-otophora,* and backcrosses of *tabacum* × *tabacum-otophora* (6*x*). The coral allele *co* was introduced into the hybrids and derivatives by the *tabacum* parent, and the Co^v allele by the *otophora* parent (Gerstel and Burns, 1966).

The hybrids and many of the hybrid derivatives carried the Co^v allele and were heterozygotes of Co^v and *co*. The corolla color was not solid carmine, however, but variegated carmine-coral, and

Figure 32. Variegated corolla in a hybrid derivative of *Nicotiana tabacum* ×
N. otophora with the constitution Co^V/co. The color pattern consists of light
coral streaks on a darker carmine background. Flower approximately natural
size. (From Gerstel and Burns, 1966; photograph courtesy of D. U. Gerstel.)

the variegation was associated with the introduction of the Co^v
allele into the hybrid genotypes (Gerstel and Burns, 1966).

Various possible explanations of this case of variegation can be
ruled out by one line of evidence or another. The variegation is not
accompanied by chromosome rings. Mutation from carmine to coral
is not likely here. Variegated position effect was at first seriously considered but later had to be rejected as a cause (Gerstel and Burns,
1966; Burns and Gerstel, 1967).

The evidence points strongly to a connection between the
variegation and heterochromatin. The chromosomes of the parental
species have very different distributions of heterochromatic segments
(Gerstel and Burns, 1966). Furthermore, the Co^v allele in *N.
otophora* is closely linked to a particular heterochromatic block in that
species (Burns and Gerstel, 1967).

Backcross derivatives of *tabacum* × *tabacum-otophora* (6x) had
Co^v from *N. otophora* along with the linked heterochromatic block
(HB). These hybrids in the advanced backcross generations segregated for coral and variegated, and also for presence or absence of
HB. The two characters were found to be correlated in their segregations. The variegated plants all had HB in the cells of the corolla
and other somatic tissues, while the solid coral plants generally
lacked HB (Fig. 33) (Burns and Gerstel, 1967).

The *otophora* chromosome containing the heterochromatic
block in question is subject to spontaneous breakage in somatic
cells. The breakage leads to loss of some or all of the HB region and

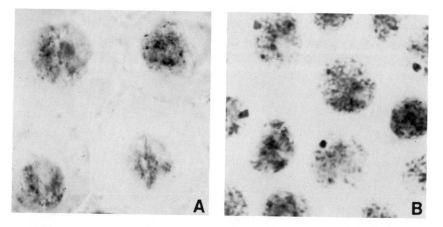

Figure 33. Epidermal cells of corollas in different individual plants derived from *Nicotiana tabacum* × *N. otophora*. The photographs show absence and presence of heterochromatic blocks in interphase nuclei. (*A*) Coral plant, lacking heterochromatic blocks. (*B*) Variegated plant, with heterochromatic blocks in several nuclei. Approximately 1000×. (From Burns and Gerstel, 1967; photographs courtesy of D. U. Gerstel.)

of the linked Co^v allele in some mitotic divisions and in the resulting cell sectors. The loss of the dominant Co^v allele uncovers the recessive *co* allele. This permits the formation of coral streaks on the carmine background (Burns and Gerstel, 1967).

Additional evidence indicates that the HB region in the *N. otophora* genome is greatly stabilized and breakage is much reduced by other genetic units in the same genome. In the hybrids, however, HB becomes separated from the influence of these stabilizing factors, and undergoes frequent spontaneous breakage. The phenotypic result is variegation (Gerstel and Burns, 1970; Burns and Gerstel, 1971).

Variegated flowers appear in the F_2 and later generations derived from the cross of *Nicotiana langsdorffii* × *N. sanderae*. The corolla of *N. sanderae* is solid red. The variegated flowers are red and white. The color pattern varies in different flowers, from red speckles on a white background, or white speckles on red, to broad sectors of red and white tissue (Smith and Sand, 1957).

Repeated backcrossing of variegated hybrids to *N. sanderae* has placed the determinant of the variegated condition in a *sanderae* genotype. The heterozygous backcross types then segregate solid-colored and variegated individuals in ratios which indicate that the segregating character is determined by a single locus (V) on chromosome 2. *Nicotiana sanderae* is homozygous for the dominant allele, V, which determines solid color. Variegation is due to two slightly different recessive v alleles governing the different color patterns (Smith and Sand, 1957).

The variegated alleles (v) are not present, or at least are not present in any substantial frequency, in the parental species, which are inbred. The variegated condition appears to arise de novo in the early hybrid generations. Two possibilities exist. The variegation could be due to frequent mutation from V to v in the hybrid genotypes. Or exchange of segments between *sanderae* and *langsdorffii* chromosomes in the hybrids could place the V allele of *sanderae* in a new spatial relation to heterochromatin and thus alter its expression. In other words, the variegation in this case could be due to a mutable gene or to the variegated type of position effect. The available evidence does not distinguish between these possibilities (Smith and Sand, 1957).

The last example to be considered briefly concerns flower variegation in hybrids of *N. tabacum* × *N. plumbaginifolia*. The parental types are both white-flowered. The hybrids have variegated corollas with white streaks on a purple background. Complementary factors for flower color thus evidently exist in the two white-flowered parents, but have an unstable expression in the hybrids (Ar-Rushdi, 1957). Chromosome bridges have been observed to undergo unequal breakage in some cells in the hybrids. Ar-Rushdi suggested that this unequal breakage leads to different cell lines which either possess or lack a color gene (1957). An alternative suggestion in line with other cytological observations is that whole *plumbaginifolia* chromosomes containing factors for corolla pigmentation are lost in some cell sectors in the hybrids (Moav and Cameron, 1960; Moav, 1961).

TYPES OF MOSAICS IN ANIMALS Stern (1968) has recently reviewed the subject of mosaicism in animals. It is of interest to compare the situation in plants with that in animals, and Stern's survey provides the basis for such a comparison. The following paragraphs are condensed and rearranged from Stern's review. I list in synoptical form the main types of mosaics in animals, give short descriptions of them, and mention typical examples.

1. *Sex mosaics (gynanders, gynandromorphs)*. One part of the body is female, another part male. The differences are sometimes between halves of the body and sometimes between small sectors. The condition has various causes, including polyspermy, abnormal numbers of sex chromosomes (see type 2), and others. Examples are found in man, other vertebrates, bees, Drosophila, and other insects.

2. *Sex-chromosome mosaics*. Cells have different chromosome numbers due to different and sometimes abnormal combinations of sex chromosomes. For example, a human male might have 46-chromosome cells and 47-chromosome cells with sex-chromosome constitutions XY and XXY, respectively. Human females may be mosaics of XX and XXX cells. Many other possible combinations are known. The combination of XY and XO cells or sectors gives a gynander (type 1). Examples in man, Drosophila.

3. *Autosomal mosaics*. Some cells of the body have abnormal dosages of certain chromosomes. Thus, in humans, chromosome 21 may be present in trisomic condition in some cells and in the normal duplex condition in others. Examples in man.

4. *Body mosaics for sex-linked genes, I*. Some sectors of the individual have cells of the constitution XX for the sex chromosomes, and other sectors are XO, due to elimination of an X chromosome. Sex-linked marker genes then yield phenotypically different cell sectors. Example of eye color in Drosophila. This can be considered a special case of type 1.

5. *Body mosaics for sex-linked genes, II*. Occur in female mammals with a normal sex-chromosome constitution XX and a heterozygous condition for sex-linked genes. In some cells one of the X chromosomes may fail to function. The gene expression of the heterozygous sex-linked allele pair will vary accordingly in different cells and sectors. Example of fur color mosaics in mice.

6. *Body mosaics due to somatic crossing-over*. Small areas of aberrant cells occur on a larger background of normal phenotype. Somatic crossing-over in a heterozygote gives some cell sectors in

which a recessive allele is exposed to expression. Example of body bristles in Drosophila.

7. *Twin-spot mosaics.* Paired patches of contrasting aberrant phenotypes are superimposed on a background of normal body tissue. The condition results from somatic crossing-over and recombination in a double heterozygote in which the two mutant alleles are in the trans position. Such crossing-over leads to complementary recombination types appearing as pairs of aberrant spots. Example of body bristles in Drosophila. A special case of type 6.

8. *Mosaics due to double fertilization.* Fertilization of two eggs by two sperm gives a single individual with mosaic condition. Examples in mice, man, insects.

9. *Position-effect variegation.* Tissues or organs have a variegated appearance with different colored streaks occurring intermixed. The condition is due to a gene being transposed to a site close to heterochromatin where that gene does not function properly or regularly. The result is adjacent sectors of different phenotype representing different states of gene action or inaction. Example of eye-color mosaics in Drosophila.

10. *Antibody variation.* A single individual produces a great variety of protein types, each of which acts as an antibody to a particular antigenic substance, and the antibody types are distributed in mosaic fashion throughout the body. Several possible causes have been suggested whch involve permutations of gene specification of proteins in different body parts. Examples in mammals, birds.

11. *Mutational mosaics.* A mutable gene mutates frequently, giving sectors of mutant tissue on a normal background. Example of fur color in mice.

There are both similarities and differences between the two kingdoms as regards mosaic phenomena. Some types of mosaicism in animals, such as mutational mosaicism and position-effect variegation, have known analogs in plants. Somatic crossing-over, on the other hand, though known in Drosophila and also in fungi, has not been established in plants.

A prominent place in animal mosaicism is occupied by mosaics with differences in sex characters, sex chromosomes, or sex-linked genes. Parallel types of mosaics may yet be found in higher plants but cannot be very common, since dioecism based on sex chromosomes is relatively unusual in plants.

CONCLUSIONS Mosaicism is a heterogeneous array of phenomena with different underlying genetic causes. Some of these mosaic phenomena have long been known. But they were also formerly regarded as rare and freakish accidents of development. An increasing number of cases have been documented in both plants and animals, making it difficult to hold to the older viewpoint. Mosaicism in one form or another is fairly common in higher organisms.

PART III : LINKAGE SYSTEMS

CHAPTER SIXTEEN

THE CHROMOSOME SET

INTRODUCTION

THE KARYOTYPE

SIGNIFICANCE OF KARYOTYPIC CHARACTERS

SOMATIC VARIATION IN CHROMOSOME NUMBER

VARIATION IN DNA CONTENT

INTRODUCTION It is necessary to bring the chromosomes back into the picture as a prelude to the discussion of linkage systems in the next several chapters. The present chapter highlights selected aspects of the chromosome set in higher plants; no attempt is made to give a systematic account. More complete treatments are available in various other works (Darlington, 1963, 1965; Lewis and John, 1963; John and Lewis, 1965; Rieger and Michaelis, 1967; Stebbins, 1971; Brown, 1972).

THE KARYOTYPE A somatic chromosome set has certain numerical, morphological, and what may be called anatomical characteristics. These characteristics taken together constitute the karyotype of a group of organisms. Jackson (1971) defines the karyotype as "the phenotypic appearance of the somatic chromosomes in contrast to their genic content."

A particular combination of chromosomal features, that is, a given karyotype, often characterizes a plant group. The group may be a large one, a section or genus or higher taxonomic unit; or it may be a small one, a species or race. In any case, the karyotype is a part of the total character combination of the group, and more especially a part of its endophenotype, and shows taxonomic variation comparable to that in the external gross morphology of the organisms.

The principal karyotypic features which are subject to taxonomic variation are listed here synoptically.

1. *Chromosome number.* (a) Basic number. (b) Aneuploid variation. (c) Presence of polyploidy. (d) Presence of supernumerary B chromosomes.

2. *Chromosome morphology,* usually as seen in mitotic metaphase. (a) Chromosome size; often expressed as length of mitotic chromosome in microns. (b) Size of complement; the cumulative total length or volume of all chromosomes in the set. (c) Position of centromere; expressed as arm ratio, that is, as the relative lengths of the two chromosome arms. (d) Symmetry or asymmetry in arm lengths. Are the chromosomes equal-armed (metacentric), or unequal-armed (subtelocentric, etc.), or intermediate? (e) Symmetry or asymmetry in chromosome size. Are the chromosomes of the complement approximately equal in length, or are there distinct classes of large and small chromosomes, or is there an intergrading series of sizes? (f) Number of chromosomes with satellites.

3. *"Anatomical" features.* (a) Amount of heterochromatin. (b) Distribution of heterochromatin. Is it in blocks near the centromeres, or distributed throughout euchromatic regions as narrow bands with characteristic banding patterns, and so on? (c) Other banding patterns as revealed by special methods of preparation. (d) Type of centromere activity; whether localized in single standard centromeres, or diffused in polycentric chromosomes. (e) Strandedness; number of DNA strands per chromosome.

A fourth type of character has been added in recent years, which, while not karyotypic in the strict sense, does show interesting group-to-group variation, and must be mentioned here.

4. *Molecular features.* (a) Total amount of DNA per chromosome, per set, or per nucleus.

The numerical and morphological characters of a basic haploid chromosome complement can be shown diagrammatically in the

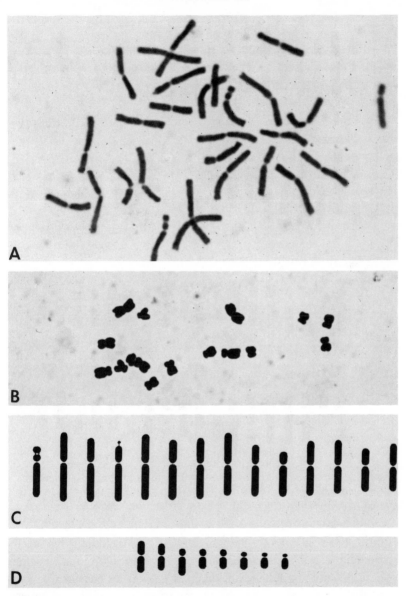

Figure 34. Somatic chromosome complements of two species of New Zealand angiosperms. (*A, B*) Photomicrographs of mitotic metaphase. (*C, D*) Idiograms. All figures to same scale. (*A, C*) *Cockaynea gracilis* (Gramineae), 2*n* = 28. (*B, D*) *Myosurus novae-zelandiae* (Ranunculaceae), 2*n* = 16. (Rearranged from Hair, Beuzenberg and Pearson, 1967; photomicrographs courtesy of J. B. Hair.)

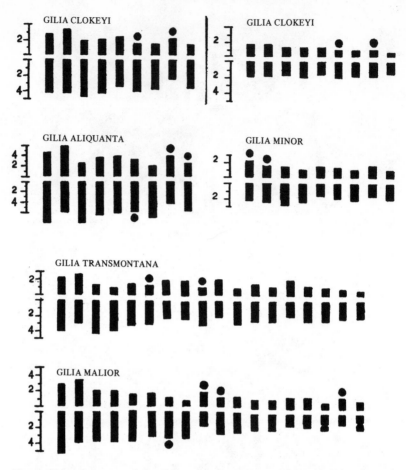

Figure 35. Idiograms of five interrelated species of Gilia (Polemoniaceae), x = 9. Two races of *G. clokeyi* are shown: Utah race (*left*) and eastern California race (*right*). Scales are in microns. (From Day, 1965.)

form of an idiogram. The amount and distribution of hetero-chromatin can also be represented diagrammatically on an idiogram. The differentiating features of different karyotypes can then be compared directly and visually by means of their respective idiograms. Examples of karyotypes and idiograms are shown in Figures 34 and 35.

Great differences exist between plant groups with respect to the karyotypic characters listed above. Take number and size, for

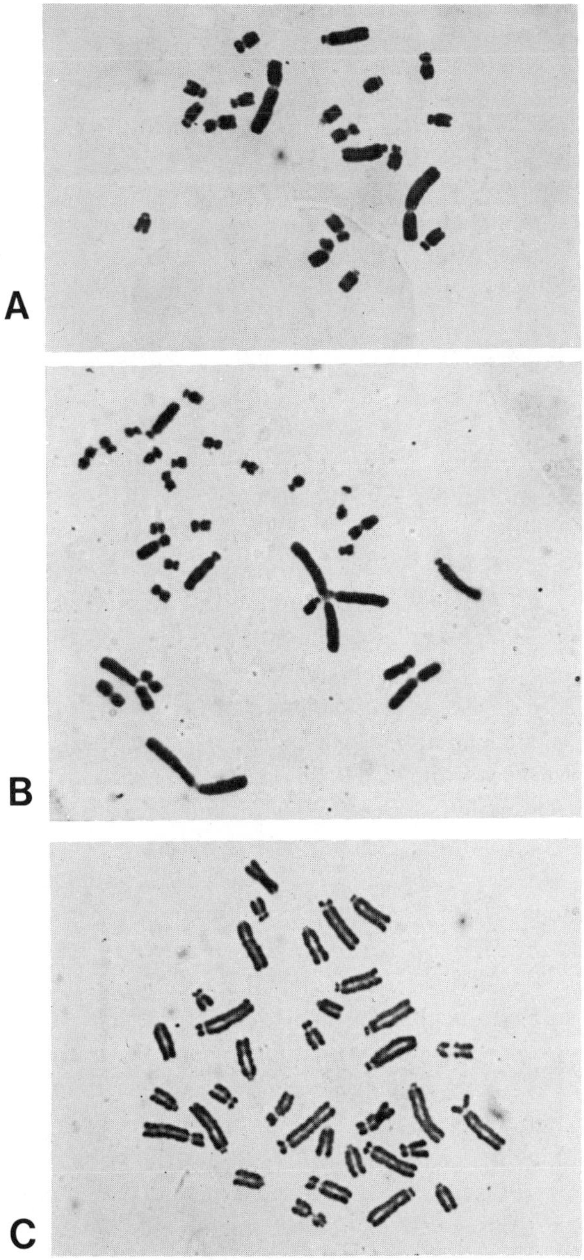

Figure 36. Asymmetrical karyotypes. Photomicrographs of mitotic metaphase. All figures to same scale. (*A*) *Luzuriaga parviflora* (Philesiaceae), $2n = 20$. (*B*) *Ripogonum scandens* (Smilacaceae), $2n = 30$. (*C*) *Gunnera albocarpa* (Haloragaceae), $2n = 34$. (From Beuzenberg and Hair, 1963; photomicrographs courtesy of J. B. Hair.)

example. Known chromosome numbers in the angiosperms range from $2n = 4$ in *Haplopappus gracilis* (Compositae) to $2n =$ ca. 540 in *Graptopetalum pachyphyllum* (Crassulaceae) (Jackson, 1957; Uhl, 1970). Single chromosomes of *Haemanthus kalbreyeri* (Amaryllidaceae) are larger in length and in volume than the whole chromosome complement of *Scirpus maritimus* (Cyperaceae) (Müntzing, 1961, p. 285). Degree of asymmetry, amount and distribution of heterochromatin, and other features also show wide ranges of variation. Some of this variation is shown in Figures 34 to 36.

Figure 34 shows the karyotypes of two species belonging to distantly related families, the Ranunculaceae and Gramineae. Their chromosomes differ strikingly in number, size, arm ratio, and satellites. The Myosurus species (Fig. 34B, D) has a diploid complement $(2n = 16)$ of small chromosomes. The grass Cockaynea (Fig. 34A, C) has a tetraploid complement $(2n = 4x = 28)$ of long-armed chromosomes (Hair, Beuzenberg and Pearson, 1967).

In Figure 35 we see the idiograms of five closely related species of Gilia (Polemoniaceae) with the same basic number, $x = 9$. Three of the species are diploids $(2n = 18)$ and two are tetraploids $(2n = 36)$. The tetraploids are allotetraploid derivatives of the diploids, with the following genomic constitutions: *G. transmontana = G. minor + clokeyi; G. malior = G. minor + aliquanta.* Differences in size, symmetry, and satellites are apparent between the diploid species, and, in one case, between races of the same species. On the other hand, resemblances are found between the genomes in the tetraploid species and those in their diploid ancestors (Day, 1965).

Two kinds of karyotype asymmetry are illustrated in Figure 36. Complements containing large and small chromosomes are shown in Figure 36A and B. The size classes are especially distinct in the Ripogonum (= Rhipogonum) (Fig. 36B). The chromosome set of the Gunnera species (Fig. 36C) consists of chromosomes with very unequal arms; many of these chromosomes are subtelocentric (Beuzenberg and Hair, 1963).

SIGNIFICANCE OF KARYOTYPIC CHARACTERS It is probably safe to assume that all regular karyotypic characters have some genetic significance; but just what that significance is, and what the genetic role of a given karyotypic feature is, is unclear. We do not understand the significance of some karyotype characteristics at all. With regard to other characteristics we have plausible suggestions but insufficient evidence. Still other aspects of karyotype variation in plants are becoming fairly well understood.

Changes in arm ratio result from unequal reciprocal translocations, unequal periclinal inversions, inter-arm transpositions, and to a lesser extent from other structural rearrangements. Differences in length between nonhomologous chromosomes also result from unequal reciprocal translocations. Changes in basic chromosome number are another result of unequal translocations. These structural changes alter the linkage relations of the genes on the rearranged chromosomes. Therefore karyotype variation in basic number and in many aspects of symmetry, within an interrelated group of plants, is a visible manifestation of changes in gene linkage.

Big differences in chromosome size exist in the angiosperms, and often occur between members of the same family. Thus in the Leguminosae the mean length of the somatic chromosomes is 1.8μ in *Lotus tenuis* and 14.8μ in *Vicia faba* (Stebbins, 1971, pp. 12 53). The meaning of such size differences is not clear.

A clue is provided by a correlation in some groups between chromosome size and climate. In the Gramineae, the tropical and warm temperate genera have small or medium-sized chromosomes, whereas the cold temperate genera mostly have large chromosomes (Avdulov, 1931; Stebbins, 1966). A similar correlation of large chromosomes with cool climates is found in the Liliales and to some extent in the Leguminosae (Stebbins, 1966). We have observed a size gradient from small chromosomes in tropical genera to large chromosomes in cold temperate genera in the Polemoniaceae (Grant, 1959) (see Fig. 37). A systematic collation of the available cytotaxonomic evidence is desirable in order to determine how widespread this pattern of size variation is.

Stebbins (1966) has attempted to explain these correlations by the hypothesis that physiological processes require more complex means of regulation in cool climates. Plants of warm climates can grow and photosynthesize more or less continuously during the growing season. These same processes are subject to frequent and

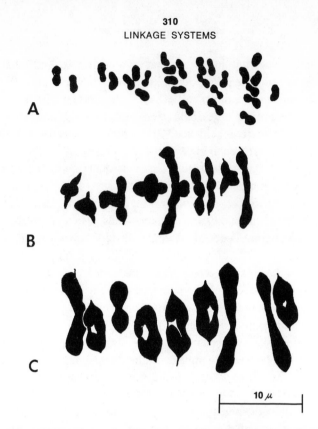

A

B

C

|_____ 10 μ _____|

Figure 37. Differences in chromosome size between representative members of three tribes of the family Polemoniaceae inhabiting different climatic zones. Chromosomes paired at metaphase I in pollen mother cells. All figures to same scale. (A) *Cantua candelilla* (2n = 54), Cantua tribe, tropical. (B) *Navarretia atractyloides* (2n = 18), Gilia tribe, warm temperate. (C) *Polemonium foliosissimum* (2n = 18), Polemonium tribe, cold temperate.

erratic interruptions in plants inhabiting cool climates. The plants of cold temperate regions require a more elaborate system of controlling genes to switch gene-controlled metabolic processes on and off, in adjustment with variable weather conditions, and this increase in the number of controlling genes is expressed visibly in larger chromosomes.

The role of heterochromatin, despite extensive discussion, is still rather poorly understood. In previous chapters we have discussed heterochromatin in relation to gene regulation (in Chapter 3), controlling elements (Chapter 8), variegated position-effect (Chapters

8 and 15), and polygenic systems (Chapter 11). The tenuousness of the evidence was indicated in these earlier discussions.

If we assume as a working hypothesis that heterochromatin is in fact involved in gene regulation, what then is the significance of karyotype differences in the amount and distribution of the heterochromatin? The answer to this question, although necessarily speculative at present, could be related to the differences in chromosome size.

In the Gramineae, heterochromatin has a different distribution in groups with small chromosomes and those with large chromosomes. Species of grass with small chromosomes have their heterochromatin concentrated in regions around the centromeres. Grass species with large chromosomes have the heterochromatin dispersed in a series of small segments throughout the chromosomes (Avdulov, 1931; Stebbins, 1966). Stebbins (1966) states that this correlation between heterochromatin distribution and chromosome size holds true in other plant families. He then suggests that the dispersed heterochromatin of large chromosomes may be regulating the activity of numerous gene sites in plants growing in cool climates. Much more evidence is needed to verify this interesting idea.

In this chapter we can only point out some of the unsolved and incompletely solved problems of karyotype variation. In subsequent chapters we shall devote much attention to linkage systems in higher plants. Changes in linkage systems will be related to chromosomal rearrangements in the mechanical sense in Chapter 17 and in the evolutionary sense in the final chapters of the book. The aspects of the karyotype which are connected with linkage are becoming better understood and are emphasized.

SOMATIC VARIATION IN CHROMOSOME NUMBER The orthodox view of development holds that an individual receives its genetic endowment at zygote formation, that this genetic constitution is replicated exactly by mitosis during the course of development, and that differentiation is due to variations in the expression of the constant somatic genotype. This view has long been known to be only partly correct. Several aspects of genetic mosaicism having external morphological expressions were discussed in the Chapter 15. In this section we mention a few examples of mosaicism with respect to chromosome numbers.

The shoot apex in *Datura stramonium*, the Jimson weed, normally consists of three concentric germ layers (L1, L2, L3) and a central core of cells behind these layers. In normal plants these cell layers are all diploid. The Datura cytogeneticists have found, however, a number of periclinal chimeras consisting of diploid layers and polyploid layers. The cell layer involved and the degree of polyploidy vary from case to case. For example, the L1 layer may be tetraploid and the other layers diploid; or L2 may be tetraploid; or L2 and L3 are both tetraploid; or a given layer is octoploid. In the growth of some of these periclinal chimeras, segregation of the cytologically different cell layers occurred, and resulted in all-diploid branches and all-tetraploid branches (Avery, Satina and Rietsema, 1959, ch. 8).

Endopolyploidy is a normal condition in some tissues of the plant body which have a nutritive function. The tapetum of the anthers and the endosperm of the developing seeds are usually polyploid in otherwise diploid plants. Parts of the embryo sac may also be polyploid. The multiplication of gene loci which results from endopolyploidy in such nutritive tissues probably facilitates their intense metabolic activity.

Small supernumerary or B chromosomes, which are not part of the regular complement, occur in varying numbers in many plant species. When present, they sometimes occur in more or less constant numbers in all parts of the plant body. In other cases, they vary in number in different parts of an individual plant. The B chromosomes are sometimes present in the shoot and flower buds but absent in the root, due to elimination during root development.

An early case of this latter sort was reported in *Xanthisma texana* (Compositae) (Berger and Witkus, 1954). The roots of this plant have the diploid number $2n = 8$. The cells of the shoot have higher numbers varying in the range $2n = 8-12$. The extra chromosomes in the shoot are B chromosomes which manage to perpetuate themselves here but are eliminated in the root because they lag during mitotic divisions (Berger and Witkus, 1954).

Claytonia virginica (Portulacaceae) has $2n = 28$ chromosomes in the regular complement, plus some B chromosomes. The B chromosomes occur only rarely in the roots but are fairly common in the stems and flower buds. One individual was found to have 28 chromosomes in root cells and 30 chromosomes in stem cells and pollen mother cells; another individual had 28 chromosomes in the root and 29 in pollen mother cells. Such differences in chromosome number between root and shoot were found in 23 of 34 plants from a single population in Texas (Lewis, 1970; Lewis, Oliver and Luikart, 1971).

These somatic differences pass over into individual variation in chromosome numbers within a population. The Texas population of *Claytonia virginica* was found to contain great individual variation in chromosome numbers as determined from counts of pollen mother cells. The range of variation extends from $2n = 24$ to $2n = 58$. Most individuals have $2n = 28$ or 29. But many other individuals have more chromosomes than this mode and some have fewer. Furthermore, the frequency distribution of chromosome numbers shifts perceptibly to a higher or lower range in different years. The annual variations may be correlated with weather conditions (Lewis, 1970). This interesting situation is still under investigation.

We cannot cover the large subject of B chromosomes in plants as a whole. For reviews of the subject see Müntzing (1959), Darlington (1963, pp. 20 ff.), and Brown (1972, ch. 9).

VARIATION IN DNA CONTENT Modern techniques of molecular genetics have made it possible to measure the amount of DNA per chromosome, per nucleus, or per cell, and to express this amount in various arbitrary or physical units. These techniques have been applied widely, with the result that an extensive backlog of information is now available on the DNA content of many and diverse species of organisms. The information is summarized in a series of tables by Sparrow, Price and Underbrink (1972).

This line of investigation has yielded a mixture of expected and unexpected results. An expected finding is that great differences in DNA content occur between different major groups and that these differences are roughly correlated with the degree of complexity of the organism. Thus the number of nucleotides per haploid cell ranges as follows in a series of species belonging to the groups listed:

	Nucleotides		
Viruses	1.3×10^3	to	5.3×10^5
Bacteria	1.2×10^6	to	5.7×10^7
Fungi	4.8×10^7	to	3.8×10^8
Bryophytes	9.0×10^8	to	8.6×10^9
Gymnosperms	8.4×10^9	to	1.4×10^{11}

The reader may refer to the account of Sparrow and co-workers (1972) for the full details and a discussion of some of the anomalies.

A surprising result which concerns us here is the discovery of rather large differences in DNA content between related species of angiosperms with identical or similar chromosome numbers. Such interspecific differences have been found in Lathyrus (Rees, Cameron, Hazarika and Jones, 1966; Rees, 1972), Allium (Jones and Rees, 1968), Vicia (Chooi, 1971a, 1971b), and Sorghum (Paroda and Rees, 1971).

Three species of Allium with the same chromosome number ($2n = 18$) show the following differences in amount of nuclear DNA as measured in arbitrary units (Jones and Rees, 1968):

A. zebdanense	25 DNA units
A. triquetrum	36
A. karataviense	45

Two closely related species of Sorghum with $2n = 20$ differ in nuclear DNA as follows (Paroda and Rees, 1971):

S. roxburghii	4.3 DNA units
S. durra	7.2

Three related species of Vicia with $2n = 12$ or 14 differ in average DNA content per chromosome as follows (Chooi, 1971a):

V. bithynica	4.9 DNA units
V. narbonensis	7.8
V. faba	16.7

Differences in nuclear DNA content are positively correlated with differences in chromosome size. This is true within the genera Allium and Lathyrus (Jones and Rees, 1968; Rees, 1972) and in a larger sample of 30 distantly related species of angiosperms (Baetcke, Sparrow, Nauman and Schwemmer, 1967).

The meaning of the variation in DNA content within a related group, particularly within a genus, is not clear. The main possibilities in comparisons between related diploid or homoploid species are differences in number of strands, duplicate segments, or B chromosomes. The evidence is not conclusive. Some evidence suggests that the interspecific differences in DNA content in Vicia might be due to the amount of duplication (Chooi, 1971a, 1971b). We discussed extra DNA in relation to duplications and polygenes in Chapter 11. The possibility that extra DNA functions as controlling genes must also be borne in mind.

CHAPTER SEVENTEEN

CHROMOSOMAL REARRANGEMENTS AND LINKAGE

TYPES OF CHROMOSOMAL REARRANGEMENTS
HOMOZYGOUS REARRANGEMENTS AND LINKAGE CHANGES
HETEROZYGOUS INVERSIONS
SMALL RECIPROCAL TRANSLOCATIONS
LARGE TRANSLOCATIONS

TYPES OF CHROMOSOMAL REARRANGEMENT Chromosomes occasionally break, and the broken ends rejoin in new ways, to produce a new segmental arrangement and hence a new gene order. These structural rearrangements set up new linkage relations between the genes in the rearranged segments or close to the breakage points.

This chapter reviews the mechanics of linkage as affected by various types of structural rearrangement in either homozygous or heterozygous condition. The subject is a classic one. The cytological behavior, sterility effects, and linkage effects of chromosomal rearrangements are variously treated in genetics and cytogenetics texts. We follow here an earlier treatment (Grant, 1964, chs. 8 and 9 in part) which focuses on the linkage effects and is suitable for our present purposes.

The number and position of the breaks determine the type of rearrangement. The main types are listed and diagramed in Figure 38. Rearrangements in a single chromosome are deficiencies, duplications, inversions, and transpositions. Rearrangements of segments belonging to two or more nonhomologous chromosomes are translocations.

CHROMOSOMAL REARRANGEMENTS

1. Terminal deficiency

Std.　A B C D　E F G H I

Rear.　A B C D　E F G

2. Interstitial deletion

Std.　A B C D　E F G H I

Rear.　A B C D　E F I

3. Duplication

Std.　A B C D　E F G H I

Rear.　A B C D　E F G H G H I

4. Paracentric inversion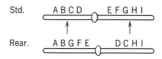

Std.　A B C D　E F G H I

Rear.　A B C D　E H G F I

5. Pericentric inversion

Std.　A B C D　E F G H I

Rear.　A B G F E　D C H I

6. Transposition

Std.　A B C D　E F G H I

Rear.　A D　E F B C G H I

7. Small reciprocal translocation

Std.　A B C D　E F G H I　M N O P　Q R S T

Rear.　A B C D　E F G H S T　M N O P　Q R I

8. Whole arm translocation

Std.　A B C D　E F G H I　M N O P　Q R S T

Rear.　A B C D　Q R S T　M N O P　E F G H I

9. Successive translocations

Std.　A B C D　E F G H I　M N O P　Q R S T　U V W　X Y Z

1st Rear.　A B C D　Q R S T　M N O P　E F G H I　U V W　X Y Z

2nd Rear.　A B C D　Q R S T　M N O P　X Y Z　U V W　E F G H I

Figure 38. Types of chromosomal rearrangement. The structurally altered chromosome (labeled *Rear.*) is compared in each case with one and the same standard arrangement (labeled *Std.*). The small arrows mark the breakage points on the standard chromosome which give rise to the structural rearrangement. The centromere is shown as a median or submedial oval. (From Grant, 1964, reproduced by permission of John Wiley and Sons.)

Duplications and deficiencies may arise from unequal crossing-over. Two homologous chromosomes with the standard arrangement could be paired homologously throughout most of their length at meiosis, but with one homologous segment displaced, so that crossovers occur at different points in the two chromosomes. Thus in the arrangement shown below the *GH* segments are displaced and the crossover points occur to the left of *GH* in one chromosome and to the right of *GH* in the other.

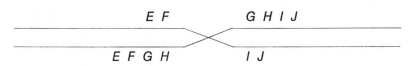

One daughter crossover chromosome will now carry the duplication for *GH,* being *EFGHGHIJ.* The other crossover chromosome will have the complementary deficiency, being *EF–IJ.*

Inversions can result from breaks and reunions at corresponding points in different gyres in a coiled chromosome. If the broken ends of two different gyres reunite in a new way, the region between the breaks will be reversed in direction, and will appear as an inversion. Translocations require breaks in two or more chromosomes and interchanges of their segments or arms.

HOMOZYGOUS REARRANGEMENTS AND LINKAGE CHANGES

With the standard gene order as the starting point, any given new linkage relation can be established by the appropriate structural rearrangement. The new linkage can then be perpetuated in homozygous condition in a derivative diploid strain or population. A comparison of the genes on the standard and rearranged chromosomes in the different parts of Figure 38 shows various new linkage relations in the rearranged chromosomes.

The deletion (Fig. 38.2) brings the genes *F* and *I*, formerly located some distance apart, into neighboring loci. The paracentric inversion (Fig. 38.4) brings the genes *E* and *H* into close linkage. A pericentric inversion brings about new close linkages between genes formerly located on different chromosome arms (*B* and *G* in Fig. 38.5). Transpositions can also set up close linkage between genes formerly on opposite arms (Fig. 38.6). Reciprocal translocations, finally, detach certain genes from one chromosome and place them in a new linkage group. Thus in Figure 38.7 the genes *ST* are on the second chromosome in the standard genome but on the first chromosome with *ABC-* in the rearranged genome.

It is of historical interest that inversions were first detected in Drosophila by the observation of differences in the linear order of genes in the related species *D. melanogaster* and *D. simulans*. The gene order in a region of chromosome III in *D. melanogaster* is scarlet-peach-delta. In *D. simulans* Sturtevant observed that the sequence of the homologous genes is scarlet-deltoid-peach. This difference could be accounted for by the rotation of the segment containing the genes peach and delta or deltoid (Sturtevant, 1921). The predicted phenomenon of inversions was confirmed later by direct cytological evidence. Note that the genes scarlet and peach are closely linked in *D. melanogaster,* lying 3 map units apart, but are more widely separated—at least 45 units apart—in *D. simulans* (Sturtevant, 1921).

HETEROZYGOUS INVERSIONS Inversions in heterozygous condition have an additional effect on linkage. The genes in the inverted region are inherited as a linked block in the inversion heterozygote. This effect is shown with the aid of diagrams in Figure 39.

Figure 39 shows pairing in a heterozygote for the paracentric inversion diagramed in Figure 38.4. The inverted region is *FGH* and *HGF* in the standard and rearranged chromosomes, respectively. Only one of the two strands of each chromosome is shown in Figure 39. Only one pair of strands need be considered at a time, and this simplification of the diagram makes the presentation easier to follow. Crossovers are indicated in the three diagrams at different points outside and inside the inverted region.

The paired chromosomes form a loop configuration in the inverted region at meiosis in the inversion heterozygote (Fig. 39). Crossing-over within the inversion leads to the formation of crossover chromosomes which carry deficiencies for some segments and duplications for others (see Fig. 39.2). Furthermore, the first crossover strand has two centromeres and the second has none. The meiotic

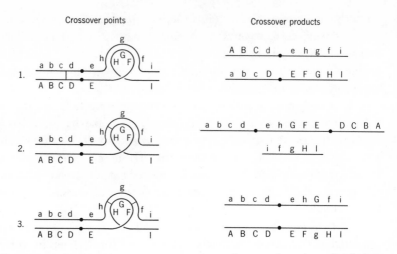

Figure 39. The genetic consequences of crossing-over in an inversion heterozygote. For the sake of clarity only two of the four strands are shown. (*1*) Crossover outside the inversion. (*2*) Single crossover within the inversion. (*3*) Compensating double crossover within the inversion. (From Grant, 1964, reproduced by permission of John Wiley and Sons.

products which carry the deficiency-duplication chromosomes with two or no centromeres are generally nonfunctional and are eliminated (Sturtevant and Beadle, 1936; Carson, 1946).

The functional gametes produced by this inversion heterozygote are mainly those in which no crossing-over occurred within the inverted region (as in Fig. 39.1).

Gene recombination in the inversion heterozygote is therefore possible between any two genes in the homologous segments (e.g., *C-D*), and between two genes lying in a homologous segment and in the inverted region (*C-F*) (Fig. 39.1). But the genes in the inversion (*F-G-H*) do not ordinarily form recombination types in a viable product of meiosis. Therefore the allele combinations *FGH* and *fgh* are usually inherited as closely linked blocks in the inversion heterozygote.

It can be shown that the net genetic effects of crossing-over within a heterozygous pericentric inversion are the same in principle as those for a paracentric inversion. The cytological behavior of a heterozygous pericentric inversion is different, in that the crossover chromosomes possess one centromere each, and not two or none as in the paracentric inversion heterozygote. But otherwise the cross-over products of a heterozygote for a pericentric inversion carry deficiencies and duplications for particular segments and are consequently eliminated.

The stage of elimination varies in different classes of organisms. In plants, as exemplified by *Zea mays,* the deficiency-duplication types are eliminated in the gametophyte stage, and most stringently in the pollen (Russell and Burnham, 1950; Rhoades and Dempsey, 1953). In Drosophila the deficiency-duplication gametes may function, but then they produce inviable zygotes (Sturtevant and Beadle, 1939, pp. 182–83; Swanson, 1957, p. 176).

An exception to the rule of tight linkage between genes within an inversion occurs when there is compensating, double crossing-over in the inverted region (as in Fig. 39.3). Such double crossovers produce recombination types (e.g. between *F* and *G*) in balanced crossover chromosomes which can enter into functional gametes or zygotes. However, compensating double crossing-over within an inversion, particularly a short one, will occur only rarely, since one crossover tends to suppress the formation of others in the immediate vicinity.

Generally, whether in paracentric or pericentric inversions, and whether in plants or animals, the genes in an inverted region constitute a closely linked block or supergene in the inversion heterozygote. Effective recombination between these genes is restricted to the exceptional products of compensating double crossing-over inside the inversion loop in the structural heterozygote.

In *Zea mays* the recombination value between two genes, *Lg,* and *A,* on chromosome 3, is 28% in normal stocks. That is, the heterozygote *Lg A/lg a* normally produces the recombination-type gametes (*Lg a* and *lg A*) in the average frequency of 28%. But where the genes *Lg* and *A* are included in an inversion, and the inversion heterozygote is simultaneously a genic heterozygote, with the constitution

$$\frac{inv\ Lg\ A}{std\ lg\ a}$$

the recombination between the two genes in the pollen mother cells is only 0.5%. This small amount of recombination in the inversion heterozygote is attributed to occasional double crossovers within the inverted segment (Rhoades and Dempsey, 1953).

SMALL RECIPROCAL TRANSLOCATIONS Translocations, like inversions, bring about linkage of particular blocks of genes when in heterozygous condition. The mechanisms are somewhat different in the case of heterozygous translocations, as is explained in this and the next section. For a comprehensive review of the older work on translocations in plants see Burnham (1956); for an up-to-date review see Brown (1972, ch. 13).

We first consider the breeding behavior of a heterozygote for a small reciprocal translocation. The translocation shown in Figure 38.7 is illustrated again in Figure 40. Only one strand of each chromosome is shown, for simplicity. The standard and rearranged chromosomes are homologous through most of their length but differ by translocations in the short terminal segments (*ST* and *I*).

At meiosis in the heterozygote, the two chromosomes, being predominantly homologous, will usually form bivalents (as in Fig. 40). Crossing-over occurs at various points along the main homologous portion of the two chromosomes. But the small homologous *ST* segments on different chromosomes do not pair. And the nonhomologous *ST* and *I* segments on the otherwise homologous chromosomes do not cross over. This precludes recombination between *S* and *T,* and between *ST* and *I*.

The restriction on recombination between *ST* and *I* can be demonstrated from a different standpoint. The independent assortment of the two pairs of chromosomes in the translocation heterozygote leads to four classes of gametes. Two of these classes are parental types and two are recombination types with respect to the translocation. The constitution of the meiotic products as regards the segmental rearrangement can be seen most clearly in Figure 40.1, where the simplifying assumption of no crossing-over is made, but it can also be seen in the other cases in Figure 40.

One type of gamete resulting from the assortment of noncross-

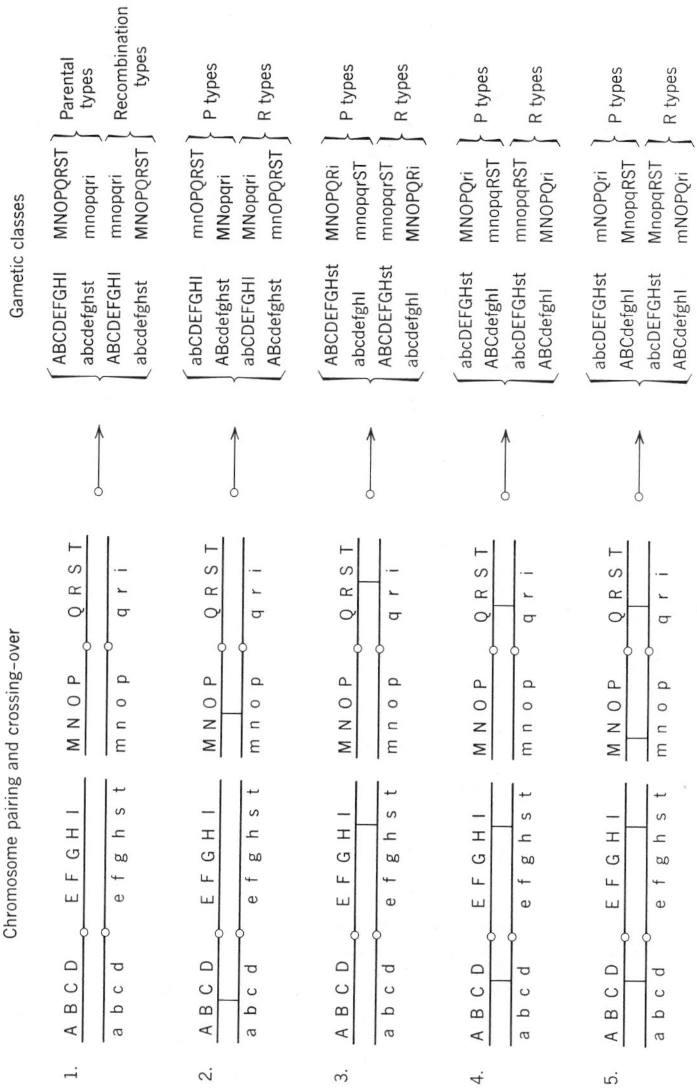

Figure 40. The products of meiosis in an individual that is heterozygous for a small translocation under various conditions of crossing-over. Only two of the four strands are shown. The gametic products are classified as parental or recombination types with respect to the translocation. (From Grant, 1964, reproduced by permission of John Wiley and Sons.)

over chromosomes (in Fig. 40.1) contains the standard arrangement in both chromosomes. A second gametic class has two translocation chromosomes. The third and fourth classes of meiotic products are recombinations of standard and translocation chromosomes. One of these recombination types carries a deficiency for the segment *ST* and a duplication for *I*. The other recombination type is deficient for *I* and duplicate for *ST* (Fig. 40.1). Crossing-over in the homologous chromosome pairs does not change this result so far as deficiencies and duplications are concerned. The recombination-type gametes containing crossover chromosomes still have deficiencies and duplications for the translocated segments *ST* and *I* (see Fig. 40, diagrams 2–5).

In angiosperms where the immediate products of meiosis must develop into functioning gametophytes, pollen or embryo sac, the recombination nuclei which are deficient in some genes and duplicate in others are normally eliminated before the stage of fertilization. In Drosophila the deficiency-duplication gametes may function in fertilization but then the resulting zygotes are inviable. In one stage or another, therefore, the recombinations of the two parental chromosomes, and of the genes borne on the translocated segments of these chromosomes, are eliminated.

The genes in the short translocation segments are thus associated with deficiencies and duplications in chromosomal recombination products. These genes cannot be recombined in viable products of meiosis. Therefore they are effectively linked.

In terms of our model (in Fig. 40), the genes *S* and *T* are linked together and are likewise linked with *I* on a different chromosome. The translocation heterozygote produces only the parental allele combinations in viable products of meiosis. The allele combinations *ST-I* and *st-i* are transmitted as two alternative linked blocks of genes.

It can be shown that the genes in a transposed segment (as in Fig. 38.6) will be linked in a similar manner in a transposition heterozygote.

LARGE TRANSLOCATIONS   The mechanism of linkage is different in a heterozygote for a large translocation. Let us take as a model

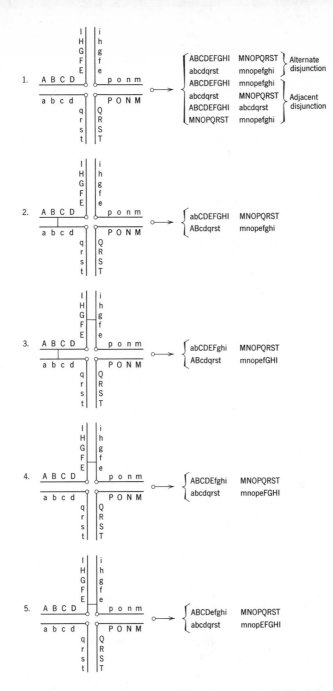

Figure 41. The products of alternate disjunction in an individual that is heterozygous for a whole arm translocation under various conditions of crossing-over. The products of adjacent disjunction are also shown in diagram 1. (From Grant, 1964, reproduced by permission of John Wiley and Sons.)

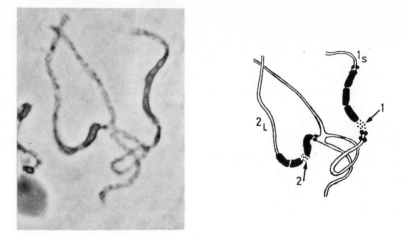

Figure 42. Chromosome pairing in a heterozygote for a large translocation in *Plantago insularis*. The translocation involves chromosomes 1 and 2 of a small complement (2*n* = 8.) The paired chromosome arms are shown at pachytene of meiosis in pollen mother cells. Photomicrograph (*left*) and interpretive drawing (*right*). Centromere regions denoted by arrows in drawing. (From Whittingham and Stebbins, 1969; photomicrograph courtesy of A. D. Whittingham.)

the whole-arm translocation shown in Figure 38.8. Meiosis in the heterozygote for this translocation is diagramed in Figure 41.

In a heterozygote for a large translocation (in contrast to the case described in the preceding section), the homologous segments on separate chromosomes *can* pair at meiosis. Homologous pairing arm-by-arm yields an association of four chromosomes. This association has a four-armed shape in early stages of meiosis when pairing and crossing-over take place (Figs. 41, 42). It later opens out to form a ring or chain of four chromosomes in preparation for the movement of the chromosomes to the poles.

The separation of the chromosomes to the daughter nuclei can occur in various ways. If adjacent chromosomes on the ring go to the same pole, the two daughter nuclei both inherit a deficiency for one large segment and a duplication for another. There are two modes of adjacent disjunction, depending on the orientation of the chromosome ring or chain prior to chromosome movement, but either mode leads to large deficiencies and duplications in the daughter nuclei, as shown in Figure 41.1. The deficiency-duplication nuclei give rise to inviable gametophytes, gametes, or zygotes. The only balanced products of meiosis are those resulting from the migration of alternate chromosomes in the ring or chain to the same pole, or, in other words, from alternate disjunction (see again Fig. 41.1).

The types of gene recombinations recovered in the progeny of this translocation heterozygote depend on the points of crossing-over between chromosomes that become distributed to the poles by alternate disjunction. The allele combinations resulting from crossing-over at various points, and from alternate disjunction of the crossover chromosomes, are shown in Figure 41 (diagrams 2–4). Genes located in many parts of the homologous segments of the four chromosomes can form recombination types in viable products of alternate disjunction. The parental allele combinations in the translocated chromosomes, in most possible cases, are capable of recombining successfully in the progeny of the translocation heterozygote.

Mechanical difficulties interfere, however, with pairing between chromosome regions lying close to the translocation breakage points. These breakage points happen to be near the centromeres in our model (Fig. 38.8). Crossing-over in the neighborhood of the breaks is therefore also reduced or inhibited. The genes in such regions consequently tend to be linked in the translocation heterozygote (Dobzhansky, 1931; Darlington, 1936, 1937).

Consider the genes D and E which occur close to and on opposite sides of a translocation breakage point in our model. These genes can be recombined only by crossovers at a point like that indicated in Figure 41.5. Pairing and crossing-over may occur only rarely or not at all in this region. Then the parental allele combinations DE and de will be linked to the same degree in the translocation heterozygote. Furthermore, the alleles d and e which occur on separate chromosomes are as tightly linked in the heterozygote as the alleles D and E which occupy neighboring loci on the same chromosome.

The linkage system can be extended to other chromosomes by means of successive overlapping translocations. A model of two successive translocations involving three chromosomes is shown in Figure 38.9 and again in Figure 43. Pairing and crossing-over between the homologous arms will occur as diagrammed in Figure 43.2. Alternate disjunction of the crossover chromosomes is shown

1.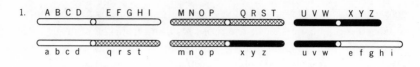

2.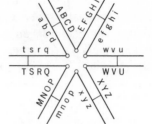

3.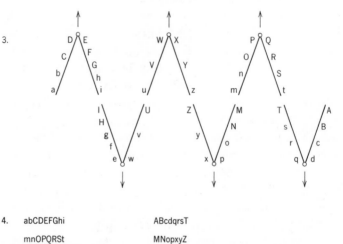

4. abcDEFGhi   ABcdqrsT

    mnOPQRSt   MNopxyZ

    uVWXYz   UvwefgHI

Figure 43. The course of meiosis in an individual heterozygous for two successive translocations. (1) Chromosomal and genic constitution of the heterozygote. (2) Chromosome pairing and crossing-over at meiosis. (3) Alternate disjunction at meiosis. (4) Genic constitution of the two balanced products of meiosis. (From Grant, 1964, reproduced by permission of John Wiley and Sons.)

in Figure 43.3. This alternate disjunction yields the two balanced and viable daughter nuclei indicated in Figure 43.4.

These diagrams show that genes located on the chromosome arms at some distance from the breakage points can cross over and recombine without any special inhibitions in the translocation heterozygote. But genes located close to the breakage points can recombine only rarely or not at all in the structural heterozygote. Consequently, the alleles of such genes which entered the heterozygote from one parent, on three separate chromosomes but near the translocation breaks on these chromosomes, are transmitted by the heterozygote as a linked group.

The genes E, Q, and X in our model, for example, occur on three separate chromosomes in each parental type. They are located near the breakage points. One parental genome carries the allele combination EQX, the other the combination eqx. Despite the fact that the three genes occur on different chromosomes, the parental allele combinations are normally linked in inheritance in the translocation heterozygote, as indicated in Figure 43 (diagrams 3 and 4).

CHAPTER EIGHTEEN

MULTIFACTORIAL LINKAGE

INTRODUCTION
MULTIFACTORIAL LINKAGE FOR A SINGLE CHARACTER
MULTIFACTORIAL LINKAGE FOR TWO OR MORE CHARACTERS
GENETIC COHERENCE IN POTENTILLA AND MIMULUS
DISCUSSION

INTRODUCTION Quantitative characters and complex characters in general are determined by multiple gene systems of one sort or another, as noted in earlier chapters. The several or many genes comprising a multiple gene system or polygenic system are distributed among a limited number of chromosomes in diploid plants. Under these conditions it is to be expected that some of the members of the gene system will occur in the same linkage groups.

The result is a loose linkage system known as multifactorial linkage (Grant, 1956a, 1964). Multifactorial linkage in two or more quantitative characters results in loose character correlations which have been aptly designated genetic coherence (Clausen and Hiesey, 1958, 1960).

A special type of multifactorial linkage, the balanced polygenic system, was discussed in Chapter 11. A phenomenon related to multifactorial linkage, the so-called M-V linkage, is described in Chapter 19.

MULTIFACTORIAL LINKAGE FOR A SINGLE CHARACTER An F_2 progeny segregating for a series of multiple genes with additive

MULTIFACTORIAL LINKAGE

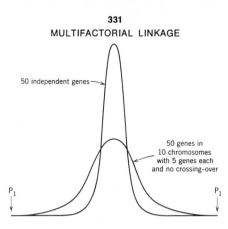

50 independent genes

50 genes in
10 chromosomes
with 5 genes each
and no crossing-over

P_1

P_1

Figure 44. Frequency distribution of phenotypes in an F_2 generation derived from a cross between two parents differing in 50 multiple genes with equal and additive effects and no dominance. The two curves describe the expected frequency distributions under two conditions of linkage. (From Anderson, 1949, reproduced by permission of John Wiley and Sons.)

effects and no dominance will form a more or less continuous array of phenotypes with a high frequency of intermediate classes and a low frequency of extreme classes. Moreover, as we saw earlier, the greater the number of such multiple genes, the higher the frequency of intermediate phenotypic classes and the lower the frequency of parental types in the F_2 generation.

We must now consider the effect of another factor. For any large number of such multiple genes segregating in F_2, linkage reduces the frequency of the intermediate types and increases the frequency of the parental types.

Consider two parental types differing with respect to 50 independent multiple genes or polygenes with equal and additive effects and no dominance. The series of 50 polymeric genes or polygenes, however we wish to refer to them, will segregate in F_2 to produce a preponderance of intermediate phenotypes and a low proportion of extreme phenotypes. If the 50 segregating genes are unlinked, the frequency of the intermediate phenotypic classes is very high, and that of the parental types is very low. The frequency distribution curve has a high peak in the intermediate range, as in Figure 44. But, as Anderson (1949) has shown, if the same 50 genes are equally distributed among 10 chromosomes, with no crossing-over between the 5 genes on each chromosome, the frequency distribution curve of phenotypes will have a much lower peak in the intermediate range and will slope off much more gradually toward the extremes (Fig. 44).

The general effect of linkage on segregation for multiple factors is thus to increase greatly the proportion of F_2 individuals which resemble one or the other parent. And some linkage, rather than its complete absence, is the normal condition when large numbers of genes are involved (Anderson, 1949).

Diploid plants have much fewer than 50 linkage groups. Therefore, in the hypothetical case under discussion here, the assumption of independent assortment of 50 multiple genes and the frequency curve of phenotypic classes drawn from that assumption are unrealistic. The alternative assumption that the 50 multiple genes are grouped in 10 blocks of 5 genes each, while oversimplified, comes closer to representing a real situation in diploid plants.

MULTIFACTORIAL LINKAGE FOR TWO OR MORE CHARACTERS

Let us next consider the effect of multifactorial linkage in an F_2 progeny segregating simultaneously for two or more different multiple-gene characters. The distribution of numerous genes among relatively few chromosomes brings about stringent restrictions on recombination between the different characters (Anderson, 1939a, 1939b, 1949; Dempster, 1949).

Anderson (1939b) has given a vivid demonstration of the restriction on recombination between different quantitative characters in the F_2 of *Nicotiana alata* × *N. langsdorfii*. The characters involved are three floral features: length of corolla tube, diameter of corolla limb, and depth of clefts between corolla lobes (Fig. 45). The parental character combination in *N. alata* is long tube, broad limb, and deeply cleft lobes (Fig. 45A). *Nicotiana langsdorfii* has the contrasting character combination (Fig. 45B).

If these three characters were capable of recombining freely in the F_2 progeny of the interspecific cross, one would expect to obtain the extreme recombination types depicted in Figure 46A. The most extreme recombination types observed in an actual F_2 progeny of 347 plants are shown in Figure 46B. It is clear from a comparison of the two series that the actual extremes of recombination between floral characters did not even come close to the ideal extremes.

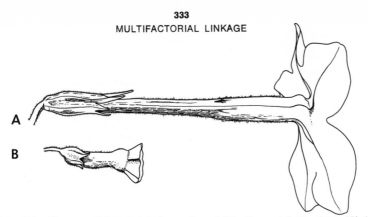

Figure 45. Flowers of two related species of Nicotiana. (*A*) *N. alata.* (*B*) *N. langsdorffii.* (From Anderson, 1939b.)

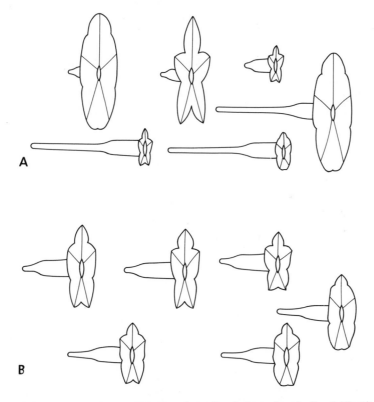

Figure 46. Recombination between three floral characters in F_2 of *Nicotiana alata* × *N. langsdorffii.* (*A*) Extreme recombination types expected with completely free recombination. (*B*) Extreme recombination types actually obtained in a hybrid progeny. (From Anderson, 1939b.)

TABLE 34
WEAK CORRELATION BETWEEN INTERNODE LENGTH AND INTERNODE NUMBER IN THE F_2 PROGENY OF A CROSS BETWEEN TALL AND SHORT PLANTS OF *PISUM SATIVUM* (LAMPRECHT, 1962)

No of F_2 plants with a given no. of internodes	No. of F_2 plants with internodes of a given average length															
	18	22	25	28	31	34	37	40	43	46	49	52	55	58	61	64 mm
30	—	—	—	—	—	—	—	—	—	1	—	—	—	—	—	—
	—	—	—	—	—	—	1	1	—	2	1	—	—	—	—	—
25	—	—	—	—	—	1	2	1	1	1	3	1	—	—	—	—
	—	—	1	—	3	3	8	3	4	—	2	2	—	1	—	—
20	—	—	6	6	18	15	16	24	13	8	4	—	3	—	—	—
	—	2	5	10	10	10	25	14	11	8	7	—	2	1	1	1
15	1	10	8	11	17	15	16	23	14	11	4	5	1	2	1	—
	1	—	—	—	5	5	8	1	5	2	3	—	—	—	—	—
10	2	—	1	—	2	1	2	2	—	1	—	1	—	1	1	—

The restriction on recombination in the Nicotiana hybrid is due partly but not exclusively to genetic linkage. Other restrictive factors are also involved. Pleiotropy is one contributing factor, and selection against the extreme types may be another (Anderson, 1939b). The exact relative contribution of multifactorial linkage to the observed results remains to be determined.

Lamprecht's (1962) data on inheritance of stem length in *Pisum sativum* provide a clearer insight into the linkage factor in character correlations. Lamprecht resolved the difference in stature between tall and short varieties of the garden pea into two characters: length of internodes and number of internodes. Both characters are multifactorial.

The F_2 progeny of a cross between tall and short pea plants were scored for both internode characters, with the results summarized in Table 34. Most F_2 plants are in the intermediate range for both internode length and internode number, as would be expected in multifactorial characters. A certain degree of independence between the two characters is also evident. This independence is indicated by some recombinants with few but long internodes (Lamprecht, 1962). The opposite recombination type is absent, however, and the trend of the data suggests a weak correlation between the two characters (see Table 34).

The observed results are understandable in the light of the

TABLE 35

LINKAGE RELATIONS OF TWO SERIES OF GENES CONTROLLING STEM GROWTH

IN *PISUM SATIVUM* (LAMPRECHT, 1962)

Gene system	Chromosome						
	I	*II*	*III*	*IV*	*V*	*VI*	*VII*
Internode length	Cot	Coe			Coh	Cona	
Internode number	Mie				Miu		Min

known linkage relations of the genes determining internode length and internode number. Internode length is controlled by the *Co* series of genes; four genes in this series are located in as many linkage groups, as shown in Table 35. Internode number is determined by three genes in the *Mi* series, with the locations indicated in Table 35. Several members of the two gene series are independent in inheritance. But certain members of these series occur together in linkage groups I and V (Lamprecht, 1962). The two characters would therefore be expected to show partial correlation in a segregating F_2 progeny, and they do.

GENETIC COHERENCE IN POTENTILLA AND MIMULUS   The most extensive and thorough studies of multifactorial linkage are those of Hiesey and his colleagues in *Potentilla glandulosa* and in the *Mimulus cardinalis* group (Clausen and Hiesey, 1958; Hiesey, Nobs and Björkman, 1971).

Coastal and alpine races of *Potentilla glandulosa* in California, as represented by populations from Santa Barbara and Upper Monarch Lake respectively, differ in numerous physiological and morphological characters, as listed in Table 36. Most of these character differences are determined by two or more genes each.

Assume, for the purpose of obtaining an overall numerical estimate, that in Table 36 "many" can be equated with 5 genes. Then at least 51 genes in the aggregate, and probably many more, determine the interracial differences in the 14 characters. *Potentilla glandulosa* has seven pairs of chromosomes. Therefore some linkage is expected to exist among the 51 genes and consequently among the 14 characters (Clausen and Hiesey, 1958).

TABLE 36

ESTIMATED MINIMUM NUMBER OF GENES CONTROLLING 14 CHARACTER DIF-
FERENCES BETWEEN A COASTAL RACE (SANTA BARBARA) AND AN ALPINE
RACE (UPPER MONARCH LAKE) OF *POTENTILLA GLANDULOSA* (COMPILED
FROM CLAUSEN AND HIESEY, 1958, pp. 54–108)

Character	No. of genes	Character	No. of genes
Winter dormancy	3	Leaf length	many
Flowering time	many	Leaflet number	1
Density of		Seed weight	6
inflorescence	1	Seed color	4
Glandular		Sepal length	5
pubescence	5	Petal width	2
Anthocyanin	4	Petal length	4
Stem length	many	Petal color	1

Clausen and Hiesey (1958) grew an F_2 generation of 992 plants from the interracial cross. The F_2 progeny segregated simultaneously for the 14 characters listed in Table 36. Almost every plant was scored for each character. The correlation coefficients were then computed between the various pairs of characters. There are 91 combinations of the 14 characters taken in pairs.

The 91 correlation coefficients ranged in magnitude from below 0.08 to above 0.7. Perfect correlation would of course be represented by $r = 1.0$ and no correlation by $r = 0.0$. In a sample of almost a thousand individuals, a correlation as low as $r = 0.08$ has a probability less than 1% of being due to chance alone, and is therefore significant statistically. Of the 91 pairs of characters in the segregating Potentilla hybrid, 67 showed statistically significant correlation (Clausen and Hiesey, 1958).

The distribution of the values of r is interesting. The bulk of the statistically significant correlation coefficients, 35 in all, fell in the range $r = 0.20$–0.40, which represents partial and incomplete correlation. Another 28 character pairs had correlation coefficients in the low range, $r = 0.09$–0.20, indicating weak but significant correlation. Only 4 character pairs showed fairly strong correlation, $r = 0.40$–0.80. The significant correlation coefficients were nearly all positive, pointing in the direction of the parental character combinations (Clausen and Hiesey, 1958, p. 116). The degrees of cor-

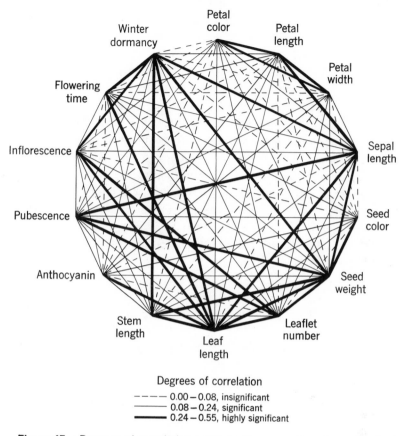

Degrees of correlation

---- 0.00 – 0.08, insignificant
———— 0.08 – 0.24, significant
━━━━ 0.24 – 0.55, highly significant

Figure 47. Degrees of correlation between 14 segregating characters in the F_2 progeny of an interracial cross in *Potentilla glandulosa*. (From Clausen and Hiesey, 1958.)

relation among the 14 characters are summarized graphically in Figure 47.

Correlations in the direction of the parental character combinations could be due to pleiotropy or linkage or both. Pleiotropy would be expected to play a part, and probably a large part, in the correlations between paired characters such as petal length and sepal length, which are the end products of common or similar growth processes. The correlations resulting from pleiotropy are also likely to be high. The partial correlations observed between characters which have separate courses of development, like seed weight and petal color, are best explained as the result of incomplete multifactorial linkage.

Figure 47 shows that no character in the interracial F_2 generation of *Potentilla glandulosa* segregates independently of all the other characters. Each character is partially correlated and probably partially linked with many or most of the other 13 characters. Yet each character is uncorrelated and may be unlinked with some of the other characters. Moreover, the various characters do not fall into mutually exclusive linkage groups, but rather into overlapping groups. Thus two characters (like sepal length and seed color), which are not correlated with each other, may both be linked independently of one another to a third character (pubescence), and the latter in turn is linked to a fourth character (anthocyanin) which segregates independently of the first (sepal length).

These results were confirmed in a subsequent study of the F_2 generation of *Mimulus lewisii* × *M. cardinalis* (Hiesey, Nobs and Björkman, 1971). The two parental species are diploid, with $2n = 16$, and are highly interfertile. The F_2 generation segregates for numerous morphological and physiological characters, most of which are multifactorial. These characters were scored in a large F_2 population, and the correlation coefficients between them were computed.

Figure 48 shows the degree of correlation between 14 characters. One character (dry weight) is recorded three times in the chart to show its behavior in different environments. Most of the characters listed are self-explanatory. The character called star is a red color pattern in the corolla throat, and rose is the density of anthocyanin pigmentation in the corolla lobes.

Here again, as in Potentilla, the vast majority of the correlation coefficients are statistically significant. And most of the significant correlation coefficients are in the low middle range, $r = 0.25–0.45$, reflecting partial correlation. Relatively few pairs of characters show independence in inheritance. The observed character correlations can be attributed partly or mainly to linkage (Hiesey, Nobs and Björkman, 1971).

The same general pattern of overlapping character correlations was found in an early study of the diploid species *Gilia capitata*

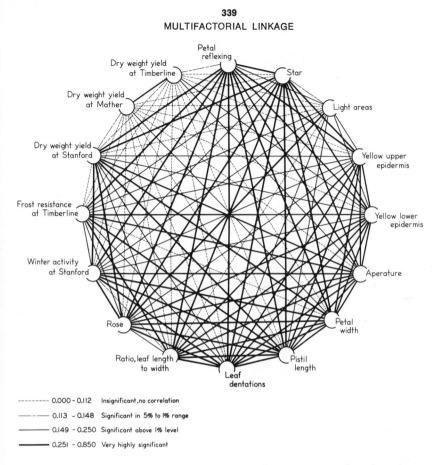

-------- 0.000 - 0.112   Insignificant, no correlation

— - — 0.113 - 0.148   Significant in 5% to 1% range

———— 0.149 - 0.250   Significant above 1% level

━━━━ 0.251 - 0.850   Very highly significant

Figure 48. Degrees of correlation between different segregating characters in F_2 of *Mimulus lewisii* × *M. cardinalis*. (From Hiesey, Nobs and Björkman, 1971; photograph courtesy of M. A. Nobs.)

($2n = 18$). Segregating F_2 progenies of interracial crosses were scored for eight characters selected to minimize the chances of pleiotropy. A large proportion of the paired combinations of these eight characters exhibited partial correlation (Grant, 1950b).

DISCUSSION   A tendency for parental character combinations to remain associated in segregating hybrid progenies is demonstrated by experimental results in diploid species of Gilia, Mimulus, Nicotiana, Pisum, and Potentilla. The character association or genetic coherence is fairly strong but not absolute. Recombination between different quantitative characters remains possible and occurs at a low rate but is generally opposed by multifactorial linkage.

The overlapping system of multifactorial linkages reflects the following underlying condition in the genotype: Genes responsible for a particular quantitative character difference are located on different chromosomes. Some of the genes controlling character A are linked with some of the genes for character B. But other genes for character A segregate independently of other genes determining character B. Therefore characters A and B are only partially correlated in inheritance. The genes for character A that are not linked with any genes for character B may occur in the same linkage group or groups with some of the genes controlling character C. Characters B and C segregate independently of one another. But B and C are separately and incompletely linked with A.

The extension of the system of interlocking linkages to characters D, E, and so on, eventually ties all the characters together, directly or indirectly, into a loose but coherent complex. Any character that displays continuous variation in F_2 is also likely to display correlations with other such characters. Conversely, no multifactorial character is transmitted in inheritance as a unit separate from the character complex. Recombination between different characters in this character complex is not prevented, to be sure, but is strongly restricted by the interlocking multifactorial linkages (Grant, 1950b, 1956a; Clausen and Hiesey, 1958, 1960; Hiesey, Nobs and Björkman, 1971).

The degree of genetic coherence within a group of multifactorial characters is proportional to the basic chromosome number, which varies widely in different groups of organisms.

The most common basic numbers in diploid angiosperms lie in the range $x = 7$–14, and lower basic numbers are not uncommon (see Grant, 1971, pp. 226 ff. for details). These basic numbers, especially the lower ones, are low enough to bring about a considerable amount of genetic coherence.

Low basic numbers also occur in various groups in the animal kingdom. Thus the haploid numbers range from $n = 3$ to $n = 9$ in many species of Diptera and from $n = 6$ to $n = 12$ in marsupials (Makino, 1951). Such groups of animals would be expected to show genetic coherence.

Much higher basic numbers prevail in other animal groups such as placental mammals, birds, and Lepidoptera. For example, gametic numbers of $n = 20$–33 are common in the Lepidoptera, and a few diploid species in this order have 50 or more chromosome pairs. Typical gametic numbers in ungulates are $n = 24$–36 (Makino, 1951). It remains to be seen how much genetic coherence is present in variable animal populations with such high basic chromosome numbers.

The consideration of basic chromosome number in relation to multifactorial linkage and genetic coherence has implications for both minor systematics and numerical taxonomy, implications which have scarcely been explored.

Botanists studying geographical variation in diploid plant species find that separate morphological characters usually vary in a correlated fashion, as is expected if genetic coherence is an effective factor (see Grant, 1971, ch. 12, for review). Animal systematists, on the other hand, often seem to regard the variation pattern of an animal species as a series of separate clines crisscrossing one another. Mayr, for example, in a discussion of geographical variation in animals, states that "The various characters of a species may and usually do vary independently. Neighboring populations agree, therefore, in some characters and differ in others" (1963, p. 333). Such independence of characters is not found in many groups of plants and would not be expected in low-number animal groups either. Here we may have an interesting difference in variation pattern between major groups.

Numerical taxonomists have assumed that an index value based on the statistical or computer treatment of a very large number of characters gives a better measure of phenotypic affinities between populations than the older methods of systematics, which employ a selected ensemble of taxonomic characters combined with that hard-to-define quantity, taxonomic judgment. The assumption made in numerical taxonomy does not take the phenomenon of genetic coherence into consideration.

It may be possible in practice to ignore genetic coherence in some high-number animal groups such as the Lepidoptera. Here one might conceivably pick out numerous taxonomic characters which happen to be independent of one another in variation. As applied in many other animal and plant groups with relatively low chromosome numbers, however, the numerical taxonomic method may often be inefficient in comparison with conventional methods. The use of additional quantitative characters beyond a certain point, in a low-number group, is tantamount to measuring variation in different facets of one and the same coherent character combination. A good taxonomist of the traditional school, approaching the same group of organisms, would select a few indicative characters intuitively, and reach the same or similar conclusions.

CHAPTER NINETEEN

LINKAGE BETWEEN MORPHOLOGY
AND VIABILITY

SKEWED SEGREGATION RATIOS IN MIMULUS   Two closely related species, *Mimulus lewisii* and *M. cardinalis* (Scrophulariaceae), occur over a wide area in western North America. *Mimulus lewisii* ($2n = 16$) occurs in the mountains and far north and is adapted to cold climates. *Mimulus cardinalis* ($2n = 16$) inhabits coastal and foothill zones with milder and warmer climates. The two species are interfertile and, in some racial combinations, are highly interfertile (Hiesey, Nobs and Björkman, 1971).

One of the conspicuous interspecific character differences is flower color: pink in *M. lewisii* and bright orange in *M. cardinalis*. This character difference is due to a single allele pair, with pink dominant over orange. Apparently the gene controls the presence or absence of a yellow carotenoid pigment in the epidermal cells of the corolla. The presence of this yellow pigment in combination with other underlying pigments brings about the orange color of the flowers in *M. cardinalis*. *Mimulus lewisii* has a dominant suppressor of carotenoid pigment production in the corolla. Accordingly it is pink-flowered and so are its F_1 hybrids with *M. cardinalis* (Vickery and Olson, 1956; Hiesey, Nobs and Björkman, 1971).

We may note parenthetically that other flower color differences exist between these Mimulus species and have a more complex genetic basis (see Hiesey, Nobs and Björkman, 1971).

The species also differ in various physiological characteristics fitting them for growth in their respective climatic zones. Some of these physiological differences are expressed in the early seedling stage. Nobs and Hiesey compared the seedling mortality of the two species at two experimental stations in California with contrasting climatic conditions, Stanford and Timberline. A coastal race of *M. cardinalis* showed almost complete seeding survival at Stanford (in the coastal zone), but had 100% seedling mortality at Timberline (in the alpine zone). Conversely, a high-mountain race of *M. lewisii* showed 60% seedling mortality at Timberline, but 90% mortality at Stanford (Nobs and Hiesey, 1957).

The cross of pink-flowered *M. lewisii* × orange-flowered *M. cardinalis* yields a pink-flowered F_1. The F_1 hybrid, being fertile, produces an abundance of good seeds from which large F_2 and F_3 progenies can be grown. F_2 generations and segregating F_3 families, when grown in optimum greenhouse environments, give approximately 3 : 1 segregation ratios for flower color (Vickery and Olson, 1956; Hiesey, Nobs and Björkman, 1971). One large F_3 family, for example, had a ratio of pink to orange of 3.1 : 1.0, and another a ratio of 2.9 : 1.0 (from data of Hiesey, Nobs and Björkman, 1971, p. 53).

When the F_2 seedlings are exposed to the normal climatic conditions at Stanford or Timberline, however, the segregation ratios for the adult floral character are strongly distorted, as shown in Table 37. The ratio in the F_2 population grown at the Timberline station is strongly skewed toward the pink, or *M. lewisii,* parental type. The ratio in a parallel sample grown in summer at Stanford is skewed in the opposite direction, toward the orange, or *M. cardinalis* parental type. A winter-grown F_2 at Stanford shows skewness in the direction of the *M. lewisii* type again. Finally, an F_2 generation grown in the intermediate environment at Mather has an intermediate ratio approximating 3 : 1 (Nobs and Hiesey, 1958; Hiesey, Nobs and Björkman, 1971).

These results are understandable in the light of the known bias in the seedling mortality of the parental species in the foreign trans-

TABLE 37

SEGREGATION FOR FLOWER COLOR IN FOUR F_2 PROGENIES OF *MIMULUS CARDINALIS* × *M. LEWISII* GROWN IN DIFFERENT ENVIRONMENTS (NOBS AND HIESEY, 1958)

Climate for seed germination and seedling growth	Station	Seedling mortality, %	Ratio of pink to orange flowers
Coastal, 150 ft. elevation, seeds sown in summer	Stanford, California	89	1.7 : 1.0
Coastal, as above, seeds sown in winter	Stanford, California	71	4.4 : 1.0
Montane, 4,600 ft. elevation, summer	Mather, Sierra Nevada	71	2.5 : 1.0
Alpine, 10,000 ft. elevation, summer	Timberline, Sierra Nevada	78	8.7 : 1.0

plant stations. Each parental type is at a strong selective disadvantage in the seedling stage in the foreign climate. Environmental selection in that stage is clearly responsible for the skewness in the segregation ratio of an adult phenotypic character toward the indigenous parental type.

The corollary is that the conspicuous morphological character of flower color must be linked with physiological characters determining survival ability of the seedlings in the contrasting coastal and alpine climates (Nobs and Hiesey, 1957, 1958).

THE GENERALIZED *M-V* LINKAGE   The Mimulus story furnishes a particularly good example, but only one among many, of a widespread phenomenon in higher plants. Underlying the phenomenon is a common genetic condition, which we can refer to as the *M-V* linkage, generalizing from a broad array of examples (Grant, 1967). It is convenient to describe the *M-V* linkage in general terms here so that it can serve as a frame of reference for considering the empirical evidence later.

The *M-V* linkage system is the linkage between genes determining particular morphological characters and other genes affecting viability. By morphology-determining genes we mean simply any genes determining visible exo-phenotypic traits including coloration of plant parts. The viability genes are taken to be genes affecting growth and vigor.

It is apparent that so-called viability genes may have pleiotropic effects on morphology, and vice versa. Albino seedlings which appear as a simple Mendelian recessive in many plant species, for example, are the expression of a viability gene which has an associated phenotypic effect. It is essential, therefore, to distinguish between pleiotropic morphological effects of viability genes, on the one hand, and M-V linkage on the other. Phenotypic traits which are first expressed in the flowers or fruits, and not detectably in the vegetative stage, and which are correlated with differences in viability, suggest linkage between separate morphological and viability genes rather than pleiotropy, since the two classes of characters are not closely related developmentally. Such correlated character combinations exist. Mimulus is one such case, and others are cited later.

The next postulate is that races and species of higher plants carry different alleles of the linked M and V genes. The different alleles of the V gene(s) are supposed to have different environmental optima for growth.

The simplest case would be that of two linked allele pairs without significant gene interactions. One parental race or species has the constitution M_1V_1/M_1V_1; the other is M_2V_2/M_2V_2. Then the progeny of the hybrid, when grown in an environment favoring one of the V alleles, will give an altered monofactorial segregation ratio for the morphological marker M.

More complex arrangements can be expected to occur commonly. Thus viability differences between the parental races or species might be determined by combinations of two or more V genes. One parental combination $(V_1 + V_{1'})$ would be viable; and so would the other $(V_2 + V_{2'})$; but recombinations like $V_1 + V_2$ would be subvital. The difference in a particular morphological character is also likely to be determined by two or more M genes. Furthermore, the different V genes may be independent or linked; and the M genes may also be independent or linked between themselves. But if at least one M gene is linked with at least one V gene, we will observe a correlation between viability and a physiologically unrelated morphological character in the segregating hybrid progeny.

We have assumed up to this point that the linked *M-V* genes are borne on homologous chromosomes or chromosome segments in parental types which form fertile or semifertile species. In the simplest case this situation is realized and the *M-V* linkage system is indeed borne on homologous segments. But another important complexity is introduced by segmental rearrangements in plant groups consisting of species with different structural karyotypes. Evidence to be examined later indicates that *M-V* linkage systems are associated with rearranged chromosome segments as well as with homologous segments.

M-V LINKAGE ON HOMOLOGOUS CHROMOSOME SEGMENTS

Genetic ratios for morphological markers are commonly distorted by environmental selection in segregating hybrid progenies in plants. Most experimental plant geneticists have probably encountered examples in their own plant groups. The collective experience of plant geneticists and breeders is not adequately reflected in the literature, however, for an obvious reason.

The experimental work is usually focused on the loci or the gene systems governing the morphological markers themselves. Cases of distorted segregation represent unsuccessful attempts as far as the primary purpose of such experiments is concerned and are assigned to the status of pilot experiments. The work is then repeated under more optimal environmental conditions in which all genotypes show good survival. The latter results are published; the aberrant preliminary observations are not.

In this chapter, following an earlier review on the subject (Grant, 1967), we wish to focus attention on this important but neglected complementary aspect of hybridization experiments; for it is this aspect which furnishes evidence for a widespread and probably general occurrence of *M-V* linkage in higher plants.

We will first review the evidence for *M-V* linkage in plant groups which produce hybrids with complete chromosome pairing at meiosis. The normal chromosome pairing in the hybrids suggests that the *M-V* linkage systems are on homologous segments in such groups. This conclusion is safest in groups like Potentilla and Mimulus, where the hybrids are also highly fertile. The presumption of homology is reasonable, and supported by cytological evidence, in other cases such as Lycopersicon that have sterile hybrids.

The best example of M-V linkage in homologous chromosomes is that of Mimulus, already presented, since all aspects of the problem are documented. The studies to be considered next were carried out with other objectives in mind, and are consequently incomplete for our present purposes, but nevertheless provide suggestive evidence for M-V linkages on homologous segments.

A cross between two interfertile races of *Potentilla glandulosa* (Rosaceae), a California foothill race and a subalpine race from the Sierra Nevada, was studied by Clausen and Hiesey (1958). The 509 individuals of the F_2 population were cloned and grown in replicate cultures at three transplant stations. The Timberline station has nearly the same climate as the natural habitat of the subalpine parent. The Stanford and Mather stations approximate the climatic conditions of the foothill parental race.

The F_2 progeny at each transplant station were scored for 12 character differences. These were nearly all multifactorial characters. About half of them pertained to vegetative parts and half to reproductive features, such as color, shape, and orientation of petals, and color and size of achenes. The scores for the 12 characters were pooled to give aggregate index values expressing the overall resemblance of each F_2 plant to one or the other parental race.

A record was also kept of the percentage of the F_2 plants which survived for five or more years at the different transplant stations. Survival was found to be correlated with the phenotypic index values. The F_2 clones which approached the foothill race in 12 characters combined had 71% long-term survival at Stanford, and 76% long-term survival at Mather, but only 13% survival at Timberline. Conversely, the F_2 clones resembling the subalpine race had 76% survival at Timberline, compared with 49% at Mather and 18% at Stanford (Clausen and Hiesey, 1958).

Environmental selection is undoubtedly responsible for the skewed distributions of phenotypes in *Potentilla glandulosa,* as Clausen and Hiesey (1958) suggest. The selective elimination apparently occurred mainly in intermediate growth stages or at least in post-seedling stages. Natural selection may well have acted directly on some of the segregating vegetative characters scored, which would then reflect viability genes with pleiotropic morphological effects. The altered ratios of the floral characters can be attributed with greater likelihood to *M-V* linkage.

The F_1 hybrid of *Lycopersicon esculentum* × *L. chilense* has visibly normal chromosome pairing but is highly sterile as to seeds, having only about 1% of the seed output of normal parental plants (Rick, 1963).

Rick has shown that the hybrid also gives altered segregation ratios for quantitative characters and marker genes in F_2 and B_1. The F_2 frequency curve for several quantitative characters separating the two species does not center on the F_1 mean, but is skewed toward *L. chilense.* This trend is confirmed by a more detailed analysis of the inheritance of six marker genes governing various vegetative characters. The *L. esculentum* parent contributed recessive and dominant markers on four or five chromosomes. In F_2 and in B_1 × *L. chilense* there was a consistent deficiency of *esculentum* types; in B_1 × *L. esculentum* a consistent excess of *esculentum* types occurred (Rick, 1963).

Gamete elimination is a known cause of skewed ratios in tomato hybrids (see Chapter 13), but is ruled out in this particular case. The skewness of the segregation is toward the recurrent parent and is not independent of the other parental type, as it should be if the determining factor lies in the gametophytes themselves (Rick, 1963). The most probable cause of the altered ratios is, as Rick suggests, a selective elimination of zygotes in F_2 or B_1.

The marker genes showing altered segregation all affect vegetative characters and are expressed in the seedling stage (Rick, 1963). Some of them, for example, dwarf stature, could have viability effects of their own. Others, such as the three genes for leaf shape, may not have important viability effects in the embryo stage; if this is the case, their behavior would point to *M-V* linkage.

Among the differences between *Phaseolus vulgaris* and *P. multiflorus* (= *P. coccineus*) are three floral characters and one seed character. Their F_1 hybrid is intermediate. In the F_2 Lamprecht (1941, 1944) observed a preponderance of *vulgaris* types and a marked deficiency of *multiflorus* types for each of the four characters. Wall and York (1957) have confirmed this observation for a seedling character in F_2, F_3, and F_4. Lamprecht's (1941) suggestion is that the *P. multiflorus* characters in question are not transmitted successfully through the pollen in the interspecific hybrid. An alternative explanation which is at least equally plausible is that the floral and seed characters are linked with viability differences. This explanation is in line with Lamprecht's (1941) observation of hybrid breakdown in the F_2 generation, with the segregation of dwarfs and chlorophyll-defective plants.

The F_1 hybrid of *Rubus idaeus* × *R. arcticus* is intermediate in several vegetative and fruit characters. The F_2 and F_3 populations derived from open-pollinated seeds are, however, skewed toward the *R. idaeus* parent in morphology (Vaarama, 1954). Vaarama attributes this result, with good reason, to differences in ease of crossing between the hybrids and the two parental species. It is quite possible, however, that viability differences linked with morphological character differences are also involved here.

M-V LINKAGE ON REARRANGED SEGMENTS The genetic and functional role of rearranged chromosome segments in the life of species which differ by segmental rearrangements has been a very difficult problem to tackle experimentally. The difficulty stems from the fact that the hybrids of such species exhibit chromosomal sterility. The hybrid sterility makes it impossible to carry out a standard factorial analysis. As a result, our knowledge of the genic contents of rearranged segments is meager.

Two older hypotheses have influenced thinking on this subject for many years. The first is that the rearranged segments determine the basic physiological characteristics of the species by means of numerous position effects or one overall pattern effect (Goldschmidt, 1940). Position effect probably is part of the story, but not the whole story. We will return to this aspect in Chapter 20.

The second older hypothesis is that the chromosomal rearrangements are neutral in their effects on morphological characters; they are independent of the morphological and physiological differences between the species (Stebbins, 1950, pp. 230–32, 324). In the light of later findings we would have to say that this view is also an overgeneralization. What the older evidence entitles us to conclude is merely that certain particular segmental rearrangements are independent of certain particular phenotypic traits.

Experimental evidence in several plant groups points to the occurrence of viability genes or morphological genes, or both, on rearranged segments. The evidence comes from experiments in four plant groups—Clarkia, Gilia, Gossypium, and Zea—and is described in the sections that follow.

This body of evidence leads to a third generalization. Rearranged segments normally contain genes affecting growth and vigor and/or genes determining morphological differences between the parental species (Grant, 1966b). There does not seem to be any important difference between the M-V linkage systems found in homologous chromosomes and those inferred to exist in rearranged segments (Grant, 1967).

PETAL SPOT IN CLARKIA   This section deals with the location of the gene for petal spot in two related species of Clarkia (Onagraceae), formerly placed in the segregate genus Godetia. The older and modern names of the species are: *Godetia amoena* (= *Clarkia rubicunda*) and *G. whitneyi* (= *C. amoena*) (Lewis and Lewis, 1955). The research was performed under the older nomenclature. In reviewing this work here we use the older names so our account will correlate with the original research papers.

Godetia amoena and *G. whitneyi* have the same chromosome number ($2n = 14$), but differ in segmental rearrangements including at least one translocation. Their F_1 hybrid is fairly sterile, with 6% good pollen, 3% seed fertility, and some meiotic irregularities. The usual configuration at meiosis is six bivalents plus two univalents; but a translocation chain appears occasionally (Hiorth, 1942; Håkansson, 1947).

The two species also differ in several morphological characters. One character difference involves the petal spot, which is basal in *G. amoena* and central in *G. whitneyi*. This difference is determined by a gene F represented by the allele F_b in *G. amoena* and by the allele F_x in *G. whitneyi*. The F_b/F_x heterozygote has both basal and central spots (Hiorth, 1940).

Some characters of *G. amoena* can be introduced readily into *G. whitneyi* by backcrossing the hybrid to the latter in successive generations. Petal spot was an exception, however. The F_x types in B_1 and B_2 were fertile but the F_b types were sterile (Hiorth, unpublished data; Håkansson, 1947).

In B_3 an exceptional individual appeared which carried the F_b allele and yet was partly fertile. It had six bivalents and a chain of three chromosomes. The bivalents were composed of *whitneyi* chromosomes. The chain had the following chromosomal composition:

amoena–whitneyi–amoena

3–2   2–1   1–6

The F_x allele is in the 1–2 chromosome of *G. whitneyi* and the F_b allele in the 1–6 chromosome of *G. amoena* (Håkansson, 1947). Therefore the two chromosomes having the number 1 arm in common can be rewritten as:

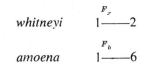

This B_3 plant gave rise to a fertile F_b/F_x heterozygote in the next (B_4) generation. Evidently the F_b allele crossed over from the *amoena* 1–6 chromosome to the *whitneyi* 1–2 chromosome. On selfing, the B_4 plant produced the following types of F_6 progeny (Håkansson, 1947):

67 F_b/F_b

213 F_b/F_x

115 F_x/F_x

These Godetias (or Clarkias) thus furnish us with an example of a morphological gene that is borne on a translocated segment and differentiates two species. The deficiency of F_b/F_b types in the F_6 generation suggests, moreover, that the F_b segment may reduce the viability of the backcross products which carry it. This is turn would suggest linkage between the morphological gene and the viability factors on the rearranged segment.

BLOCK INHERITANCE IN GOSSYPIUM   The two tetraploid cotton species, *Gossypium hirsutum* and *G. barbadense,* cross easily to produce vigorous and fertile F_1 hybrids. Chromosome pairing is regular at meiosis in the hybrids. Hybrid breakdown appears in the F_2 generation, which includes many inviable and subvital individuals.

These two species possess a number of homologous genes, such as that for petal spot, but they carry different sets of modifier genes. Harland's classic study of the differentiating modifier complexes in the two cotton species was presented in Chapter 7 (Harland, 1936). A heterozygous allele pair, like S/s for petal spot, segregates in clear-cut Mendelian ratios in intraspecific F_2 progenies, but gives a semicontinuous array of phenotypes in interspecific F_2 and F_3 progenies. If an allele (for example, S) is transferred from one species to the other by repeated backcrossing, however, the heterozygote (S/s) containing the alien allele in the new genetic background will again segregate in Mendelian fashion.

Harland's (1936) interpretation of this breeding behavior was that the species of cotton differ mainly in their modifier genes, each species being homozygous for a different set of alleles of the modifier complex. Harland assumed that the various genes composing each modifier complex were independently assorting and unlinked. Furthermore, the complexes were assumed to be independent of any structural rearrangements. In fact, such rearrangements were not thought to be present, in view of the regular chromosome pairing in the interspecific F_1 hybrid.

Stephens later showed that Harland's hypothesis of sets of unlinked modifier genes did not agree with all of the experimental data. The situation was more complex than Harland had realized. Stephens's alternative hypothesis contains three elements. First, the modifier genes are linked together in blocks. Second, these blocks are coextensive with segmental rearrangements in the two species. And third, the rearranged segments are small, too small to interfere visibly with chromosome pairing at meiosis in the interspecific hybrid, but large enough to bring about linkage effects in the structural heterozygotes. The net result is that the modifier complexes of the species are not broken up by recombination in interspecific progenies but are inherited as linked blocks of genes (Stephens, 1949, 1950).

There is some cytogenetic evidence for small rearrangements between the closely related cotton species. The strongest evidence is the occurrence of preferential pairing in the polyploid derivatives of the interspecific hybrids. These amphiploids form more bivalents and fewer quadrivalents at meiosis than do the autopolyploid derivatives of the parental species themselves (Stephens, 1950). A second line of evidence is the reduction of crossover values between certain linked genes in interspecific backcrosses as compared with intraspecific backcrosses (Stephens, 1950). Finally, translocation chains or rings are found occasionally in the cotton hybrids, *G. hirsutum* $\times$ *G. barbadense* and *G. arboreum* $\times$ *G. herbaceum* (Gerstel, 1953; Gerstel and Sarvella, 1956).

Three main lines of evidence indicate that the modifier genes are grouped in linked blocks: the rate of reversion from continuous to Mendelian segregation in successive interspecific backcross generations; the nonrandom nature of the hybrid breakdown in interspecific F_2s; skewed segregation ratios in interspecific B_1 progenies.

Stephens argues that if the cotton species differ in respect to numerous independently assorting modifier genes with minute individual effects, the interspecific backcrosses should exhibit a slow and gradual change from continuous variation to clear-cut segregations in successive generations. Actually, sharp segregation into Mendelian ratios appears after only two or three generations of interspecific backcrossing. Such a rapid reversion from continuous to noncontinuous variation in the backcrosses cannot be explained on the hypothesis of many freely assorting modifiers. A rapid reversion from blurred to sharp segregation would be expected, however, if the modifiers were grouped in blocks corresponding to chromosome segments which had a different segmental arrangement in the two parental species (Stephens, 1949, 1950).

It was noted earlier that, whereas the F_1 hybrid of *G. hirsutum* $\times$ *G. barbadense* is vigorous and fertile, the F_2 generation shows much depression of vigor. The same is true in the cross *G. arboreum* $\times$ *G. herbaceum*. In both crosses the F_2 types that are viable are those which approach the parental species or the F_1 hybrids. The recombination types in F_2 are inviable. In F_3 and later generations only the types resembling the parents become established.

The selective nature of the hybrid breakdown is consistent with Stephens's hypothesis. Crossing-over and segregation of only partially homologous chromosomes in the interspecific F_1 hybrids lead to the production of gametes or F_2 zygotes that are deficient or unbalanced for particular segments. The most vigorous F_2 segregates would be those carrying the most noncrossover chromosomes and hence those which most closely resemble either parent or the F_1 in phenotype. In later generations, further crossing-over would tend to eliminate the types resembling the F_1 and increase the number of segregates approaching one or the other parental species (Stephens, 1950).

Perhaps the most interesting and suggestive evidence is that of skewed segregation ratios in interspecific B_1 progenies. Stephens utilized marker genes on several chromosomes of *G. hirsutum* and *G. barbadense*. He produced the interspecific F_1 hybrid, which was then heterozygous for these independently assorting markers, backcrossed the hybrid reciprocally to the two parental types, and compared the segregation ratios in the reciprocal backcross progenies (Stephens, 1949). The results are summarized in Table 38.

The table shows that the observed ratio differs in nearly every case, and often significantly, from a theoretical 1 : 1 backcross ratio. In the backcross to *G. hirsutum,* furthermore, there is a consistent excess of *hirsutum* types and deficiency of *barbadense* types for five of the seven marker genes tested. Conversely, in the backcross to *G. barbadense,* there is a substantial deficiency of *hirsutum* types for three of the five genes tested. Furthermore, the ratios for the gene *L* are skewed in opposite directions in the reciprocal backcrosses. The allele of *L* derived originally from the *barbadense* parent is represented in far less than the expected 50% of the progeny in the backcross to *G. hirsutum,* whereas the alternative allele of *L* introduced from the *hirsutum* parent attains a lower frequency in the reciprocal backcross to *G. barbadense.* The skewness of the segregation ratios is reversed when the direction of backcrossing is reversed (Stephens, 1949).

Let us compare the observed and expected results for all seven independent marker genes taken together. On the basis of a 1 : 1 ratio for each gene, and random recombination between genes, we would expect the frequency of individuals in B_1 with any given

TABLE 38

SEGREGATION RATIOS FOR SEVERAL INDEPENDENT MARKER GENES IN BACK-
CROSSES OF *GOSSYPIUM HIRSUTUM* × *G. BARBADENSE* TO THE PARENTAL
SPECIES (STEPHENS, 1949)

| Direction of backcross | Gene | No. of individuals in B_1 carrying | |
		hirsutum allele	barbadense allele
B_1 (F$_1$ × *hirsutum*)	R_1	88	70
	R_2	89	50
	P	78	69
	Y	80	67
	K	47	98
	N	69	77
	L	103	54
B_1 (F$_1$ × *barbadense*)	R_1	40	37
	K	27	42
	N	33	36
	L	19	58
	Cr	26	51

number of marker alleles from one parental species to form a symmetrical distribution. The expected frequency distributions for the
two reciprocal backcrosses are indicated by the heavy dark lines
superimposed on the two histograms in Figure 49. The actual frequency distributions are shown by the shaded areas of the histograms.

The graphs in Figure 49 show clearly that the distribution of
types in B_1 progenies carrying various numbers of *hirsutum* alleles
is not symmetrical. The frequency distribution in the backcross to
G. hirsutum is skewed toward *G. hirsutum*. That in the reciprocal
backcross to *G. barbadense* is skewed in the opposite direction. In
general, the frequency distributions of types in the interspecific B_1
generations are skewed away from the donor parent and toward
the recurrent parent. The skewness of the backcross ratios toward
the recurrent parent is evidence for a selective elimination of the
donor parent genotype (Stephens, 1949).

Some of the marker genes utilized in this experiment affect
vegetative characters and therefore could have pleiotropic effects
on viability in themselves. This explanation can scarcely hold for
other marker genes which govern petal color and pollen color and

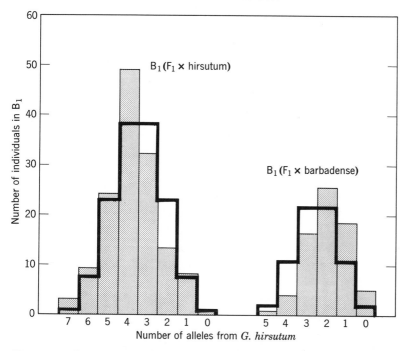

Figure 49. Frequency distribution of plants carrying alleles of *Gossypium hirsutum* for various numbers of independent marker genes in backcross progenies of the hybrid *G. hirsutum* × *G. barbadense*. In each histogram, the expected frequency is outlined by a heavy black line, while the observed frequency is shown as a shaded area. (From Stephens, 1949.)

also exhibit altered ratios. Such genes are likely to be separate from but linked with genes affecting viability.

STAGE OF SELECTIVE ELIMINATION IN GOSSYPIUM Selective elimination of the nonrecurrent parental genotype occurs in interspecific backcrosses in Gossypium. The question therefore arises, Where does this selective elimination take place in the individual life cycle and in the course of generations? There is additional evidence bearing on this question.

In later backcross generations the deviations from the normal 1 : 1 ratio tend to disappear. In successive backcross generations of the F_1 hybrid to *G. hirsutum,* the ratio between individuals carrying the allele derived from the *hirsutum* parent and those carrying the *barbadense* allele was as follows:

hirsutum : *barbadense*

B_1	166 : 98
B_2	27 : 24
B_3	186 : 172
B_4	9 : 10

Only in the B_1 generation is the class containing the allele from the donor parent significantly deficient. This result narrows the process of selective elimination mainly to the reproductive units produced by the F_1 hybrid, either the gametes or the zygotes (Stephens, 1950).

It is difficult to determine whether the selective process involves gamete elimination or zygote elimination. Stephens (1949, 1950) argues for the former and Rick (1963) for the latter. Rick's argument is that the skewness of the segregation would not be expected to be consistently toward the recurrent parent if the cause of the elimination resided in the gametes themselves; and therefore the elimination is probably zygotic. Stephens (1949), however, has presented evidence suggesting gamete elimination.

The backcross of the F_1 hybrid to one parental species can be made by using either the parental species or the hybrid as female parent. When the parental species is used as the female for the backcross, only a negligible proportion of abortive seeds is produced. The proportion of abortive seeds is about the same as that produced normally when the same species is selfed. Therefore, little or no selective elimination is apparent in the B_1 zygotes. On the other hand, when the F_1 hybrid is used as the female parent in making a backcross, around 8–10% abortive seeds are formed, which may be due to the elimination of some female gametes produced by the hybrid (Stephens, 1949).

The skewness of the backcross ratios was compared in reciprocal B_1 progenies. The comparisons were between B_1 (*G. hirsutum* ♀ × F_1) and B_1 (F_1 ♀ × *G. hirsutum*), and again between B_1 (*G. barbadense* ♀ × F_1) and B_1 (F_1 ♀ × *G. barbadense*). The deviations from 1 : 1 ratios for all genes tested were generally the same, independently of the direction in which the backcross was made (Stephens, 1949).

The deficiency of nonrecurrent parental genotypes in B_1 (F_1 ♀ × species ♂) agrees with the evidence of ovule abortion for selective elimination of female gametes in the F_1. The similar deficiency in the reciprocal backcross, B_1 (species $♀ \times F_1 ♂$), indicates a selective elimination of F_1 male gametes in similar proportions (Stephens, 1949).

In conclusion, the selective elimination of the nonrecurrent parental genotype in interspecific backcrosses in cotton appears mainly in the B_1 generation. Some considerations point to zygote elimination and others to gamete elimination in this generation. It would not be surprising if both modes of selective elimination were operating simultaneously. The gametes and zygotes selected against the cotton backcross hybrids are probably those carrying one or more blocks of genes introduced from the donor parental species.

BLOCK INHERITANCE IN ZEA   Insight of a different sort into the problem of *M-V* linkage on rearranged segments is provided by Mangelsdorf's (1958) studies of mutable loci in maize-teosinte hybrid derivatives.

Zea mays and its close relative *Zea mexicana* (teosinte) differ in a number of morphological characters. The two species can be crossed easily to produce a highly fertile hybrid. The chromosomes pair regularly in the hybrid ($2n = 10$ bivalents). Crossing-over between particular marker genes also occurs at a normal rate in the hybrid. Nevertheless, there are grounds for suspecting that the maize and teosinte chromosomes differ by small rearrangements in some parts of their respective genomes (Mangelsdorf, 1958; Ting, 1965). The maize-teosinte cross thus provides favorable material for studying the genetics of character differences between plant species.

Findings relevant to our topic have resulted from the introduction of individual teosinte chromosomes into inbred strains of maize. Mangelsdorf (1958) backcrossed maize-teosinte hybrids to inbred maize for three successive generations, selecting for certain teosinte characters, particularly those of the cob, in each generation. He then self-pollinated selected B_3 plants in order to fix the teosinte chromosome or segment in homozygous condition in at least some of the derivatives. The genotypes of the backcross derivatives thus consisted of inbred maize in most of the 10 chromosome pairs and of substituted segments of chromosomes of teosinte, present either in homozygous or heterozygous condition, in other parts of the genome.

The backcross derivatives carrying introduced teosinte chromosomes or segments turned out to be highly mutable, producing mutant gametes at the rate of 1.8% or higher. Inbred strains of pure corn, by comparison, are very stable. The mutations in introgressive maize affected the endosperm, plant stature, and various other characters, usually deleteriously. Certain endosperm loci were observed to mutate repeatedly. Some of the mutations were stable, others unstable. Unequal crossing-over between small rearrangements is implicated in both types of mutations (Mangelsdorf, 1958).

The stable mutations, which are usually deleterious, arise most frequently in plants that are heterozygous for the chromosome pair bearing the mutant locus. This suggests that the mutation may be a product of rare crossing-over and probably of unequal crossing-over, as in the case of Bar eye in Drosophila. If the chromosomes of maize and teosinte differ slightly in gene order, the plants can breed true so long as crossing-over does not occur in the small nonhomologous region. Rare crossovers within the rearranged segment, however, would lead to some minute deficiencies and duplications, which, on becoming homozygous, would express themselves as stable deleterious mutations (Mangelsdorf, 1958).

The unstable mutants sometimes appear in strains inbred for many generations and hence highly homozygous. Unstable mutations for inhibited development of endosperm show different degrees of inhibition when crossing-over has occurred, as revealed by marker genes. Mangelsdorf suggests that the mutable locus may not be a single gene but rather a block of multiple genes or polygenes introduced from teosinte into maize. The block causes poor endosperm development because the teosinte genes can substitute to some extent but not completely for the maize genes (Mangelsdorf, 1958).

BLOCK INHERITANCE IN GILIA Our final case history involves two species of Gilia (Polemoniaceae), which are known to differ by numerous segmental rearrangements, and their homoploid hybrid derivatives. The parental species are *Gilia modocensis* ($2n = 36$) and *G. malior* ($2n = 36$). The hybrid derivatives with which we are particularly concerned here are designated as Branch II ($2n = 36$) (see Fig. 50).

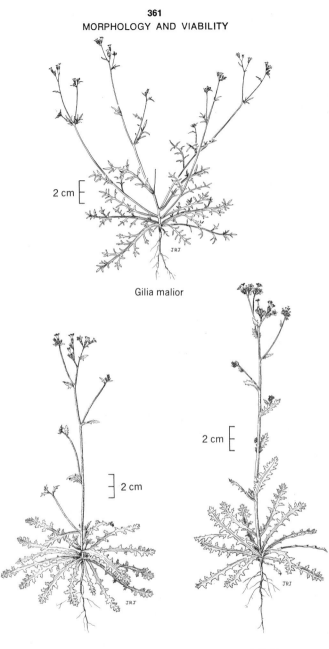

Gilia malior

Gilia modocensis

Branch II, F$_{11}$

Figure 50. Two intersterile species of Gilia and a hybrid derivative which closely resembles and is interfertile with one parental species. (From Grant, 1967.)

The F_1 hybrid of *G. modocensis* × *G. malior* was highly sterile, with a low degree of chromosome pairing due mainly to extensive segmental rearrangements on most or all of the chromosomes. In the F_2 to F_6 generations, inviable seeds and subvital seedlings were preponderant, and sterile or semisterile adult plants were also frequent (Grant, 1966a).

Selection for viability was carried out in a series of inbred lines in the early generations (F_2 to F_6) under environmental conditions favorable to the types approaching the *G. modocensis* parent. The artificial selection was for viability and not for any morphological characteristic as such. Nevertheless, the inbred lines which became wholly vigorous in response to this selection resembled the *G. modocensis* parent in most of their vegetative and floral characters, and were intermediate in others, but did not resemble *G. malior* in any character (Fig. 50). Some morphological traits of *G. modocensis,* therefore, are apparently linked with viability genes in that species.

Selection for fertility and meiotic regularity was also carried out in the viable fraction of the early segregating generations, and was successful in bringing about the recovery of full fertility and regular bivalent pairing in three independent sets of lines, referred to as branches. The genomic relationships of these branches with the parental species were tested by backcrosses. Branch II, which resembled *G. modocensis* morphologically, proved to be segmentally homologous and interfertile with *G. modocensis,* and chromosomally intersterile with *G. malior.* A second fertile branch was closely related genomically to *G. modocensis* (Grant, 1966a).

It is highly unlikely that a set of independent rearrangements belonging to one parental species could be broken up by hybridization and then reassembled in the fertile hybrid derivatives by chance alone in one or two out of three trials. The chief anti-chance factor in this experiment was selection for viability. The selection for viability in an environment favorable to *modocensis* types is correlated with the restoration of a *modocensis* set of segmental rearrangements. This correlation indicates that some of the viability genes and morphological genes of *G. modocensis* are located on the rearranged segments of that species (Grant, 1966a, 1966b).

To put it in another way, the results of the Gilia experiment are inexplicable if the rearranged segments differentiating the parental species are neutral in their morphological and viability effects. The observed results can readily be explained, however, on the hypothesis that the rearranged segments carry linked blocks of *M-V* genes.

CONCLUSIONS In higher plants, genes determining morphological characters are generally linked with genes controlling growth and vigor. Related races or species differ allelically with respect to the linked morphological and viability genes, as indicated by the evidence of hybridization experiments in various plant groups.

The *M-V* linkage system is found on homologous chromosome segments in related species or races of Lycopersicon, Mimulus, Phaseolus, Potentilla, and probably Rubus.

The evidence regarding the genic contents of rearranged segments in plant groups with structurally differentiated chromosomes is necessarily more indirect and less certain, due to the complicating factor of chromsomal hybrid sterility. Nevertheless, there are indications that linked *M-V* genes occur also on rearranged segments in chromosomally differentiated plant groups.

The evidence is derived from the analysis of interspecific hybridization experiments in Clarkia, Gilia, Gossypium, and Zea. The experiments differ greatly in detail but point to similar general conclusions. In Clarkia a simple morphological gene is located on a simple rearrangement, with the additional possibility of linked viability genes. In Zea, morphological genes and apparently also viability factors are located on probable rearranged segments. In Gossypium known blocks of morphological and viability genes occur on what appear to be rearranged segments. In Gilia, finally, there is the complementary case of known rearranged segments carrying linked morphological and viability factors.

As far as the available evidence goes, there does not appear to be any essential difference between the *M-V* linkage systems on homologous segments and those on rearranged segments. The situation in plant groups with chromosomal rearrangements requires much further study.

CHAPTER TWENTY

LINKAGE OF
FUNCTIONAL GENE SYSTEMS

GENE CLUSTERING

PETAL SPOT IN GOSSYPIUM

LINKED SERIAL GENE SYSTEMS

THE ROLE OF DUPLICATIONS

THE CASE OF *CLARKIA LINGULATA*

ENZYME LOCI IN TRITICUM

GENE CLUSTERING   The linkage maps of genetically well studied multicellular organisms such as Drosophila and corn reveal, at first glance, a spatial distribution of genes that is apparently random as regards gene function. Neighboring genes on the same chromosome often control different characters. Separate genes controlling a single character are often located on different chromosomes or chromosome arms.

Undoubtedly there is some dispersion of functionally related genes throughout the chromosome set. This, however, is not the whole story. Many cases have been found in which different genes controlling the same or similar characters are clustered together. The spatial grouping in such cases is probably not a matter of chance.

In *Drosophila melanogaster,* all of the known mutations for extra wing venation occur in chromosome II, and all of the homeotic mutations, which transform wings or antennae into legs and other body parts, occur in a small region of chromosome III (Dubinin, 1948). In the mouse, five genes governing the characteristics of the tail lie close together in a segment 8 crossover units long on one chromosome (Dunn and Caspari, 1945).

Similar cases have been found in higher plants. The chromosomal distribution of 15 seedling mutations in *Lycopersicon esculentum* has both random and nonrandom aspects. Nine of these genes are distributed among 7 of the 12 chromosomes, illustrating the random aspect. But 6 seedling mutations are located on chromosome 2. This represents a high concentration of seedling genes on one chromosome in the complement of the tomato (Rick, 1959).

Several examples of compound loci are known in *Zea mays.* The locus consists of two adjacent but separable and recombinable genes or subgenes controlling the same general character. This is the case in the *A* and *R* loci for pigmentation, the tunicate (*Tu*) locus for type of ear, and the *Rpl* locus for rust resistance (Laughnan, 1948, 1952; Rhoades, 1954; Mangelsdorf and Galinat, 1964; Hooker and Saxena, 1971). In these cases we have close linkage between functionally related genic units.

An interesting situation in Gossypium involving close linkage between two genes governing petal spots is described in the next section. Many cases of gene clustering are known in bacteria and other microorganisms (see Chapter 4).

A related and possibly complementary phenomenon is the occurrence of so-called silent regions in the chromosomes. Large sections of the linkage maps of tomato and maize lack any known mutant genes; such sections are genetically "silent." In both maize and tomato the silent regions occur on different chromosomes of the complement and are most common in the distal or terminal regions. The corollary is that the known Mendelian genes tend to be concentrated in the proximal or centromeric regions. The significance of this aspect of nonrandom gene distribution is still not clear (Rick, 1971b).

PETAL SPOT IN GOSSYPIUM   The yellow or white petals of the asiatic diploid cotton species, *Gossypium arboreum* and *G. herbaceum,* may have a red spot at the base or may be spotless. The difference between spotted and spotless is controlled by at least two adjacent genes, G and S ($S = R_2$). The spotted genotype is *GS* and the usual spotless genotype is *gs* (Fig. 51) (Stephens, 1948).

The cross of spotted × spotless yields an F_1 of the constitution *GS/gs,* which exhibits the dominant condition of petal spots, and segregates in F_2 in a 3 : 1 ratio of spotted to spotless. The spotless F_2 plants mostly breed true to type. The spotless condition is thus usually inherited as though it was determined by a simple Mendelian recessive allele (Stephens, 1948).

Occasionally, however, the intercrossing of two spotless plants yields a spotted progeny. The exceptional spotted progeny of spotless parents is the result of rare crossing-over between *G* and *S*. Crossing-over between *G* and *S* in the double heterozygote *GS/gs* leads to the formation of the recombination types *Gs/Gs* and *gS/gS*, which are both spotless (see Fig. 51). Intercrossing of two such spotless types gives the complementary product, *Gs/gS*, which has petal spots. These exceptional types show that the presence or absence of petal spots is not controlled by a single gene, contrary to the indication of the 3 : 1 ratio in most segregating progenies, but by two closely linked genes (Stephens, 1948).

The genes *G* and *S* are separable but closely linked. They both control petal-spot formation. The two homozygous recessive genotypes are both spotless, whereas the double heterozygote in the trans arrangement (*Gs/gS*) is spotted (Fig. 51). Therefore *G* and *S* are separate but linked complementary genes (Stephens, 1948).

The positional and functional relations of *G* and *S* suggested the plausible hypothesis that these two genes control successive biochemical steps in pigment formation on the petal base. *G* and *S* could be compared with two steps on an assembly line (Stephens, 1948). It was not possible to confirm this hypothesis from a biochemical standpoint. Apparently the formation of red pigment in the petal bases is more complex than a two-step process. There is evidence that a third linked gene (*M*) is involved peripherally in this process, the probable gene order being *M-G-S* (Yu and Chang, 1948; Stephens, 1951a).

LINKED SERIAL GENE SYSTEMS   The phenomenon of pseudo-allelism was first discovered and worked out in Drosophila, corn, and fungi in the period 1945–1955, and further studies in multicellular organisms have of course been made since then (see Chapter 4). The related phenomenon of gene clustering was also de-

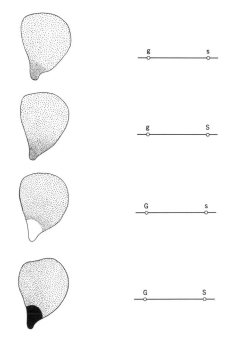

Figure 51. Types of flower petals in asiatic cotton (*Gossypium arboreum* and *G. herbaceum*). Phenotypes (*left*) and corresponding genotypes (*right*). (Rearranged from Stephens, 1948.)

tected at approximately the same period in Mus, Drosophila, Gossypium, and Lycopersicon. Subsequently, a similar clustering of functionally related genic units was found in bacteria and viruses.

In representatives of several kingdoms of organisms, therefore, we find evidence of close linkage between subgenes or genes with related biochemical and developmental functions and often with complementary effects. This situation poses difficult problems for the definition of the gene. We discussed these problems at some length in Chapter 4. Our interest here is not in the gene concept, however, but in the significance of the linkage relations themselves.

In some of the cases analyzed in Drosophila, corn, and cotton, the linear order of the contiguous genes seems to correspond to a developmental sequence. Several workers have therefore compared a section of a chromosome with an assembly line. The suggestion was put forward that functionally related genes are often arranged on a chromosome in assembly-line fashion to control the successive steps in a biochemical reaction chain (Stephens, 1948, 1951a; Lewis, 1951, 1955; Demerec, Blomstrand and Demerec, 1955; Demerec, 1956; Beadle, 1957a).

With this hypothesis in mind, I later proposed to call the chromosome section containing the functionally related genes a linked serial gene system (Grant, 1964, ch. 6). The more widely known term, gene cluster, carries the same connotation and refers to the same genetic unit.

The relatively simple organization of bacteria and viruses makes these organisms especially suitable subjects for analysis of genetic fine structure. The mapping of gene clusters has been carried out in the greatest detail in Salmonella, bacteriophage, and other prokaryotic forms. For a review of these studies see Hayes (1968, ch. 8). Some aspects were also introduced here in Chapter 4.

The now classic case of tryptophan synthesis in *Salmonella typhimurium* is relevant to our present discussion. The synthesis of tryptophan takes place in the following sequence of steps: precursor → anthranilic acid → indoleglycerol phosphate → indole → tryptophan. Let the four steps or arrows be designated A, B, C, and D. A group of 10 different tryptophan-deficient mutants in Salmonella was found to fall into four classes—A, B, C, and D—corresponding to the four steps in tryptophan synthesis. Thus the mutant type *Try-8* belongs to the A class and blocks step A; *Try-2* and *Try-4* are in the B class and block step B; and so on (Demerec, 1956; Hayes, 1968).

The position of the mutations in the gene-string was then mapped. It turns out that the classes of tryptophan mutations occur in a sequence on the gene-string which corresponds to the sequence of biochemical steps. The linear order of genic sites in the tryptophan region is: *Try-A, Try-B, Try-C, Try-D* (Demerec, 1956; Hayes, 1968).

The section of the gene-string in Salmonella which is concerned with tryptophan synthesis is thus organized in the manner of an assembly line for production. The assembly-line hypothesis, which is probably correct as applied to multicellular organisms, is fully verified in this and other similar cases in bacteria. I feel that we can extrapolate from bacteria to eukaryotic organisms in this instance without necessarily assuming homologies in other features of their genetic systems.

The widespread occurrence of linked serial gene systems provides an explanation for a common type of position effect. Position effects are phenotypic changes, and usually deleterious changes, resulting not from gene mutations but from changes in the spatial relations of the genes (for reviews see Goldschmidt, 1938; Lewis, 1950; see also the brief discussion in Chapter 4). Some chromosomal rearrangements, particularly in Drosophila and other animals, bring about position effects. Deleterious position effects would be expected to result from a fraction of segmental rearrangements if the genes in some segments were lined up like the parts of an assembly line; for in such cases the chromosome break would disrupt a linked serial gene system and hence block the normal production.

THE ROLE OF DUPLICATIONS   In the course of evolution toward greater complexity, organisms must frequently require the addition of new functions. Gene mutations alter an original gene function but do not bring about a new and additional function. A gene in a duplicated segment, however, is relieved of the task of performing its original function and is thus free to diverge and acquire a new function. Duplications could enable an organism to gain new gene-controlled functions without giving up preexisting ones.

Duplications are therefore looked upon by various geneticists as a source of new genes in evolution. Tandem duplications, in particular, have been regarded as the possible starting point of gene clusters or linked serial gene systems, in which a linear series of related genes performs related sequential functions.

These ideas have had advocates in three disjunct periods of genetics. The general idea of duplications as sources of new genetic material was first put forward by Bridges (1935, and earlier). The specific idea of an association between tandem duplications and gene clusters was explicitly suggested by various workers in the early period of pseudoallelism studies (Stephens, 1948, 1951a, 1955; Komai, 1950, Lewis, 1951; Laughnan, 1952, 1955; Demerec, Blomstrand and Demerec, 1955; and others). This plausible suggestion was widely accepted. It was stated in genetics books which reached a wider audience than did the original research papers (e.g., Swanson, 1957, ch. 15; Sinnott, Dunn and Dobzhansky, 1958, ch. 28; Grant, 1964, ch. 8). Then, after another lapse of time, the genetical and evolutionary potential of duplications was rediscovered by Ohno (1970), who does not refer to most of the relevant earlier studies. A recent, balanced view of the whole problem of duplications in evolution, with a fairly complete bibliography, is given by Mayo (1970).

There is some cytological evidence in *Drosophila melanogaster* for an association of gene clusters with segmental duplications. The two adjacent genic units, star and asteroid, lie in a doublet or double band of the salivary gland chromosome II (Lewis, 1945). The subgenes W and W^a for eye color in Drosophila also appear to be located in a doublet of a salivary chromosome (Lewis, 1952). Two components of the gene forked, governing a body bristle character, occur in or near a doublet in the X chromosome (Green, 1955). These doublets are believed to be duplications.

THE CASE OF *CLARKIA LINGULATA* Duplicate genetic material can become established in the genotype by mechanisms other than tandem duplication. One such alternative is polyploidy. Another is the fixation of extra chromosomes or chromosome arms. Either mechanism adds extra genetic material on a large scale. The additional genes, moreover, are not linked with the original ancestral genes.

The effects of an extra chromosome are illustrated by a case in Clarkia, described in this section. An example involving polyploid duplication is presented in the next section.

Clarkia biloba and *C. lingulata* (Onagraceae) are a pair of genomically closely related species in the Sierra Nevada of California. *Clarkia biloba* has a wide geographical range in the central and northern Sierras, where it occurs in relatively moist environments. Its chromosome number ($2n = 16$) is close to the original basic number ($x = 7$) of the genus. *Clarkia lingulata* ($2n = 18$) is endemic in a small area on the arid southern periphery of the range of *C. biloba;* it represents a derived condition in both ecology and chromosome number. The whole pattern of relationships suggests strongly that *C. lingulata* has been derived rapidly as an offshoot from *C. biloba* (Lewis and Roberts, 1956).

The main morphological difference between the two species is the lobing of the flower petals. The petals are cleft at the tip or heart-shaped in *C. biloba* and entire or tongue-shaped in *C. lingulata.* In addition the two species differ in their ecological preferences, *C. biloba* being more mesic and *C. lingulata* more xeric (Lewis and Roberts, 1956).

At meiosis in the interspecific F_1 hybrid, the chromosomes pair to form four bivalents, one ring of four, and a chain of five. An inversion bridge appears occasionally in one of the bivalent-forming pairs. The chain of five includes the ninth chromosome of *C. lingulata* (Lewis and Roberts, 1956).

The chain of five contains the 1–2 and 3–4 chromosomes of both parental species along with the ninth chromosome of *C. lingulata.* The pairing relations of this ninth chromosome in the chain indicate that it has the end arrangement 1–4. The genomic constitution of the two species with respect to the chromosomes participating in the chain is therefore (Lewis and Roberts, 1956):

C. biloba	1–2	3–4	
C. lingulata	1–2	3–4	1–4

The genomes of the two species are thus predominantly homologous. Six chromosomes are structurally homologous with the exception of one paracentric inversion on one of the bivalent-forming pairs. Two other chromosomes in each complement differ by a reciprocal translocation. A second translocation is incorporated into the ninth chromosome of *C. lingulata.* The derivation of the nine-chromosome complement of *C. lingulata* from the ancestral eight-chromosome complement of *C. biloba* involved the fixation of one reciprocal translocation in homozygous condition and the establishment of a second translocation as a constant tetrasomic or aneuploid type (Lewis and Roberts, 1956).

The ninth chromosome in *C. lingulata* duplicates the number 1 and 4 arms of the ancestral standard complement. It thus provides a source of genetic material for the divergence of *C. lingulata* from the ancestral species in morphological and physiological characters. It is very significant in this connection that the morphological character difference of petal shape is determined by genes on the ninth chromosome. It is possible that the extra genetic material on this chromosome is also involved in the adaptation to a more xeric environment (Lewis, 1954, Lewis and Roberts, 1956).

ENZYME LOCI IN TRITICUM The possible effects of polyploidy on gene divergence have been investigated in *Triticum aestivum* (wheat), which is hexaploid with the genomic constitution AABBDD. Twelve enzyme systems in the leaves, determined by enzyme loci distributed among the different genomes, were identified by means of isozyme bands of leaf material (Brewer, Sing and Sears, 1969).

Specific enzyme loci in the A, B, and D genomes were compared by using nullisomic-tetrasomic types. A nullisomic type lacks a given chromosome in one genome and carries the corresponding homeologous chromosome in another genome in tetrasomic condition, or double dose. For example, a particular nullisomic-tetrasomic type could have four doses of chromosome 2 in the A genome and no chromosome 2 of the B genome. There are 6 possible nullisomic-tetrasomic types for each chromosome type in hexaploid wheat, and, with a basic number of $x = 7$, a total of 42 possible types for hexaploid wheat. Nearly all of the possible nullisomic-tetrasomic types (38) were available for study of the leaf enzyme systems (Brewer, Sing and Sears, 1969).

If the duplicate enzyme loci in different genomes of hexaploid wheat have diverged since the origin of the species, the divergence should be revealed by differences between related nullisomic-tetrasomic types in isozyme banding patterns for a given enzyme system.

Actually, for 11 of the 12 enzyme systems, no variation in enzymes was detected within a series of related nullisomic-tetrasomic types. The corresponding homeologous chromosomes in the different genomes do not give different isozyme banding patterns for these 11 enzymes. Consequently the homologous genes for a given enzyme on homeologous chromosomes belonging to different genomes are very similar if not identical in function (Brewer, Sing and Sears, 1969).

The twelfth enzyme system studied is alkaline phosphatase. This enzyme is controlled by genes on chromosome 4 in the different genomes. In this case the chromosome 4 of the B genome and the chromosome 4 of the D genome do produce different isozyme patterns. Different forms of the gene for alkaline phosphatase occur at the homologous loci in the B and D genomes (Brewer, Sing and Sears, 1969).

The alkaline phosphatase enzyme affects the plant's mineral nutrition for soil phosphates. The authors suggest that a diversity of gene types for this enzyme may be advantageous in permitting the plant to grow in a variety of soils with different amounts of phosphate. Selection has favored gene diversity in this case (Brewer, Sing and Sears, 1969).

With regard to the other 11 enzyme systems, however, the polyploid condition in hexaploid wheat has not, contrary to expectations, led to divergence between homologous genes. There is even evidence for some convergence between these genes during the history of hexaploid wheat. The reasons why gene similarity has been favored over gene divergence in these cases are not clear (Brewer, Sing and Sears, 1969).

CHAPTER TWENTY-ONE

LINKAGE OF
ADAPTIVE GENE COMBINATIONS

INTRODUCTION
SUPERGENES
INVERSION POLYMORPHISM
CHROMOSOMAL VARIATION IN TRILLIUM
ADAPTIVE GENE COMBINATIONS IN *AVENA BARBATA*

INTRODUCTION Most of the adaptations of higher plants are based on complex characters or character combinations. The complex adaptations are determined by combinations of genes which are more or less unrelated at the level of elementary gene action. Thus a particular floral mechanism may be the result of the combined action of separate and functionally different gene systems governing time of flowering, corolla length, stamen position, style length, corolla pigmentation, and other features.

The adaptive value of the complex character or character combination depends on the joint action of all the component genes and gene systems. Consequently the preservation of the underlying gene combination is selectively advantageous.

The main threat to the integrity of any adaptively valuable allele combination (e.g., $A_1A_1B_1B_1C_1C_1$) is sexual reproduction in a population containing alternative alleles (A_2, B_2, C_2) which do not contribute to the development of the desirable character combination. The heterozygote produced by outcrossing in such a polymorphic population will segregate a fraction of ill-adapted progeny. This fraction rises with the number of genes in the gene combination.

Any factor which reduces the disintegration of the adaptive allele combination due to segregation and recombination in the polymorphic population will itself have adaptive value. Linkage is such a factor. Certain chromosomal rearrangements are vehicles of linkage, as noted in Chapter 17. Therefore the adaptive value of the phenotypic character combination may be transferred ultimately to the inversion, translocation, or other rearrangement which holds the allele combination together.

SUPERGENES   A supergene is a group of contiguous genes linked as a block by heterozygous inversions or translocations, or localized chiasmata. The component genes are not necessarily related functionally, but they usually form an adaptive gene combination, and they segregate as a single unit in inheritance. Good general discussions of supergenes are given by Darlington and Mather (1949), Darlington (1956b), Dobzhansky (1959), and Ford (1964, 1966).

An interesting and well-known example of supergenes is found in the polymorphic mimetic butterfly, *Papilio dardanus,* in Africa. A series of alternative forms exists in the female of *P. dardanus;* these forms differ in the color pattern and shape of the wings. The characters are genetically complex but exhibit simple Mendelian inheritance. This is due to the fact that the genes involved are grouped in two supergenes which switch wing development into one polymorphic type or another (Clarke and Sheppard, 1960).

A very interesting supergene in higher plants is that governing flower type in heterostylous species of Primula. As was discussed in Chapter 14, populations of many Primula species consist of two flower forms and mating types known as pins and thrums. Pins and thrums differ in a complex of morphological and physiological characters: style length, anther height, pollen-grain size, stigma papilla size, and self-compatibility reaction. The legitimate cross-pollinations are pin × thrum and the reciprocal. The progeny of such legitimate crosses segregate into pins and thrums in a 1 : 1 ratio. The contrasting states of the character combination usually segregate as intact units, like sex differences in a dioecious species. For this reason, the pin and thrum types in Primula are judged to be due to different allelic forms of a supergene S.

Rare crossing-over within the S supergene gives rise to recombination types which reveal the structure of this gene block. The gene order within the S supergene is as follows (Ernst, 1933; Lewis, 1954a; Darlington, 1971a):

G	style length
S_p	stigma papilla length
I	self-incompatibility
P	pollen grain size
A	anther height

Pins and thrums, then, are determined by different allelic forms of the S supergene. Thrums are the heterozygous type (Ss). Rare illegitimate thrum × thrum crosses yield segregating progeny in a ratio of 3 thrums to 1 pin, indicating that the genotype of thrum is Ss. Pins are the homozygous recessive type (ss). They breed true to type in occasional illegitimate pin × pin crosses (Ernst, 1933; Lewis, 1954a; Darlington, 1971a).

Another example of a supergene in higher plants is the speltoid mutants in cultivated hexaploid wheat, *Triticum aestivum*. The speltoid mutants represent a change in a whole complex of characters in the wheat spike. The change is in effect a reversion to the spike characteristics of more primitive forms of wheat.

The wheat spike consists of a main axis (rachis) bearing the spikelets, each of which contains at maturity several grains enclosed by two bracts (glumes). In the cultivated wheats, *Triticum aestivum, T. carthlicum,* and others, the rachis is tough and the glumes are loose, so that when threshed the grains fall out freely from the head. Such types are called naked wheats. In the more primitive species and varieties of wheat, both the wild and the primitive cultivated forms, the rachis is brittle and the glumes persistent. When these wheat varieties are threshed, the rachis breaks and the grains remain enclosed within the glumes. These types are known as spelt wheats, after *T. spelta* which normally has these characteristics.

The speltoid mutational change from the normal naked type of *T. aestivum* affects simultaneously several features of the spike, including strength of rachis and nature of glumes. The characters involved mutate together, and there is no recombination between them. Furthermore, the mutation always occurs from normal naked wheat to the speltoid type, and never in the opposite direction (Frankel and Munday, 1962).

The genes involved in the speltoid mutations are clustered in a short region on the long arm of chromosome IX in hexaploid wheat. This segment is designated Q. When the Q segment is present, the phenotype is that of normal naked wheat; when Q is absent, the speltoid characters result. The speltoid mutants thus result from deficiency for the Q segment. The deletion has been observed cytologically in chromosome IX (Sears, 1944; Mackey, 1954).

The ultimate origin of the Q supergene and its phylogenetic source in *Triticum aestivum* are still matters of debate among wheat geneticists (see Frankel and Munday, 1962; Johnson, 1972). Sears pointed out that Q is present in another, more ancient, tetraploid species, *T. carthlicum,* in the Caucasus Mountains. Therefore it is possible that *T. aestivum* received Q by hybridization from *T. carthlicum* in the past (Sears, 1959). Its ultimate origin within *T. carthlicum* or some extinct ancestor of *T. carthlicum* remains obscure.

The success of modern cultivated wheats as grain cereals depends to a large extent on the complex of spike characters determining the response of the heads to threshing. The evolution of the wheat plant under domestication has involved, among other changes, the replacement of the spelt type of spike by the naked type. The Q supergene has played a central role in this change in the wheat spike and is therefore adaptively valuable for wheat under the conditions of domestication (Frankel and Munday, 1962).

INVERSION POLYMORPHISM Polymorphism for inversions is well known in Drosophila and other Diptera. The inversion types carry different genes affecting the adaptedness of the flies. These genes are held together in alternative linked blocks in the inversion heterozygotes, by the mechanism explained in Chapter 17, and the inversions thus behave as supergenes. Heterozygotes and homozygotes for the different inversions or supergenes have characteristic adaptive values in different environments. The inversion polymorphism is then a means by which the fly population maximizes its adaptive fitness in an environment which varies seasonally or microgeographically, or both (for reviews see Da Cunha, 1955, 1960; Dobzhansky, 1970).

It was only natural to look for parallel phenomena in plants. Early studies in a number of plant species did seem to indicate the presence of inversions. Chromosome bridge configurations at meiosis, deemed to be indicative of inversion heterozygosity, were found in many individuals in natural populations of *Paris quadrifolia* (Liliaceae), Clematis spp. (Ranunculaceae), Polygonatum spp. (Liliaceae), and other plant groups (Geitler, 1937, 1938; Meurman and Therman, 1939; Suomalainen, 1947). More recent investigations have made it necessary to revise the older conclusions.

The problem lies in the realm of cytological observation and inference. Inversion heterozygosity in Drosophila can be readily seen and reliably identified in paired loop configurations in giant salivary-gland chromosomes. In plants the inversion loops can be seen, but only with considerable difficulty, in the pachytene stage of meiosis. This stage is usually difficult to work with in plants on the extensive scale required for population analysis. It is much more feasible to record chromosome bridges and fragments at anaphase. The early studies of population cytology in Paris, Polygonatum, Clematis, and other groups did make use of chromosome bridges and fragments as indicators of inversion heterozygosity. Bridge-fragment configurations certainly result from heterozygosity for paracentric inversions (see Chapter 17). But bridges also result from chromosome breakage and refusion, as demonstrated in Zea, Bromus, and other plant groups, and this source of bridges seems to be fairly common (Haga, 1953; J. L. Walters, 1956; M. Walters, 1957; Lewis and John, 1966). The interpretation of inversions in Paris and other plants is therefore uncertain.

A key study of this problem was made by Newman (1966, 1967). Newman examined both pachytene and anaphase stages in

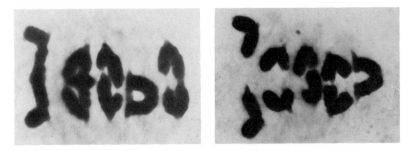

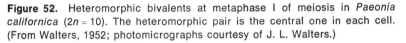

Figure 52. Heteromorphic bivalents at metaphase I of meiosis in *Paeonia californica* (2n = 10). The heteromorphic pair is the central one in each cell. (From Walters, 1952; photomicrographs courtesy of J. L. Walters.)

individual plants of *Podophyllum peltatum* (2n = 12; Berberidaceae). He found anaphase bridges in individuals that did not exhibit inversion loops at the pachytene stage (Newman, 1966). There was, to be sure, some valid evidence for inversions in *Podophyllum peltatum,* but these were not as frequent as would be inferred from the bridge configurations alone (Newman, 1967). Newman reviewed the older literature in the light of these findings and concluded that the frequency of inversions in plant populations has been overestimated (Newman, 1966).

The best valid example of inversion polymorphism in plants at present appears to be in *Paeonia californica* (2n = 10). Translocation polymorphism in this species is discussed in the next chapter. Anaphase bridges have long been known in the California peonies but are disregarded here. More convincing evidence of inversions is the occurrence of heteromorphic bivalents in which one centromere is shifted or displaced relative to the other (Fig. 52). This displacement is probably, though not certainly, due to heterozygosity for a pericentric inversion (Walters, 1952).

Two types of heteromorphic bivalents were found in *Paeonia californica.* They probably correspond to different pericentric inversions. One of these is rare (Fig. 52, left), the other common (Fig. 52, right). Plants heterozygous for the common type of pericentric inversion were found in 10 or more populations throughout the northern range of the species. The structural heterozygote is common, whereas the corresponding structural homozygotes are very rare in nature. The high frequency and wide geographical distribution of the inversion heterozygote suggest that it has a selective advantage in the natural populations (Walters, 1952).

CHROMOSOMAL VARIATION IN TRILLIUM Pretreatment of chromosomes by means of cold relaxes the coiling in some segments but not in others. Staining of such pretreated chromosomes then reveals a marked linear differentiation between coiled and uncoiled regions which is not brought out by ordinary cytological methods. (The terms nucleic acid starvation, differential segment, and allocycly are variously applied by different authors to the artificially uncoiled condition.)

Fine details of structure can be made out in the uncoiled chromosome regions, thus furnishing a tool for the analysis of chromosomal variation in plant populations. By this method it is sometimes possible to detect structural differences between homologous chromosomes, in individual plants and in populations, which could not be seen otherwise. Polymorphism for fine structural features has been revealed by cold treatment in species of Trillium, Paris, Fritillaria, Tulbaghia (Liliaceae), and Cestrum (Solanaceae) (Haga and Kurabayashi, 1954; Kurabayashi, 1958; Darlington and Shaw, 1959; Dyer, 1963).

Chromosomal variation of this sort has been extensively studied in *Trillium kamtschaticum* (2n = 10) in Japan. The five chromosomes in the haploid complement are designated A, B, C, D, E. A series of structurally different types is found for each chromosome in the basic complement, and these types are designated A1, A2, A3, and so on. Nine types of chromosome C are illustrated in Figure 53. Differences are apparent among them in the length of the uncoiled region, the number and position of beadlike knobs, and the position of the centromere (Haga and Kurabayashi, 1954).

These structural differences are apparently due to translocations, inversions, and deletions. There is direct cytological evidence for translocations in *T. kamtschaticum,* and indirect cytological evidence for inversions and deletions (Haga and Kurabayashi, 1954).

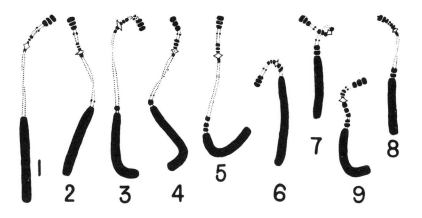

Figure 53. Nine structurally different types of chromosome C in *Trillium kamtschaticum.* Somatic chromosomes pretreated with cold and stained with Feulgen stain. Centromere shown as diamond; standard coiled region is black; uncoiled regions indicated by dotted lines; satellites and beadlike knobs in the uncoiled regions shown in black. (From Haga and Kurabayashi, 1954.)

The number of variant forms of each chromosome in *Trillium kamtschaticum* is considerable. As of 1958 the known number of types for each chromosome was as follows (Kurabayashi, 1958):

Chromosome	No. of Types
A	71
B	19
C	15
D	16
E	18

Populations of *T. kamtschaticum* at scattered localities throughout northern Japan were sampled and analyzed cytologically. Some chromosome types exhibited geographical variation. Some exhibited polymorphic variation. In general, chromosomal polymorphism was found in large Trillium populations, while small isolated populations were often monomorphic or nearly so (Haga and Kurabayashi, 1954).

For example, consider two large populations (Samani and Sizunai) on the coast of Hokkaido Island. Chromosome A was represented in the two populations by the following polymorphic types in samples of 50 plants:

	Types
Samani	1, 2, 3, 4, 5, 6, 7, 8, 9, 10, 13, 14, 15, 16, 17, 18, 19
Sizunai	1, 2, 3, 4, 5, 8, 9, 12, 14, 16, 18, 19

Similar variation occurs in other chromosomes of the complement in these populations (Haga and Kurabayashi, 1954). Conversely, a small population at Kusakai on Honshu Island which was also well sampled turned out to be monomorphic for chromosomes A to E.

Most of the plants in the large populations and in some of the small ones are structural heterozygotes for one or more chromosomes in the basic complement. That is, they have the chromosomal constitution A1/A2 B3/B3 . . . , or A1/A2 B2/B4 . . . , or are heterozygous in all five chromosomes. Structural homozygotes are uncommon, especially in the large populations (Haga and Kurabayashi, 1954).

Structural homozygotes are more common in some of the small Trillium populations. The small Kusakai population consists entirely of structural homozygotes with the chromosomal constitution A13/A13 B12/B12 C6/C6 D3/D3 E7/E7. Different small populations also tend to show fixation or high frequency of different chromosome types, suggesting that their composition has been determined by a combination of natural selection and genetic drift (Haga and Kurabayashi, 1954; Kurabayashi, 1958).

A comparable pattern of chromosomal variation occurs in *Trillium ovatum* ($2n = 10$) in western North America. Here again the chromosomes of the basic complement are represented by variant forms, as revealed by cold treatment (A1 to A9, and so on). There is both geographical and polymorphic variation in these chromosome types. A series of populations of *T. ovatum* on the Pacific coast are monomorphic for chromosome A and either monomorphic or dimorphic for chromosome E. Populations in the Rocky Mountains, by contrast, are highly polymorphic for both the A and

E chromosomes (I. Fukuda, unpublished data; Stebbins, 1971, pp. 65–66).

Chromosomal variation of this sort appears to be general in the genus Trillium. Darlington and Shaw surveyed 10 named species in North America, all diploid ($2n = 10$) with basic A to E chromosomes. They found morphological variation between homologs of each basic chromosome. Chromosomal heterozygotes were common in *T. grandiflora, T. sessile,* and other species (Darlington and Shaw, 1959).

Trillium, like Paeonia, is a good subject for cytological work but a poor one for genetic work. There is, to be sure, some exophenotypic variation between and within populations of *T. kamtschaticum* which correlates with the chromosomal constitution of the populations. But the genic contents of these chromosomes are otherwise unknown.

The cytological evidence alone, however, points to the conclusion that the different chromosome types play a role in adaptation. We have noted that the variant forms of the homologous chromosomes apparently differ by structural rearrangements, which would promote gene linkages in heterozygous condition. The high frequency of chromosomal heterozygotes in some Trillium populations, and the less than expected frequency of the corresponding homozygous segregates, suggest strongly that the heterozygotes possess an adaptive superiority (Haga and Kurabayashi, 1954).

ADAPTIVE GENE COMBINATIONS IN *AVENA BARBATA* Populations of the predominantly self-fertilizing wild oat, *Avena barbata,* in California are monomorphic or polymorphic for six enzyme loci which can be assayed by electrophoretic methods. Four of these loci govern esterase (E_1, E_4, E_9, E_{10}), one governs phosphatase (P_5), and one anodal peroxidase (APX_5). Three of the loci are linked (E_{10}-APX_5-P_5). The APX_5-P_5 linkage is tight, with 4% recombination, but the linkage of E_{10} with these loci is loose (Marshall and Allard, 1969).

Different alleles of the enzyme genes, both the linked and unlinked ones, reach high frequencies or complete fixation in different environments on a spectrum from cool moist to warm arid. The correlation between the environmental conditions and the predominant types of alleles holds up on a geographical scale, between different climatic regions of California, and also on a microgeographical scale, between different facies of the same heterogeneous habitat (Clegg and Allard, 1972; Hamrick and Allard, 1972).

Monomorphism for one set of alleles tends to develop in the warm arid regions. Polymorphism for the same alleles is common in the cool mesic regions. But polymorphic populations in such climatic regions sometimes show a relatively high frequency of the "arid alleles" in the more arid parts of the habitat and a high frequency of the opposite alleles in the more mesic subdivisions. These geographical and microgeographical patterns in the distribution of the allele frequencies indicate that they are controlled mainly by environmental selection (Allard, Babbel, Clegg and Kahler, 1972; Clegg and Allard, 1972; Hamrick and Allard, 1972).

A population at Calistoga in northern California contains two alleles of each of the following five enzyme genes: E_4, E_9, E_{10}, P_5, APX_5. The two alleles of each gene are designated number 1 and number 2 according to the electrophoretic motility of their enzymes. Thus, for example, the E_4 gene is represented by alleles $E_{4\text{-}1}$ and $E_{4\text{-}2}$. The allele combinations for the five genes, taken in order, can be written in abbreviated form as

$$11111 \quad (= E_{4\text{-}1} \ E_{9\text{-}1} \ E_{10\text{-}1} \ P_{5\text{-}1} \ APX_{5\text{-}1})$$
$$21111 \quad (= E_{4\text{-}2} \ E_{9\text{-}1} \ E_{10\text{-}1} \ P_{5\text{-}1} \ APX_{5\text{-}1})$$

and so on.

There are 32 possible allele combinations for the five genes, and hence 32 possible types of gametes. Two of these allele combinations predominate in the gamete pool of the Calistoga population, *21112* and *12221*. The gametic types *12211* and *21121* attain frequencies of 4% each. Other possible gametic types are rare or absent. The associations of alleles actually found in the Calistoga gamete pool are thus highly nonrandom (Allard, Babbel, Clegg and Kahler, 1972).

The common allele combinations are closely correlated with environmental conditions in the ecologically variable Calistoga habitat. The *21112* combination attains a high frequency (91%) in the more mesic subdivision of the population and grades off in increasingly xeric subdivisions. The *12221* combination shows the opposite gradient, from a high frequency in xeric subdivisions to a low frequency in mesic ones. These microgeographical trends in one small locality are paralleled to a certain extent by geographical variation in the frequencies of the same alleles over a broader area in California. Evidently the allele combinations *21112* and *12221* are adaptive in relation to mesic and xeric environments, respectively (Allard et al., 1972).

Consider next the classes and frequencies of genotypes in the Calistoga population. The homozygous genotypes *21112/21112* and *12221/12221* are the most frequent, and are present in large excess over expectation, whereas the other classes of homozygotes, with two exceptions, are rare. The observed excess of the two homozygous genotypes above indicates that they are favored by selection. Environmental correlations indicate that *21112/21112* has a selective advantage in mesic and *12221/12221* in xeric microhabitats (Allard et al., 1972).

Heterozygous genotypes as a group also show an excess over expectation in the Calistoga population. Evidently some heterozygous types have a selective advantage. However, the available data do not permit identification of the particular heterozygous genotypes which possess a marked heterozygote superiority (Allard et al., 1972).

The population variation is thus generally polarized into two or three main modes occupying different adaptive peaks: two homozygous genotypes in different environmental niches and some superior heterozygotes. The segregation of the population variation into alternative allele combinations is not complete, however, for there is a restricted generation of recombination types which occupy intermediate niches and are available for the colonization of new niches (Allard et al., 1972).

A second population near Geyserville, also in northern California, differs from the Calistoga population in the details of its gene pool. The Geyserville population occupies a relatively arid habitat. It is polymorphic for E_1, E_4, E_9, and E_{10}. The alleles E_{4-2}, E_{9-2}, and E_{10-2} occur in high frequency here as they do in the arid facies of the Calistoga habitat. Also the corresponding allele combination, *222*, has a high frequency (56%) in the gamete pool of the Geyserville population, whereas other gametic types are uncommon or rare. The Geyserville population is monomorphic for P_5 and APX_5, containing the alleles P_{5-2} and APX_{5-3} in fixed condition. Thus the principal adaptive allele combination at Geyserville is *22223*. This gene combination is partly the same as the prevalent one in the arid facies of the Calistoga population (Allard et al., 1972).

The adaptively valuable gene combinations in *Avena barbata* are held together partly by linkage (E_{10}, P_5, and APX_5) and partly by inbreeding. But these are only contributing factors. The primary factor is natural selection (Allard et al., 1972).

CHAPTER TWENTY-TWO

TRANSLOCATION HETEROZYGOTES

INTRODUCTION Translocations are of widespread occurrence in higher plants. Burnham (1956) in his review of the subject described translocations in 49 genera of dicotyledons, 47 genera of monocotyledons, and 2 genera of Taxaceae; many more examples have been added since then.

Chromosome rings or chains indicative of heterozygosity for translocations are found in four main situations in plants: (1) new chromosomal mutations, often induced by x-rays or other agents [*Zea mays, Sorghum versicolor, Plantago insularis* (Fig. 42)]; (2) artificial interracial hybrids (*Datura stramonium*) or interspecific hybrids (Datura, annual species; *Nicotiana alata* group); (3) in natural populations which are polymorphic for translocations (Campanula, Clarkia, Datura, Oenothera, Paeonia, Trillium); (4) in natural populations as permanent structural hybrids (Oenothera, Isotoma, Rhoeo).

Translocation heterozygosity sets up linkage of genes near the translocation breakage points in all of the foregoing situations, according to the mechanism explained in Chapter 17. Here we are concerned primarily with modes 3 and 4, where the translocation heterozygotes, and hence their linkage systems, occur in natural populations.

Modes 3 and 4 require further elucidation. In mode 3 we have a plant population which is polymorphic for one or more translocations. It consists of ring-forming individuals and the corresponding bivalent-forming structural homozygotes in varying frequencies. The frequency of the ring-forming individuals may remain at a high level from generation to generation, due to selection based on the adaptive superiority of the heterozygous types; in this case we would have a balanced chromosomal polymorphism for translocations in plants comparable to that for inversions in insects. The rings or chains in translocation polymorphism (balanced or otherwise) are often small and involve as few as four chromosomes, but they may also be large.

In mode 4 the population consists of translocation heterozygotes exclusively. This pattern occurs because the translocations are associated with a balanced lethal system, which prevents the segregation of the structural homozygotes, and consequently the translocation heterozygotes are permanent and true-breeding. The plant population may be monomorphic, containing only one heterozygous type, or it may contain two or more structurally heterozygous types. The rings are usually large and sometimes include all chromosomes of the complement.

In short, mode 3 includes transient and balanced chromosomal polymorphism for translocations, whereas mode 4 consists of permanent structural hybridity for translocations.

Although modes 3 and 4 are sufficiently different in certain cases to warrant separate recognition, it must also be admitted that difficulties exist in other cases. One difficulty is the existence of transitional conditions between the two extreme situations defined above. Another is lack of knowledge about a given example. Many naturally occuring translocation heterozygotes have been studied cytologically and cytogenetically, but not from the population genetical standpoint, and thus it is not known whether they are polymorphic variants or permanent hybrids.

A further comment is in order here regarding the practical study of translocation heterozygotes. Some plant groups have beautiful chromosomes, and lend themselves well to cytological work, but are miserable objects for genetic analysis, usually because they are difficult to grow or slow-growing. Some of the best plant groups for experimental genetic work, on the other hand, are poorly suited to cytological study because of small chromosomes or other problems. The technical difficulties in making a combined cytological and genetical analysis of translocation heterozygotes have often proven to be a real limiting factor in the analysis of this subject.

The *Oenothera biennis* group is by far the most thoroughly analyzed case of natural translocation heterozygosity in terms of both cytogenetics and population genetics. We are fortunate to have this example as a standard of reference. The examples of translocation polymorphism presented in the immediately following sections are very interesting and fairly well rounded but less thoroughly investigated than the Oenotheras. Let us consider first the case of *Paeonia californica*.

TRANSLOCATION POLYMORPHISM IN PAEONIA *Paeonia californica* (Paeoniaceae) is a perennial herb which inhabits the sage scrub vegetation in the coastal plains and foothills of central and southern California and northern Baja California. The plants bloom in the cool rainy weeks of early spring when few insects are on the wing. The flowers are predominantly self-pollinating, setting seeds following self-fertilization, and the plants reproduce sexually by seeds. Seed formation and seedling reproduction are somewhat erratic.

The species is diploid $(2n = 10)$, with beautiful large chromosomes. Three chromosomes in the haploid set have median centromeres and are designated M. One chromosome (or chromosome pair) has a submedian centromere and is designated D. The fifth chromosome has a subterminal centromere and is designated E (see Fig. 54A) (Stebbins and Ellerton, 1939; Walters, 1942).

The story begins with the discovery of ring-forming plants in a natural population in the northern part of the species area. Stebbins and Ellerton (1939) examined six individuals in a population of *P. californica* and found translocation rings of various sizes in all of them. They also found rings in the closely related *P. brownii* in northern California and Oregon. Walters (1942) thereupon

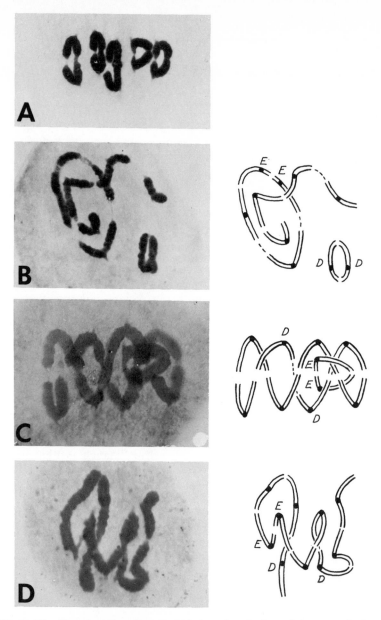

Figure 54. Translocation rings and chains at metaphase I (or prometaphase in *B*) in different plants of *Paeonia californica* (2*n* = 10). The D and E chromosomes of the complement are labeled throughout. (*A*) Structural homozygote with five bivalents. (*B*) Plant with a ring of 4, a chain of 4, and a bivalent. Interpretive diagram on right. (*C*) Plant with a ring of 8 and a bivalent. The bivalent is interlocked with the ring, as shown by interpretive diagram. (*D*) Complete translocation heterozygote with chain of 10. (Rearranged from Walters, 1942, 1952; photomicrographs courtesy of J. L. Walters.)

made a cytological survey of *P. californica* throughout the whole area of the species.

Walters found that translocation heterozygotes and also inversion heterozygotes are widespread throughout this area. He also found bivalent-forming structural homozygotes. Walter's survey revealed the existence of a diverse array of structural karyotypes in *P. californica*. The series includes all possible types of arrangements:

5 II	(Fig. 54A)
1 r4 + 3 II	
2 r4 + 1 II	(Fig. 54B)
1 r6 + 2 II	
1 r8 + 1 II	(Fig. 54C)
1 r4 + 1 r6	
1 r10	(Fig. 54D)

The total number of structural rearrangements is even greater. Among plants with a single ring of four chromosomes, the translocation ring involves two M chromosome pairs in some cases, and an M and D chromosome in others. Obviously many different translocations involving all chromosomes of the complement exist in *Paeonia californica* (Walters, 1942).

The various structural karyotypes show certain geographical trends. These trends are not well marked, inasmuch as the structural homozygotes and heterozygotes are both widely distributed, but tendencies can be discerned. The bivalent-forming homozygotes occur in the mild coastal zone in the central part of the species area (Fig. 55). Small rings also occur in this temperate central area. The complete translocation heterozygote with a ring or chain of 10 chromosomes is found in more arid interior places in the southern part of the species area. The combination of a ring of 6 plus a ring of 4 is also interior and southern in distribution (Walters, 1942).

The next step was to determine the frequency of structural homozygotes and heterozygotes within populations. This has only been done on a small scale so far. Walters (1942) reported on small samples from three populations in southern California. In

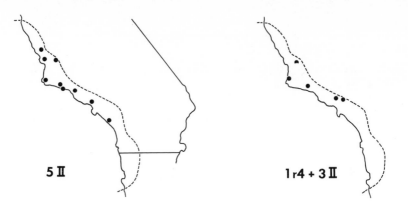

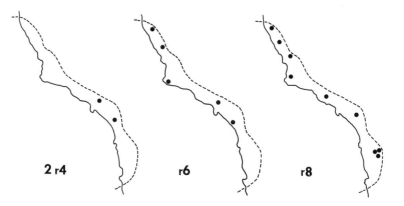

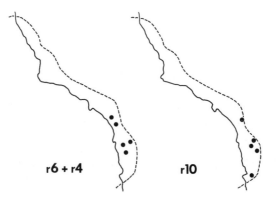

Figure 55. Geographical distribution of different types of heterozygotes and homozygotes for translocations in *Paeonia californica* in central and southern California and northern Baja California. The species area is outlined with dashed line; r = ring or chain; II = bivalent. (Redrawn from Walters, 1942, with new data of Walters added.)

TABLE 39

STRUCTURAL KARYOTYPES IN POPULATIONS OF *PAEONIA CALIFORNICA* AND *P. BROWNII* (2n = 10)

Population	Configuration *	Frequency	Sample Size	Investigator
P. californica in southern California				
Pasadena	5 II	1	10	Walters, 1942
	r4 + 3 II	6		
	r6 + 2 II	1		
	2 r4 + 1 II	2		
Altadena	2 r4 + 1 II	10	10	Walters, 1942
San Juan	5 II	4	8	Walters, 1942
Hot Springs	r6 + 2 II	3		
	2 r4 + 1 II	1		
Claremont	2 r4 + 1 II	7	24	Grant, 1956b, and
	r4 + r6	17		unpublished data
P. brownii in Sierra Nevada, California				
Calaveras	r4 + 3 II	3	53	Snow,
Big Trees	r6 + 2 II	1		unpublished data
	r7 + 1 II + 1 I	1		
	r8 + 1 II	48		

* r4 = ring or chain of 4; II = bivalent.

1955 and 1956 I scored almost all the individuals in a small colony near Claremont, southern California (Grant, 1956b, and unpublished data). Richard Snow also became interested in the problem and analyzed a larger sample from a population of the related *P. brownii* in the Sierra Nevada of northern California (Snow, personal communication, and unpublished data). The data of these independent studies are pooled in Table 39. More population sampling with statistical treatment is needed, but the preliminary results are revealing.

It is evident from Table 39 that translocation heterozygotes preponderate in the five populations examined. Structural homozygotes are either rare (as in the Pasadena colony) or apparently lacking (as in the Claremont and Calaveras populations).

Stebbins and Ellerton (1939) and Walters (1942) reported semisterility in *P. californica*. I later examined 20 plants of known chromosomal constitution in the Claremont population. The percentage of well-formed and well-stained pollen grains, taken as a measure of pollen fertility, ranged from 14 to 39% with an average of 24%. This is approximately the amount of good pollen expected in plants that are heterozygous for two independent translocation rings, which is the actual condition in the Claremont population. These plants are also semisterile as to seeds. A typical mature fruit contains one to three plump seeds and a larger number of abortive seeds (Grant, unpublished data).

The observed frequencies of translocation heterozygotes and homozygotes in several colonies of Paeonia (Table 39) obviously deviate strongly from the expected frequencies of genotypes in a population in Hardy-Weinberg equilibrium. The most probable cause of this deviation is adaptive superiority of the translocation heterozygotes. They must have advantages which outweigh their semisterility.

Seeds produced by self-pollination in known translocation heterozygotes in the Claremont colony were sown in the greenhouse. A high rate of mortality was observed in the seedlings. A similar seedling mortality occurs in nature. This mortality could well have a selective component favoring the heterozygous types. Experimental evidence bearing on this hypothesis was sought but not obtained due to technical difficulties with the plant materials. It has not been possible as yet to grow these peonies from seed to maturity in the experimental plot. This is another aspect of peony population biology warranting further study.

The case for heterozygote superiority in the California peonies rests mainly on the combined features of the breeding system and the genotype frequencies in natural populations. Here we have plants reproducing by seeds formed by self-pollination. The self-fertilization would in itself lead to a rapid decay of heterozygosity. Yet the adult populations consist predominantly or exclusively of translocation heterozygotes. The most likely source of this bias in population composition is natural selection. There is ample opportunity in the seed and seedling stages for selective elimination of the homozygous types (Grant, 1956b, 1964).

A corollary is that the genes near the translocation breakage points must form linked systems in the ring-forming peonies. And therefore linked gene combinations of adaptive value are probably associated with the translocation rings.

TRANSLOCATION POLYMORPHISM IN OTHER GROUPS Translocations are of widespread occurrence in Datura (Solanaceae), where they are mostly fixed in homozygous condition in different races and species. *Datura stramonium* contains a series of structurally homozygous races which differ by translocations. Some 11 translocation types are known in this species and a lesser number in other annual Datura species such as *D. quercifolia, D. metel,* and *D. meteloides.* The species also differ with respect to translocations (Bergner, Satina and Blakeslee, 1936; Blakeslee, Bergner and Avery, 1937; Avery, Satina and Rietsema, 1959).

Snow and Dunford (1961) have gone on to describe and analyze a polymorphism for translocations in a population of *Datura meteloides.* A large population in central California was found to contain two independent translocations, designated A and B. In a sample of 97 individuals, there were 61 structural homozygotes and 36 translocation heterozygotes. The breakdown of the heterozygous class is

> 19 heterozygotes for translocation A
>
> 15 heterozygotes for translocation B
>
> 2 heterozygotes for translocations A and B

The high frequency of the translocation heterozygotes, and particularly of the A heterozygotes, suggests that they possess adaptive superiority (Snow and Dunford, 1961).

Translocation heterozygotes are found commonly in natural populations of the normally outcrossing *Campanula persicifolia* (Campanulaceae). When the translocation heterozygotes are selfed artificially, no structural homozygotes appear in their progeny. Thus a parental plant with one ring of four chromosomes, on selfing, yielded 35 ring-forming progeny and no structural homozygotes. A plant with two r4 yielded, following self-pollination, 15 ring-forming progeny and no homozygotes. The absence of the expected structural homozygous classes is attributed to selective elimination. This in turn points to an adaptive superiority of the translocation heterozygotes in the natural populations (Darlington and La Cour, 1950; Darlington, 1956b).

Garden strains of *Chrysanthemum carinatum* consist of structural homozygotes ($2n = 9$ II) and translocation heterozygotes with one or rarely two rings of four chromosomes. In a sample of 75 plants there were 58 translocation heterozygotes and 17 homozygotes. The high frequency of translocation heterozygotes suggests that they are adaptively superior (Rana and Jain, 1965).

Populations of *Isotoma petraea* and related species (Lobeliaceae) in western Australia contain bivalent-forming individuals ($2n = 7$ II) and ring-forming individuals (r4, r6, r10) floating in varying frequencies. Complete translocation heterozygotes with r14 also occur (James, 1965).

Again, in the annual genus Clarkia (Onagraceae) translocations occur as polymorphic variants in a number of species (Lewis, 1953). Translocation heterozygotes are known in 14 of the 34 species of Clarkia, and are common in *C. amoena, C. dudleyana, C. unguiculata,* and *C. williamsonii* (Snow, 1960).

Clarkia unguiculata has five known chromosome end arrangements that differ by translocations. These are designated A to E. Translocation heterozygotes as well as some types of structural homozygotes are common throughout the range of the species in California. In some populations the translocation heterozygotes reach a frequency of 77%. These are mostly heterozygotes containing the A genome in combination with one other genome, i.e., they have the genomic constitution A/B, A/C, and so on. The structural homozygote A/A is also fairly common in *C. unguiculata.* But other classes of structural homozygotes are rare in natural populations. For example, B/B is rare in populations containing a high frequency of A/B heterozygotes. Here again a superior fitness of the translocation heterozygotes is indicated (Mooring, 1958, 1961).

Translocation heterozygotes also attain a high frequency in populations of *Clarkia dudleyana* (Snow, 1960). Chromosome rings are found in natural populations of other genera of Onagraceae such as Camissonia, Gaura, Gayophytum, Hauya, and Oenothera

(Lewis, Raven, Venkatesh and Wedberg, 1958; Raven and Lewis, 1960). In the large genus Oenothera, small rings in polymorphic condition occur in a number of subgenera, such as Anogra (the *Oe. californica* group), Chylismia, Sphaerostigma (= segregate genus Camissonia), and Euoenothera (Lewis et al., 1958; Gregory and Klein, 1960). Translocations often take the form of permanent translocation hybridity in Euoenothera, and for this genetic system the group is justly famous, but it should not be forgotten that ordinary polymorphism for small rings also occurs in *Oe. hookeri, Oe. irrigua, Oe. jamesii,* and other members of Euoenothera (Cleland, 1940, 1949, 1950).

PERMANENT TRANSLOCATION HETEROZYGOSITY IN THE *OE-NOTHERA BIENNIS* GROUP   Oenothera subgenus Euoenothera (= subgenus Oenothera) is a group of yellow-flowered herbaceous plants, often rank or weedy, of wide distribution in North and South America. In North America the group consists of 10 taxonomic species that are all diploid, with $2n = 14$. They fall into three categories on the basis of genetic system.

1. *Oenothera hookeri* in western and southwestern North America and certain other species in other areas are cross-pollinating structural homozygotes with seven bivalents at meiosis. Occasional translocations occur between or within populations in these species.

2. The second category consists of *Oe. jamesii, Oe. longissima,* and *Oe. irrigua* in the interior southwestern plains. (*Oe. irrigua* is sometimes treated as a race of *Oe. hookeri.*) These plants are cross-pollinating, like the first category, but often carry translocations in polymorphic condition. The rings are small or medium-sized (r4, r6, r8, two r4) and are not associated with balanced lethals.

3. *Oe. biennis, Oe. parviflora* (= *Oe. muricata*), and *Oe. strigosa* in the northern, central, and northeastern United States are predominantly self-pollinating and have large rings, usually r14. The rings are associated with balanced lethals. A large number of microspecies, many of them named as Linnaean species, exist in

this group. One of these is the genetically famous *Oe. lamarckiana* (its correct botanical name is *Oe. erythrosepala*), an introduced weed in Europe and Pacific North America.

We are concerned here with the third category, the North American Euoenotheras which exhibit permanent and true-breeding hybridity for large rings.

Their remarkable genetic system has been worked out by a succession of botanists: De Vries, Renner, Cleland, and many others. Renner (1917, 1925) inferred from their breeding behavior that the Oenothera plants are permanent heterozygotes for linked groups of genes which segregate as Mendelian units. Cleland (1923, 1925, 1936) then showed that successive translocations form the cytological basis of the linkage systems. The nature of the balanced lethals is discussed by Steiner (1956), and estimates of the rate of outcrossing are given by Hoff (1962). A great deal of cytogenetic information is brought together by Cleland (1950). The taxonomy of the group has been treated recently by Munz (1965).

Attention is called to reviews by Tischler (1951–1953), Cleland (1962), and Darlington (1965), which contain extensive bibliographies. I recently reviewed the phylogenetic development of the Euoenotheras, in a treatment complementary to the present one (Grant, 1971, ch. 23). At this writing a definitive monograph by Cleland has been announced (Cleland, 1972).

Oenothera lamarckiana regularly forms a ring of 12 plus one bivalent at meiosis. So do *Oe. chicaginensis* and *Oe. suaveolens*. Most microspecies have complete rings of 14 chromosomes (see Table 40). The plants are heterozygous for successive translocations on all or nearly all chromosomes of the complement.

The haploid sets are known as Renner complexes and designated by names (velans, etc.). Each type of Renner complex has a particular chromosome end arrangement. The ring-forming microspecies are thus complex heterozygotes with particular genomic constitutions (e.g., *Oe. lamarckiana* is velans/gaudens). The genomic constitutions of several microspecies are given in Table 40.

The various end arrangements are expressed in terms of an arbitrary standard arrangement. That standard is the end arrangement in the original De Vries strain of the western American *Oe. hookeri*. The Johansen strain of *Oe. hookeri* is closely related

TABLE 40
GENOMIC CONSTITUTION OF SEVERAL TYPES OF PERMANENT TRANSLOCA-
TION HETEROZYGOTES IN EUOENOTHERA (COMPILED FROM CLELAND, 1950)

Microspecies	Renner complexes	Size of ring
Oe. lamarckiana	velans/gaudens	r12
(= Oe. erythrosepala)		
Oe. chicaginensis	excellens/punctulans	r12
Oe. suaveolens	albicans/flavens	r12
Oe. muricata	rigens/curvans	r14
(= Oe. parviflora)		
Oe. grandiflora of De Vries	acuens/truncans	r14
(not Oe. grandiflora L'Her.)		
Oe. cockerellii	curtans/elongans	r14
(= Oe. strigosa in part)		
Oe. strigosa of De Vries	curtans/stringens	r14
(= Oe. strigosa in part)		

genomically to the De Vries strain, differing by one reciprocal translocation. The Johansen end arrangement is phylogenetically primitive in Euoenothera. It is widespread in natural races of *Oe. hookeri,* occurs in some other structurally homozygous species of the *Oe. hookeri* group, and is found in certain translocation heterozygotes. The heterozygous microspecies *Oe. chicaginensis* has the constitution excellens/punctulans; and the end arrangement in excellens is the same as that in *Oe. hookeri* Johansen.

The end arrangements in these and other Renner complexes are listed in Table 41. The table shows that the Renner complexes of the heterozygous microspecies, with the exception of excellens, all differ by successive translocations from the basic end arrangement in *Oe. hookeri.* Furthermore, the two Renner complexes in a hybrid microspecies differ from one another by successive translocations so as to give rings of 12 or 14 at meiosis.

The chromosomes in the ring undergo alternate disjunction at meiosis (Fig. 56). Consequently the two Renner complexes are reassembled intact in the two poles at the end of the first meiotic division. Therefore *Oe. lamarckiana* (velans/gaudens) produces two classes of gametes, velans and gaudens; *Oe. muricata* segregates rigens and curvans gametes, and so on.

Self-fertilization in a hybrid microspecies restores the parental heterozygous constitution. The velans/gaudens heterozygote pro-

TABLE 41

CHROMOSOME END ARRANGEMENTS OF VARIOUS RENNER COMPLEXES IN
EUOENOTHERA (CLELAND AND HAMMOND, 1950)

Renner complex	End arrangement of seven chromosomes						
hookeri of De Vries	1– 2	3– 4	5– 6	7– 8	9–10	11–12	13–14
hookeri of Johansen	1– 2	3– 4	5– 6	7–10	9– 8	11–12	13–14
velans	1– 2	3– 4	5– 8	7– 6	9–10	11–12	13–14
gaudens	1– 2	3–12	5– 6	7–11	9– 4	8–14	13–10
excellens	1– 2	3– 4	5– 6	7–10	9– 8	11–12	13–14
punctulans	1– 4	3– 9	5– 2	7– 8	6–12	11–10	13–14
albicans	1– 4	3– 6	5– 7	2–14	9– 8	11–10	13–12
flavens	1– 4	3– 2	5– 6	7– 8	9–10	11–12	13–14
rigens	1– 2	3– 4	5– 6	7–11	9–10	8–14	13–12
curvans	1–14	3– 2	5–13	7–12	9– 8	11–10	4– 6
acuens	1– 4	3– 2	5– 6	7–10	9– 8	11–12	13–14
truncans	1–13	3– 7	5– 2	4– 6	9–14	11–10	8–12
curtans	1– 7	3– 4	5– 8	2–10	9–11	6–12	13–14
elongans	1– 4	3– 2	5–10	7– 6	9–14	11–12	13– 8
stringens	1– 4	3– 2	5–14	7– 8	9–10	11–12	13– 6

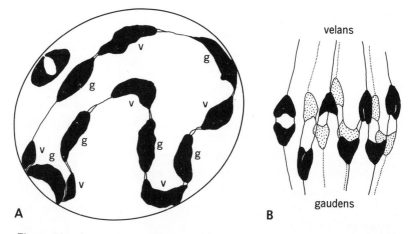

Figure 56. Chromosome configurations at meiosis in *Oenothera lamarckiana*
(velans/gaudens) ($2n = 14 = r12 + 1$ II). Chromosomes lying in upper optical
plane shown black; those in lower optical plane stippled. (*A*) Ring of 12 and
bivalent at diakinesis. Velans chromosomes labeled V, gaudens chromosomes
labeled G. (*B*) Zigzag orientation of ring at metaphase for alternate disjunction. (From Emerson, 1935.)

duces only velans/gaudens progeny, with rare exceptions, and the expected classes of homozygous segregates (velans/velans and gaudens/gaudens) are absent. The large-ring-forming types breed true for their structurally heterozygous condition.

The absence of the homozygous classes is due to the association of a balanced system of complementary lethals with the Renner complexes. The complementary lethals are l_1 and l_2. The double heterozygote $\dfrac{+ \quad l_2}{l_1 \quad +}$ is viable, but the homozygous recombination types $\dfrac{l_1 \ +}{l_1 \ +}$ and $\dfrac{+ \ l_2}{+ \ l_2}$ are lethal. The lethal factors are linked to the Renner complexes.

The velans lethal (let_{vel}) is in chromosome 5–8. The gaudens lethal apparently has not been mapped. The constitution of *Oe. lamarckiana*, taking the lethals into consideration, can be written:

$$\frac{\text{velans} \quad let_{vel} \quad +}{\text{gaudens} \quad + \quad let_{gau}}$$

The structural homozygotes, being homozygous recessive for one lethal factor or the other, are consequently suppressed and do not appear in the progeny.

The lethal factors act in the zygote stage in *Oe. lamarckiana*. In some hybrid microspecies such as *Oe. muricata*, the lethal factors operate in the gamete stage, killing one class of gametes or the other. The end result is the same.

Although a hybrid microspecies produces only one class of offspring by self-fertilization, with the exception of a low proportion of "mutant" types, it yields as many as four classes of offspring following out-crossing to another hybrid microspecies. This is a result of each hybrid microspecies producing two classes of gametes. The cross *Oe. lamarckiana* × *Oe. grandiflora*, for example, gives rise to four kinds of phenotypes in F_1. These are the four heterozygous combinations of four classes of gametes. In genomic terms, the cross and its F_1 progeny are: velans/gaudens × acuens/truncans → velans/acuens + velans/truncan + gaudens/acuens + gaudens/truncans.

The large-ring-forming Oenotheras are heterozygous for various genes affecting the visible exo-phenotype as well as for the Renner complexes. This is shown by segregation for phenotypic differences in the F_1 progeny of outcrosses between microspecies, as in the case of *Oe. lamarckiana* × *Oe. grandiflora* mentioned above, and by much other evidence. The group of microspecies known as *Oe. biennis* I in the American Middle West, for example, carries genes for broad thin smooth leaves in one of its Renner complexes and genes for thick hairy leaves in the other.

Enzyme loci in plants of *Oe. biennis* II in Connecticut have been analyzed for heterozygosity by gel electrophoresis. Nineteen genes were analyzed by this method. Heterozygosity was found at $\frac{5}{19}$ or 26% of these gene loci, a very high level of heterozygosity compared with that of most other organisms (Levin, Howland and Steiner, 1972).

The genes *R* and *P* govern the color of the leaf midribs and papillae, respectively. They are heterozygous in *Oe. lamarckiana* and *Oe. muricata*. But their linkage relations are different in the two microspecies. *Oenothera lamarckiana* has the constitution

$$\frac{\text{velans} \quad P \quad R}{\text{gaudens} \quad p \quad r}$$

and *Oe. muricata* the constitution

$$\frac{\text{rigens} \quad P \quad R}{\text{curvans} \quad p \quad r}$$

Thus both *P* and *R* are linked to the Renner complexes in the complete heterozygote, *Oe. muricata;* but in the incomplete heterozygote, *Oe. lamarckiana,* only *P* is linked while *R* is independent, being located in the bivalent-forming 1–2 chromosomes (Renner, 1928; Emerson and Sturtevant, 1932; Cleland, 1936).

HOMOZYGOUS SEGREGATES IN THE *OENOTHERA BIENNIS* GROUP

Gene recombination in ring-forming Oenotheras may occur freely, rarely, or be suppressed, depending upon the location of the genes in relation to the translocation break points.

The genes *P* and *R* can recombine freely in *Oe. lamarckiana,* since *R* is outside the Renner complexes in this microspecies.

Genes inside the Renner complexes can also recombine freely if they are located near the chromosome ends. The homologous arms of the translocated chromosomes pair and cross over in their terminal regions at meiosis, and genes in these pairing segments can consequently recombine without any special restrictions.

The brevistylis gene *Br* in *Oe. lamarckiana* determines style length, the *Br* allele giving normal styles and *br* short styles. *Br* is located near the end of the 12 arm. This arm belongs to the 11–12 chromosome in the velans set and to the 3–12 chromosome of the gaudens set (see Table 41). Pairing between the terminal regions of the 12-arms belonging to the two Renner complexes permits the alleles in the genic heterozygote *Br/br* to cross over into the opposite Renner complex with a high frequency. The short-styled brevistylis form (*br/br*) then appears as a fairly common segregate in *Oe. lamarckiana* progenies (Emerson and Sturtevant, 1932; Catcheside, 1954).

The conditions are very different in the chromosome regions close to the translocation break points, which are also near the centromeres in Oenothera, for these regions are not normally able to pair and form crossovers. Therefore it is postulated that the heterozygous gene combinations for which a large-ring-forming Oenothera breeds true must be located in these nonpairing or differential segments. The differential segments must harbor the balanced lethal factors and also the genes determining the phenotypic features characteristic of the different microspecies (Darlington, 1937).

The alleles of a gene located close to a differential segment may pair and cross over rarely. A heterozygous parent thereupon produces some exceptional homozygous progeny which deviate from type. These exceptional homozygous crossover types are one of the sources of "mutations" in Oenothera.

An example is the nanella or dwarf form in *Oe. lamarckiana*. This microspecies generally breeds true for tall stature but gives rise to a certain low frequency of dwarf forms. The nanella types have the superficial appearance of mutants but are actually rare segregates.

The gene *N* determining plant stature is located in the 4 arm and hence in the 3–4 chromosome of the velans set and in the

9–4 chromosome of the gaudens set. It is represented by the dominant allele N for normal stature and the recessive allele n for dwarfness in *Oe. lamarckiana*. Normal tall plants with the constitution N/n produce mainly tall N/n progeny. The linkage between N and the velans and gaudens lethals is not complete, however, and crossing-over occurs rarely between the two sets of factors. When such rare crossovers occur, a plant of the constitution $\dfrac{\text{velans } N}{\text{gaudens } n}$ produces exceptional progeny $\dfrac{\text{velans } n}{\text{gaudens } n}$ which exhibit the nanella phenotype and breed true for the nanella condition thereafter (Emerson and Sturtevant, 1932; Emerson, 1935).

A line may become homozygous not merely for a single gene but for a large block of genes, after the occurrence of new translocations within the chromosome set. The new translocation, by changing the end arrangement, is likely to break up a ring of 14 or 12 chromosomes into an array of smaller rings and bivalents. The same result follows from outcrossing between one translocation heterozygote and another different one. The chromosomes in some of the newly formed rings and bivalents will be separated from the balanced lethal genes. Consequently progeny can arise which are homozygous for whole chromosomes and for the genes on them.

The rare separation of a large proportion of the genes from the balanced lethal factors, by unusual crossovers or new translocations or hybridization, and the subsequent segregation of individuals which are structurally and genically homozygous to a large degree, make it possible to compare the homozygotes and the heterozygotes in Oenothera with respect to their vigor. In the following cases the homozygotes are weak compared with the heterozygotes.

A structurally homozygous, bivalent-forming derivative of *Oe. lamarckiana* known as blandina differs from normal heterozygous plants in having pale yellowish-green seedling leaves, slender rosettes, slow growth, and reduced fruit formation (De Vries, 1917; Cleland, 1950). A flavens/flavens homozygote derived from *Oe. suaveolens* (albicans/flavens) is semilethal and difficult to keep alive (Renner, 1941). The same is true of a structurally homozygous derivative of *Oe. pratincola* (Renner, 1941). Curtans/curtans homozygotes derived from *Oe. cockerellii* (curtans/elongans) also form bivalents at meiosis and are weak (Oehlkers and Harte, 1943).

Normal vigor and fertility in all these cases are associated with heterozygosity for a large block of genes, and are lost when these genes become homozygous. The heterozygous gene combination in a ring-forming Oenothera evidently has heterotic effects. By virtue of the peculiar genetic system of these plants, consisting of heterozygosity for successive translocations, balanced lethal genes, and self-pollination, any heterozygous gene combination which is adaptively valuable can be reproduced exactly and multiplied indefinitely.

A population of *Oe. biennis* II in Connecticut has recently been assayed by electrophoretic methods to detect polymorphism. The plants in this population are all highly heterozygous. Of interest to us here is the finding that most individuals in this population have the same heterozygous genotype. Preliminary indications are that this particular heterozygous genotype is established over a much wider area throughout Connecticut (Levin, Howland and Steiner, 1972, and unpublished data).

PERMANENT TRANSLOCATION HETEROZYGOTES IN OTHER GROUPS Complete translocation heterozygotes have been reported in a number of other subgenera and genera in the Onagraceae. In most cases it is not known whether the translocation heterozygosity is associated with balanced lethals.

Complete translocation heterozygotes with rings of 14 are found in the following subgenera of Oenothera: Euoenothera, in North America as already discussed and also in South America; Raimannia, in eastern North America; Lavauxia, in North America; and Hartmannia, in South America. Balanced lethals are known in Euoenothera and Raimannia (Cleland, 1950; Hagen, 1950; Hecht, 1950).

Other genera of Onagraceae exhibiting large rings are Gayophytum and Gaura. Complete rings of 14 occur in *Gayophytum heterozygum* ($2n = 14$) (Lewis, Raven, Venkatesh and Wedberg, 1958; Lewis and Szweykowski, 1964). In Gaura ($2n = 14$) small rings are common (see Lewis et al., 1958). Complete rings of 14 are found in *Gaura biennis* and *G. triangulata* (Raven and Gregory, 1972).

Outside of the Onagraceae, complete translocation rings are found in members of four other plant families: the Hypericaceae, Commelinaceae, Paeoniaceae, and Lobeliaceae. The relevant facts are summarized synoptically below.

Hypericum punctatum (Hypericaceae), northeastern United States; $2n = 16 = r16$. Related species have regular bivalent formation (Hoar, 1931; Hoar and Haertl, 1932).

Rhoeo discolor (Commelinaceae). The correct botanical name now appears to be *Rhoeo spathacea*. Central America and in cultivation; $2n = 12 = r12$. Different cultivated strains show the ring or chain of 12. Zygote lethals probably present. A population from British Honduras (*R. spathacea concolor*) has regular bivalent formation ($2n = 12 = 6$ II). This pair-forming population is sympatric with the ring-forming one. One of the two genomes in the structural heterozygote is homologous with the genome in the British Honduras entity (Sax, 1931; Anderson and Sax, 1936; Simmonds, 1945; Wimber, 1968).

Paeonia californica and *P. brownii* (Paeoniaceae), California; $2n = 10$; r10 in some individuals. The situation in Paeonia was discussed earlier; we should add here that the possibility of balanced lethals is not excluded in the case of the large-ring-forming types (Stebbins and Ellerton, 1939; Walters, 1942).

Isotoma petraea (Lobeliaceae), western Australia; $2n = 14$; r14 in some individuals. Zygote lethals present (James, 1965, 1970).

The origin of complete translocation heterozygosity, whether gradual or abrupt, is a debatable question. Probably both modes of development have taken place in different plant groups. A gradual development from bivalent-forming types through small-ring-forming types to complete heterozygotes is indicated by the cytogeographical evidence in Isotoma (James, 1965, 1970). This model may apply also to Paeonia and parts of Euoenothera. An abrupt origin by hybridization between preexisting, structurally differentiated, bivalent-forming species is suggested by morphological and cytogenetic evidence in Gayophytum (Lewis and Szweykowski,

1964). This model may be applicable to other segments of Euoeno-thera and perhaps to Hypericum.

In several groups the complete translocation heterozygotes are autogamous but have outcrossing, meiotically normal relatives. This is the case in Euoenothera, Paeonia, Isotoma, and Gayophytum. The interpretation of this correlation is not entirely clear. Whatever the explanation, a change in the breeding system often seems to accompany the development of complete translocation heterozy-gosity (for discussions see Darlington, 1958, 1963; James, 1965, 1970).

PART IV : GENETIC SYSTEMS

CHAPTER TWENTY-THREE

THE REGULATION
OF RECOMBINATION

INTRODUCTION The Mendelian mechanism works with mathematically beautiful precision to bring about segregation and recombination of genes. We described this mechanism in Chapter 1 from the classical standpoint as a more or less constant mode of hereditary transmission. The Mendelian mechanism does indeed have some universal features, which are common to all sexual organisms, but there is another more relativistic aspect.

The Mendelian mechanism is made up of components—chromosomes, crossover segments, heterozygous or homozygous loci —which are subject to variation in number, size, and frequency. Furthermore, the Mendelian mechanism operates within the context of breeding systems and breeding populations which vary greatly in their characteristics. These internal and external variables affect the release of genetic variation by the Mendelian mechanism.

The internal and external variables are themselves under genotypic control to a greater or lesser extent. Not only does the Mendelian mechanism determine the segregation and recombination of genes, therefore, but the genotype also controls reciprocally the variable components of the Mendelian mechanism. Darlington (1939, p. 70) has described this reciprocal relationship in the following way: "The genes are like the members of a legislature in being subject as individuals to the laws they enact as a body."

Recognition of the relativistic and variable aspects of the Mendelian mechanism led to the concept of the genetic system (Darlington, 1932, 1939, 1958). The genetic system is the system of hereditary variation, the set of factors determining the balance between genotypic copying and genotypic variability in reproduction. This concept has been extremely fruitful; its development at the hands of Darlington and others has furnished new insights in biology.

A diverse array of genetic systems is found in nature. This diversity makes possible a comparative approach. The various types of genetic systems exhibit certain correlations. Phylogenetically unrelated organisms which have undergone parallel evolution in morphology and life cycle are often found to possess similar types of genetic systems. In short, the genetic system displays parallel and convergent evolution. Furthermore, the particular type of genetic system is correlated with the conditions of life of the organisms possessing it. The genetic system is adaptive.

The genetic system includes various subsystems concerned with mutation, development, and other aspects. But the subsystem of central importance is the recombination system (Carson, 1957). In this chapter we discuss recombination systems in higher plants, following an earlier treatment (Grant, 1958).

TYPES OF RECOMBINATION SYSTEMS   The process of sexual reproduction has two opposing tendencies: the replication of a pre-existing gene combination and the formation of new ones. The first of these tendencies is advantageous to a population for maintenance of a state of fitness to existing environmental conditions. The second is necessary for flexibility in relation to future changes in environmental conditions (Mather, 1943).

Each population must find a point of compromise between immediate fitness and long-range flexibility. Or, more correctly, the successful living populations are those which have found a satisfactory compromise in the past. A satisfactory compromise between fitness and flexibility depends upon achieving an optimum balance between constancy and variability in reproduction. It is the main function of the recombination system to strike such an optimum balance (Mather, 1943; Stebbins, 1950).

The requirements for fitness and for flexibility differ according to the particular conditions of the organism. And actual recombination systems, in accordance with their function, do vary from those promoting maximum constancy to those promoting maximum variability.

Some restriction on recombination is universal. The genes do not float freely as individuals in the gene pool. They are always associated in chromosomes and mating groups which channelize their movements at the stages of meiosis and fertilization, respectively. Consequently, only a fraction of the potentialities of recombination is ever realized. It is this fraction of attainable recombination which varies in different species and groups (Grant, 1958).

The variation is continuous, and the differences are relative. Carson (1957) proposes to recognize three modes in this continuous spectrum. These are termed closed, restricted, and open recombination systems.

A closed recombination system is exemplified by obligate apomixis, in which the potential variation is seldom or never released. Such systems probably are nearly but not completely closed in the majority of cases in plants. A restricted recombination system imposes strong but not insurmountable barriers to the formation of new gene combinations. Predominantly autogamous plants exemplify this condition. An open recombination system generates a much larger number and range of genotypes, as in an outcrossing plant population.

The difference between open and restricted recombination systems lies not only in the amount of variability generated but also in the evenness of the flow of this variability. Plants possessing a restricted recombination system often produce new variation sporadically, in a series of bursts in a few generations separated by periods of constancy. An open recombination system, by contrast, gives rise to a slow but steady flow of variation over the course of generations.

THE REGULATORY FACTORS   The formation of a new recombination type is the end result of a chain of events in the sexual cycle. A double or multiple heterozygote may or may not produce recombination-type gametes, depending upon the course of meiosis. Let us assume that it does give rise to gametes which possess nonparental combinations of genes. Effective recombination now depends on the novel gametes uniting to form novel zygotes. Recombination can be controlled at either of the principal links in the chain, at meiosis or fertilization.

The recombination system regulates the generation of variability by restricting the types of gametes produced and the types of zygotes formed. The control of recombination at the stage of gamete formation is exercised by the chromosomal machinery, particularly by the number of chromosomes and the number and distribution of crossover segments. The distribution of sterility barriers, whether surrounding large or small assemblages of forms, will also limit the formation of recombination gametes to a certain definite range of types. Still other factors exert a channelizing influence on zygote formation. The nonrandomness of fertilization in the higher plants is affected by the breeding system, mode of pollination, dispersal potential, population size and structure, and distribution and strength of incompatibility barriers and external isolating mechanisms.

The control comes into force during each successive sexual generation. It is well known that the length of the life cycle varies widely in different organisms, whereas the environmental changes to which populations must adapt take place on a chronological scale. The amount of recombination generated per unit of time is consequently a function of the length of generations as well as of the degree of restriction in each generation. A bacterium in which recombination is strongly restricted might engender as much variability over a period of years as an oak possessing an open recombination system.

We can classify the factors regulating recombination in plants as follows (Grant, 1958):

   I. Factors controlling the amount of recombination per generation; control operative at meiosis
      1. Chromosome number
      2. Frequency of crossing-over
      3. Sterility barriers
  II. Factors controlling the amount of recombination per generation; control operative at fertilization
      4. Breeding system
      5. Pollination system
      6. Dispersal range
      7. Population size
      8. Crossability barriers and external isolating mechanisms
 III. Factor controlling the amount of recombination per unit of time
      9. Length of generation

Each of these controlling factors occurs in different conditions in different plant groups. To the extent that a particular factor varies in condition, so too does its restrictive influence on recombination. Table 42 summarizes the conditions of the various regulatory factors which determine different modes of recombination.

CHROMOSOMAL FACTORS Genes borne on separate chromosomes become combined in different ways by the independent assortment of the chromosomes at meiosis. A greater number of recombinations will arise if the genes are distributed among many chromosomes than if they are grouped on a few. An organism with the haploid number $n,$ and heterozygous for one gene pair on each chromosome pair, can produce 2^n genetically different gametes. The addition or subtraction of one chromosome pair thus doubles or halves the number of gene combinations which can arise by independent assortment, assuming that the chromosome pairs are all equally heterozygous.

Assume that the haploid number is $n = 7$; then the number of possible chromosome combinations in the gametes is 128. But if $n = 14$, the number of possible combinations is 16,384. This comparison illustrates the power of chromosome number alone to control recombination.

TABLE 42

COMPONENTS OF DIFFERENT TYPES OF RECOMBINATION SYSTEMS (GRANT, 1958)

Regulatory factor	Closed recombination	Restricted recombination	Open recombination
Length of generation	no sexual generations	long	short
Chromosome number	irrelevant, no meiosis in female line	low	high
Crossing-over	irrelevant, no meiosis in female line	low chiasma frequency; localized chiasmata; structural hybridity	high chiasma frequency; random chiasmata
Breeding system	no crossing	partially to predominantly self-fertilizing	entirely or predominantly cross-fertilizing
Pollination system	no crossing	pollinating insects with narrow foraging range; flower-constant insects	widely dispersed pollen; promiscuous pollinators
Dispersal range	irrelevant where no crossing	diaspores dispersed close to parent plant	diaspores dispersed far from parent plant
Population size	asexual clones	small colonial populations	large continuous populations
Isolating mechanisms	no crossing	related species strongly isolated	related species or larger groups incompletely isolated

Recombination of linked genes is promoted by a high frequency and random distribution of chiasmata, and by structural homozygosity. Conversely, a chromosome segment can be converted into a supergene, devoid of recombination, by localization of chiasmata or structural heterozygosity. Recombination of linked genes is slowed down by a low chiasma frequency.

Darlington (1939) has put together the two main quantifiable chromosomal factors, namely haploid number and chiasma frequency, in the so-called recombination index. The recombination index is the haploid number of chromosomes plus the average number of chiasmata per cell. This index shows much variation between plant groups.

We have compared the potential for recombination with $n = 7$ and $n = 14$. This comparison is not unrealistic, since both are common haploid numbers in the angiosperms. Chromosome numbers in the angiosperms have the following modes:

Monocotyledons	$n = 7$
Herbaceous dicotyledons	$n = 7-9$
Woody dicotyledons	$n = 11-14$

It appears that herbaceous flowering plants as a group have a smaller amount of recombination per generation and woody flowering plants have a greater amount (Stebbins, 1938; Grant, 1958, 1963).

Stebbins (1950) suggested that chiasma frequency may often be correlated with breeding system in herbaceous plant species that have the same chromosome number. The expected correlation is between low chiasma frequency and outcrossing or high chiasma frequency and inbreeding. This correlation would operate to produce a steady flow of variability in the outcrossing species and sporadic bursts of variability separated by intervals of relative constancy in the inbreeding ones (Stebbins, 1950, p. 180, also p. 169).

Data on both chiasma frequency and breeding system are available in several plant groups and were assembled some years ago to test Stebbins's hypothesis (Grant, 1958). The plant groups are Agropyron-Elymus-Sitanion (Gramineae), Collinsia (Scrophulariaceae), Crepis (Compositae), Gilia (Polemoniaceae), and Sorghum (Gramineae). In each genus there are cross-pollinating and self-pollinating species. In each case the self-pollinating species have a higher chiasma frequency than the outcrossing species. These results seem to confirm the hypothesis.

However, the recorded chiasma frequencies are mostly those at metaphase I or sometimes at diakinesis, whereas the actual process of crossing-over occurs much earlier in meiotic prophase. A reinvestigation of Gilia species and races possessing different breeding systems was therefore undertaken to determine their chiasma frequencies at the genetically significant stage of diplotene. The differences in chiasma frequency tend to disappear as one goes back from metaphase I to diplotene. The cytogenetic difference between outcrossing and inbreeding Gilias turns out to be not so much in effective chiasma frequency as in rate of terminalization of the chiasmata (Brown, 1961). The significance of these observations is not clear.

The hypothesis of a compensating relationship between chiasma frequency and breeding system with respect to the generation of variability appears to require further study.

BREEDING SYSTEM   The diversity of breeding systems in the angiosperms is very great. Outcrossing is enforced by dioecism, heterostyly, hermaphroditism with self-incompatibility, and monoecism with self-incompatibility. Outcrossing is promoted but not necessarily enforced strictly by monoecism, gynodioecism, incomplete self-incompatibility, and hermaphroditism with protandry or protogyny. Inbreeding is enforced by cleistogamy and bud autogamy, and promoted by predominant autogamy.

Between those extremes lies an intermediate zone of mixed outcrossing and inbreeding. The very frequent combination of hermaphroditism with self-compatibility leads to various mixtures of crossing and selfing. Some plants have open-pollinated and cleistogamous flowers on the same individuals, and this combination also produces mixtures of crossing and selfing.

The amount of outcrossing can be expressed quantitatively. Two units of measurement are used: the rate of natural crossing and the percentage of outcrossing. Natural crossing is defined as crossing between different genotypes, and outcrossing as crossing between different individuals.

The rate of natural crossing is the percentage of the progeny of a given individual or biotype, grown in a mixed planting and exposed to open pollination, that is derived from pollen of a different genotype. This percentage is usually inferred from the frequency of genetically marked individuals in the progeny where either the female plants or the pollen donors, or both, are carriers of some markers. The percentage of outcrossing is often larger than the rate of natural crossing. Formulas and methods of computation are given by Fryxell (1957) and Vasek (1964, 1965).

Outcrossing at the rate of 100% is presumably brought about by dioecism and complete self-incompatibility. Cleistogamy and bud autogamy presumably produce no crossing.

Predominant autogamy brings about various rates in the low range. In *Galeopsis tetrahit* the rate of natural crossing varied from 0 to 16% in different progeny tests (Müntzing, 1930). The rate of natural crossing in one population of *Clarkia exilis* was 3.6% (Vasek, 1964). Some other rates of natural crossing in predominantly autogamous plants are 1–3% in *Lactuca sativa* in California, 0.15–11.5% in *Lactuca sativa* in England, and 0.2% in *Hibiscus sabdariffa* (Thompson, Whitaker and Bohn, 1958; Watts, 1958; Sanyal and Dutta, 1954, respectively). The percentage of outcrossing is estimated to be 43–45% in *Clarkia exilis* (Vasek, 1964, 1967). Much lower amounts of outcrossing are estimated in other autogamous plants, for example, 1–2% in *Hordeum vulgare* (Jain and Allard, 1960).

A fairly large body of information has accumulated on the amounts of natural crossing and outcrossing in different plant groups. Partial summaries with references are given by Fryxell (1957) and Haskell (1961). We can mention only one good example here. A natural population of the hermaphroditic and self-compatible but nonautogamous species, *Clarkia unguiculata,* has an estimated 96% outcrossing (Vasek, 1965).

Plants with hermaphroditic, self-compatible, and partially cross-pollinated flowers sometimes show a wide range of rates of natural crossing. The rate varies with geographical locality, stage of season, position of plant in the population, and other factors. Such variation is well illustrated by cotton, *Gossypium hirsutum.*

Seed formation in cotton is by a mixture of cross- and self-fertilization. The relative amount of natural crossing in this mixture varies from more than 50% in North Carolina and Tennessee to less than 5% in some parts of Texas (Stephens, 1953; Stephens and Finkner, 1953; Simpson, 1954). Within one locality there is further variation in rate of natural crossing between early and peak flowering periods (Stephens, 1953). The rate of natural crossing is much higher between individuals standing close together than between remote individuals (Simpson and Duncan, 1956). The rate is also higher on the margins of the plant population than in the central parts (Fryxell, 1956). Finally, there are significant differences in rate of natural crossing between different flowers on an individual plant. Plants with an overall average of 11% natural crossing had some fruits with about 50% natural crossing and some with all selfed seeds (Hutchinson and Lawes, 1953).

Individual plants belonging to certain species produce open-pollinated flowers on some flowering branches and cleistogamous flowers on others, as mentioned earlier. Among the species in which this situation has been investigated are *Lamium amplexicaule* (Labiatae) and *Lespedeza cuneata* (Leguminosae) (Bernström, 1950; Fryxell, 1957). In other species the plant population is polymorphic for open-pollinated plants and cleistogamous plants. An example is *Lithospermum caroliniense* (Boraginaceae) (Levin, 1972). In either case the total seed output of the plant population is a mixture of cross-fertilized and selfed seeds. The proportions of the two components in the seed pool, and hence the relative amount of outcrossing, are influenced by various environmental factors.

Incomplete self-incompatibility in a plant with hermaphroditic flowers also leads to flexibility in the proportions of cross- and self-fertilization. Bateman (1956) found that *Cheiranthus cheiri* (Cruciferae) sets seeds freely when isolated and self-pollinated. But when the flowers of a plant are pollinated with a mixture consisting of equal amounts of its own pollen and that from another, genetically marked individual, the latter is responsible for 92% of the seed set. The foreign pollen clearly surpasses the self pollen competitively.

The mechanism underlying the competitive advantage of the foreign pollen may be similar to that discovered by Smith and

Clarkson (1956) in *Iris tenax*. These workers compared the rate of growth of pollen tubes down the style after cross-pollinations and self-pollinations. The pollen tubes derived from sib crosses reach the locules of the ovaries in 30 hours, whereas pollen tubes derived from self-pollinations require 40 hours to grow the same distance. The retarded growth rate of the self pollen constitutes a mechanism favoring outcrossing.

The localized activities of pollen-carrying insects on plants bearing numerous flowers result in a preponderance of self-pollinations. This is true in Iris, where a single individual may form a clone several feet in diameter, consisting of numerous flowering stalks. Bees working within a narrow foraging radius in such a clone will inevitably bring about more self-pollinations than cross-pollinations. Since some species of Iris are self-compatible, it would be simple to assume that the observed excess of self-pollinations leads to a high proportion of inbred progeny in the populations. Such an assumption does not take account of the factors intervening between pollination and the establishment of progeny. One of these factors, the selective inferiority of self pollen in competition with foreign pollen, may modify or reverse the relations between selfing and outcrossing established at the stage of pollination (Grant, 1958).

RELATIONSHIP BETWEEN LOW BASIC CHROMOSOME NUMBER AND BREEDING SYSTEM

The value of reduced basic chromosome numbers as a means of restricting recombination depends on the existence of a condition of genic heterozygosity. It does not make much difference in a highly homozygous line of plants whether $n = 6$ or $n = 9$, so far as the generation of recombinations is concerned. But low basic numbers are important for the restriction of recombination in outcrossing and highly heterozygous plant groups.

We would expect, therefore, to find that reduced basic numbers are prevalent in outcrossing plants which place an emphasis on restricted recombination. Predominantly self-fertilizing and homozygous plants, on the other hand, would be expected to display a variety of high and low chromosome numbers.

Around 800 species of angiosperms have basic chromosome numbers ranging from $n = 6$ down to $n = 2$. Most of these species have been studied cytologically but not genetically, and their breeding systems are consequently unknown. The evidence available at present is incomplete and inadequate for testing the expectations outlined above. However, we do find some preliminary indications.

A search in the literature disclosed some nine species of Compositae that have low numbers and that have been determined for breeding system. The numbers are in the range $2n = 4–12$. The nine low-number species are all outcrossing, and all or nearly all are self-incompatible:

Aster amethystinus ($2n = 10$)	*Haplopappus gracilis* ($2n = 4, 6$)
A. novae-angliae ($2n = 10$)	*Hypochaeris radicata* ($2n = 8$)
Crepis capillaris ($2n = 6$)	*Leontodon autumnalis* ($2n = 12$)
C. foetida ($2n = 8, 10$)	*Picris echioides* ($2n = 10$)
C. paludosa ($2n = 12$)	

Conversely, I could find no case of a self-fertilizing species of Compositae with a low number ($2n = 12$ or less), although such cases may well exist. There are numerous self-fertilizing Compositae with normal unreduced numbers, e.g., *Lactuca sativa* ($2n = 18$) and *Bellis perennis* ($2n = 18$).

The family Liliaceae also contains low-number groups. At least some of these can be identified as heterozygous on the basis of population cytological evidence. Examples are Trillium ($x = 5$) and Paris ($x = 5$).

In Amsinckia (Boraginaceae), the heterostylous and outcrossing species in the *A. furcata* group have low numbers ($n = 7$ to $n = 4$), whereas self-fertilizing species in the section Muricatae have higher numbers ($n = 8, 12, 13$, and higher) (Ray and Chisaki, 1957a, 1957b).

This desultory evidence is inconclusive, but it is suggestive and confirmatory as far as it goes. Further study of the relationship between basic chromosome number and breeding system would appear to be definitely worthwhile.

POLLINATION SYSTEM   Pollen vectors vary in two main ways which affect the rate of natural crossing and hence the amount of recombination in the plant population. The first variable is the range of pollen dispersal. The second is the extent of the deviation, if any, from random mating.

Some pollinating agents such as wind and water currents are quite promiscuous, carrying pollen to flowers or ovulate cones within their range of dispersal without regard to individual genotype or species affinity. Many relatively unspecialized flower-visiting insects, particularly among beetles and flies, are highly but not completely promiscuous pollinators. They transfer pollen more or less undiscriminatingly among a wide array of flower types in the plant community.

The opposite extreme is represented by highly specialized flower-visiting insects with behavioral habits of flower constancy. Flower-constant behavior is strongly developed in the honeybee and other social bees. Individual bees tend to remain fixed on one flower type or another, as recognized by visual and olfactory signals, during a succession of foraging visits, (see Grant, 1950a; Free, 1970; Faegri and Pijl, 1971).

Such flower-constant behavior sets up a system of assortative mating in a plant population which is polymorphic for certain floral characters. The bees will often carry out cross-pollinations preferentially between individual plants with flowers of the same color type, color pattern type, configuration, or scent. A related behavioral trait in bees—constancy for locality and position of flowers—can bring about assortative mating with respect to other relevant plant characters. Levin and Kerster (1973) have recently documented a case of assortative pollination for plant stature in *Lythrum salicaria* (Lythraceae).

Many intermediate conditions exist between promiscuous wind pollination and flower-constant bee pollination. Hawkmoths and butterflies exhibit flower constancy to a generally lesser degree than social bees. Hummingbirds are not flower constant but they are strongly territorial during the breeding season. All such moderate deviations from randomness in flower visitations presumably lead to correspondingly intermediate or slight levels of assortative mating in the plant populations.

Pollinating agents which carry pollen grains far and wide would be expected in general to bring about an open recombination system. Conversely, more sedentary pollinating agents as exemplified by various groups of small flower-visiting insects are probably associated with restricted recombination systems in their plant hosts. These correlations seem plausible enough, but much more evidence is needed to substantiate them.

DISPERSAL RANGE   The motile organs in sessile seed plants are the pollen grains, seeds (or fruits or cones as the case may be), and detached vegetative propagules (which often behave like seeds or fruits). These dispersal units are transported by a variety of agents: wind, water currents, and various kinds of animals in different ways. An interesting descriptive account of the modes of dispersal of seeds, fruits, and vegetative propagules is given by Pijl (1972).

A wide dispersal radius will tend to open up the recombination system and a narrow dispersal radius will restrict it. Undoubtedly there are large differences between plant groups in dispersal potential. Furthermore, such differences must be correlated with differences in the recombination system. But good comparative quantitative data on dispersal are difficult to find in the literature, and such data as are available have not been connected with recombination systems.

One general conclusion which has emerged from a number of independent studies is that the bulk of dispersal units comes to rest fairly close to the parent plant. A smaller fraction of the diaspores becomes dispersed to varying distances from the parental individual. The frequency curve for dispersal is leptokurtotic, with a peak close to the parent plant and a tail descending rapidly with distance from the parent. This form of the distribution curve seems to be characteristic for both pollen and seeds and for wind-borne as well as insect-borne diaspores (Bateman, 1946, 1950; Grant, 1958).

Progeny tests of wind-pollinated maize and sugar beets show that plants separated by relatively short distances are semi-isolated. Isolation distances of 60 ft in maize reduce intervarietal contamination to 1% (Bateman, 1947). Intervarietal contamination in sugar beets is only 0.3% at isolation distances of 200 m (about 650 ft) (Archimowitsch, 1949).

Progeny tests of bee-pollinated crop plants give similar results. The amount of cross-pollination between two varieties of cotton decreases from 26% when they are planted side by side to 4% when 25 feet apart (Simpson and Duncan, 1956). Isolation distances of 50 ft in bee-pollinated turnips and radishes reduce intervarietal contamination to 1%, and the next 50 ft of separation produce scarcely any additional effect (Bateman, 1946).

The localized movements of bees foraging in a plant population, which account for the vicinism described above, are supplemented by longer flights during which some widely separated plant individuals may become cross-pollinated. In cotton populations cross-pollinated by bumblebees, usually within local neighborhoods, some pollen grains or substitute dye particles have been shown to be transported at least 80 ft from the source plant (Stephens and Finkner, 1953). Abundant evidence indicates that some wind-borne pollen is dispersed far from the pollen parent.

The picture is one of much cross-pollination between neighboring plants and some crossing between more widely separated plants. The population-genetical implications of this pattern have been spelled out by Bateman (1949). The plant population is inferred to have a great deal of local inbreeding supplemented by occasional long-range dispersal of genes. There are opportunities for the formation of new recombination types as a result of long-range crossing, but equally there are opportunities to test these recombination types in inbreeding neighborhoods (Bateman, 1949).

The same pattern of dispersal seems to hold true in wind-pollinated and animal-pollinated plants. To the extent that the pattern is general, it affords no basis for predicting expansive or restrictive effects on the recombination system of the plants involved. It is at this point that our information is seriously incomplete. It is probable that wind pollination usually brings about a more open recombination system than animal pollination. The critical comparison will be the shape and scale of the tail of the frequency curve portraying the fraction of long-range pollen dispersal.

We could describe a generally parallel situation and draw

similar conclusions for seed dispersal, which is the other main vehicle of gene flow in plant populations.

A comparison which reveals the effects of long- versus short-range seed dispersal on the recombination system is that between Juniperus and Cupressus. The comparison is instructive since the two genera belong to the same family (Cupressaceae), have very similar vegetative characters, and are both wind pollinated. They differ mainly in cone characters which are related to seed dispersal. The seeds of Juniperus are contained within succulent blue berries which are extensively eaten and disseminated by thrushes, bluebirds, starlings, crows, and other birds; the cedar waxwing is named for its predilection for the berries of Juniperus. In Cupressus, on the other hand, the seeds are borne in a heavy, woody, globose cone which normally travels only as far from the parent tree as it can roll or be carried by rodents.

The North American junipers form a series of wide-ranging species: *J. virginiana* in the eastern and southern United States, *J. horizontalis* in Canada, *J. scopulorum* in the Rocky Mountain region, and others. The populations are extensive and polymorphic. The variation is clinal in some cases which have been analyzed in detail. Terpenoid compounds show variation both within and between populations of *J. virginiana,* and this variation shows clinal trends on a 1500-mile geographical transect between Washington, D.C. and Texas (Flake, von Rudloff and Turner, 1969).

The genus Cupressus in North America has a different distribution and variation pattern. Cupressus comprises a series of disjunct and more or less endemic populations ranging from Oregon through California to Texas and thence to Mexico, Guatemala, and Costa Rica. Fifteen species are recognized in this region by the latest monographer, Wolf (1948). Although some species such as *C. arizonica* form extensive stands, most of the species are represented by small numbers of groves. Eleven occur as clusters of one to seven small isolated groves per species. Morphological characters show a high degree of uniformity within a grove or endemic species. The variation in the endemic cypresses is not so much between individuals as between groves (Wolf, 1948).

In short, the variability of the endemic cypresses is channelized within an array of morphologically discrete colonies, whereas that of the wide-ranging junipers is spread out clinally over areas of subcontinental extent.

POPULATION SIZE   Population size and structure have a strong influence on the recombination system. A large population can be the repository of a much greater store of variability than a small population. The extensive, continuous populations of many dominant forest trees, plains grasses, and desert shrubs are associated in general with open recombination systems. The small, isolated, colonial populations characteristic of many subdominant or pioneering species, by contrast, correlate with restricted recombination systems. We have previously referred to the greater amount and wider diffusion of variability in the large populations of Juniperus than in the small disjunct colonies of Cupressus. This example was presented in connection with seed dispersal, but it is equally pertinent to population structure. There is interaction between the dispersal factor and the population factor.

Clinal variation is characteristic of species such as *Pinus sylvestris* in northern Europe, *Bouteloua curtipendula* in the North American plains, and *Juniperus virginiana* in the eastern and southern United States. These are species which form extensive and more or less continuous populations (Langlet, 1936; Olmsted, 1944; Flake, von Rudloff and Turner, 1969). In such species, the continual mixing of gametes drawn from adjacent neighborhoods, generation after generation, leads to gradual intergradation over large geographical areas.

The duckweeds (Lemnaceae) are small free-floating plants of ponds and streams. The aquatic habitat for any single species is relatively uniform over large geographical areas. There is, moreover, a continual interchange of individuals between the colonies as a result of transport by water birds. Consequently the breeding population must frequently embrace numerous colonies within a large territory. It is interesting, therefore, that six species of Lemna and one species of Spirodela sampled at widely spaced stations in North America and Europe and studied physiologically are only weakly differentiated into races when compared with various species of terrestrial plants inhabiting the same areas (Landolt, 1957).

A contrasting variation pattern is exhibited by plants with the colonial type of population structure. This pattern is seen in the endemic cypresses and in many groups of annual herbs. Good examples are found in the annual genera Layia, Clarkia, Gilia, and Amsinckia in the California flora (Clausen, 1951; Lewis, 1953; Grant, 1950b, 1954; Ray and Chisaki, 1957a, 1957b, respectively).

In these annuals the individual colonies may or may not be highly polymorphic. The most marked variations within the species, in any case, are those which differentiate the colonies from one another. The unit of variation is to a large extent the colony. The species presents an array of more or less discrete races or ecotypes corresponding in some cases to regional groups of colonies and in extreme cases to the individual colonies themselves.

The ability of the colonial population structure to restrict and channelize recombination is clearly manifested in those colonial species which have become contaminated by natural hybridization. The hybridized species considered as a whole exhibits the expected high degree of variability. The individual colonies, however, are frequently more or less uniform, and the variability becomes evident only on comparing the separate populations. An example of this variation pattern is provided by *Gilia achilleaefolia* (Grant, 1954).

STERILITY BARRIERS The species as a whole can be regarded as a broad, inclusive field for gene recombination (Carson, 1957). The strength and distribution of reproductive isolating mechanisms, particularly internal ones, therefore affect the extent of this field.

This factor, like all the others, shows wide variation in different plant groups. The evidence has recently been reviewed in some detail (Grant, 1971, ch. 7), and can only be touched on here.

Many north-temperate groups of trees and shrubs have only weak internal barriers between related species. Examples are Ceanothus, Pinus, and Quercus. The same condition holds in many groups of perennial herbs, as exemplified by Antirrhinum, Aquilegia, and orchids. In such groups, natural hybridization can and frequently does bring together the genes from different species or semispecies. The field for gene recombination is broad taxonomically.

A contrasting situation exists in numerous groups of autogamous annuals in such genera as Erophila, Gilia, and Vicia, and in the autogamous perennial grass, Elymus. Here internal barriers of cross-incompatibility, chromosomal sterility, and the like are strongly developed within the limits of what is considered a taxonomic species. The taxonomic species, in other words, is a cluster of morphologically similar, reproductively isolated, sibling species. The field for gene recombination is narrow taxonomically.

There are of course many intermediate conditions. For example, outcrossing annual herbs in genera like Clarkia, Gilia, and Layia have strong internal barriers surrounding related taxonomic species. The field for recombination is broader than in autogamous annuals but more restricted than in the groups of perennial plants mentioned earlier.

THE BALANCE SYSTEM OF REGULATION   The total amount of recombination generated in any population or species is the result of the combined actions of numerous regulatory factors. An interesting feature of their operation in a given organism is that often they work in opposing directions.

In comparisons between woody plants and herbaceous plants, for example, we often find that a long generation is associated with a high basic chromosome number and a short generation with a low chromosome number. If in a given group the length of generation restricts recombination, the chromosome number is apt to be such as to promote recombination; and, conversely, where the generation time is expansive, the chromosome number is often restrictive.

To take another example, the dispersal potential of the pollen is not always correlated positively with that of the seeds in a plant group. Oak trees have wind-borne pollen and heavy seeds; their dispersal range is relatively wide at the stage of cross-pollination and relatively narrow in the seed stage. Orchids have the contrasting combination of insect cross-pollination and wind-borne seeds.

Other paired combinations of regulatory factors also show opposing directions of action in particular plant groups, one factor being expansive and the other restrictive. A series of cases is listed below in synoptical form together with typical plant examples of each (from Grant, 1958).

1. Generation time and chromosome number
   a. Short generations and low chromosome number (many herbaceous plants)
   b. Long generations and high chromosome number (many woody plants)
2. Generation time and breeding system
   a. Short generations and autogamy (many annual herbs)
   b. Long generations and outcrossing (many woody plants)
3. Generation time and population size
   a. Short generations and colonial populations (many annual herbs)
   b. Long generations and large continuous populations (temperate forest trees, steppe grasses)
4. Generation time and sterility barriers
   a. Short generations and closely spaced sterility barriers (many annual herbs)
   b. Long generations and widely spaced sterility barriers (many perennial plants)
5. Chromosome number and chiasma frequency
   a. High chromosome number and low chiasma frequency (many dicotyledons)
   b. Low chromosome number and high chiasma frequency (Paeonia)
6. Chromosome number and breeding system
   a. High chromosome number and inbreeding (*Nicotiana tabacum, N. rustica,* Amsinckia section Muricatae)
   b. Low chromosome number and outcrossing (*Nicotiana alata, N. longiflora, Amsinckia furcata*)
7. Chiasma frequency and breeding system
   a. High chiasma frequency and autogamy (*Sitanion jubatum*)
   b. Low chiasma frequency and outcrossing (*Agropyron parishii*)
8. Pollination system and seed dispersal potential
   a. Promiscuous wind pollination and short-range seed dispersal (oaks, pines)
   b. Flower-constant bee pollination and long-range dispersal of seeds by wind (many orchids, many weeds)

The existence of opposing combinations of regulatory factors in recombination systems is very interesting. In biological systems generally, a fine adjustment is often attained by a balancing of opposite forces. In the genetics of individual organisms this method of control is apparent in polygenic systems composed of plus and minus polygenes. It is significant that the regulation of recombination in populations is accomplished by a similar system of checks and balances (Grant, 1958).

The net result of the combined action of all factors, however, is a more open recombination system in some species and a more restricted one in others. The regulatory factors may work in balance but they do not cancel one another out.

CHAPTER TWENTY-FOUR

THE ECOLOGY OF
RECOMBINATION SYSTEMS

INTRODUCTION The function of the recombination system is to bring about a workable compromise between the contradictory demands for fitness and flexibility in population variability. Different points of compromise are found in different organisms. A spectrum of recombination systems, ranging from open through restricted to closed, occurs in nature.

Species possessing different types of recombination systems seem to be equally successful in nature. The question then arises, What conditions favor the development of open recombination systems in some groups of organisms and restricted or closed systems in other groups? Under what conditions will each type of recombination system have a high selective value? The question posed in this form brings in ecology.

Darlington (1939), Mather (1943), and Carson 1957) considered the genetic and evolutionary, but not the ecological, aspects of recombination systems. Salisbury (1942) meanwhile investigated the ecological but not the genetic aspects of plant reproduction.

Stebbins (1950, chs. 5, 12) put the two approaches together. He showed that fitness-producing genetic systems are correlated with and favored in pioneer habitats in numerous plant groups. This opened up a new approach to the study of genetic systems.

Grant (1958) later broadened the generalization. A correlation was shown between type of recombination system and position of a species in the ecological succession from pioneering to climax community. The causal connection underlying this correlation consists of an interaction between net fecundity and stabilizing selection. Different stages in the development of biotic communities call for different reproductive strategies in the species occupying those stages.

We will elaborate on this general idea in the next several sections of this chapter. The problem has of course been discussed by a number of authors since the 1950s. We will bring in the more recent contributions as the story unfolds.

CLASSIFICATION OF HABITATS Salisbury (1942) recognized five main types of habitats which differ in ease of colonization. Salisbury's classification is basic to our later discussion and is accordingly summarized here.

Open habitats
  1. Continuously open habitats
     a. Natural: open deserts, strand
     b. Man-made: cultivated fields, roadsides
  2. Intermittently open habitats: clearings in woods, burned areas, flood plains, landslides
  3. Semi-open habitats—vegetation comparatively open and unshaded: desert grassland, arid scrub
Closed habitats
  4. Closed communities
  5. Shady habitats

Colonization is easiest in the open habitats. It is more difficult in the closed habitats because it must take place despite strong competition. The conditions of competition and colonization are intermediate in semi-open habitats. Annuals preponderate in the open habitats and perennials in the closed habitats (Salisbury, 1942).

Salisbury found characteristic trends in the seed output of species living in the various types of habitats. A sample of species was studied for seed production in each habitat; the sample size ranged from 22 to 41 species. The overall average seed output per plant in each habitat is as follows (the habitats are numbered as above):

Open habitats
1. 2,904 seeds per plant
2. 28,289 seeds per plant
3. 2,379 seeds per plant

Closed habitats
4. 1,944 seeds per plant per season
5. 280 seeds per plant per season

The figures demonstrate that the average seed output is relatively large in annual species of open habitats, which must often play a colonizing role. By far the largest seed outputs are found in species of intermittently open habitats; such species are the "opportunists of the plant world" (Salisbury, 1942, p. 232).

The average seed output is lower in species of closed habitats, which have a smaller opportunity for seedling establishment. Here perennial life forms preponderate, vegetative propagation is common, and the seeds are often furnished with abundant food reserves. These features are part of the species' equipment for coping with the generally strong competition in closed habitats (Salisbury, 1942). Salisbury's conclusions have been confirmed recently in an independent study by Gadgil and Solbrig (1972).

Seed number is thus correlated with ease of colonization. It is logical to suspect that genotypic variability in the seed output is also correlated with this factor. We are concerned with the latter aspect of seed reproduction in this chapter.

SPECIES OF CLOSED HABITATS　The hereditary mechanism in an open recombination system generates many new gene combinations as well as the parental ones in each generation. A large proportion of the new gene combinations will be ill adapted and eliminated by stabilizing selection. The production of a wide diversity of genotypes entails a certain wastage of zygotes. But this loss of zygotes is not disadvantageous to the same degree for all populations in all species.

Some species normally produce a great excess of zygotes relative to their opportunities for establishment. Since an extensive elimination of progeny is inevitable, it may as well be a selective elimination, involving a variable progeny. It is no more expensive for a species with a large excess fecundity to lose a given number of recombination-type zygotes than to lose the same number of parental-type zygotes. Such a species has nothing to lose and something to gain by gambling with its zygotes. It can afford an open recombination system (Grant, 1958).

A capacity for increase that is greatly in excess of the chances of establishment is the normal condition in plant species belonging to closed communities. In a climax temperate forest, for example, the trees produce their crops of seeds year after year; yet the seedling reproduction is limited to the replacement of older trees as they die and leave an opening in the community. It is a common observation that a forest opening is quickly populated with hundreds or thousands of seedlings of the same species, only a small fraction of which will survive to maturity. Under these conditions, strong restrictions on recombination are unnecessary for the preservation of constancy in reproduction. This can safely be left to stabilizing selection. And there is always the chance that some of the new recombination types may be better adapted than the preexisting types.

The regulatory factors are typically such as to promote open recombination in many north-temperate woody plants. The populations are often large. Internal barriers to natural interspecific hybridization may be weakly developed, making the potential breeding group still larger. Outcrossing breeding systems prevail. Pollen and seeds are often comparatively widely dispersed by wind. Basic chromosome numbers are relatively high (see Chapter 23; see also Grant, 1971).

Some of the same factors for open recombination are present in plains grasses and many perennial herbaceous dicotyledons. The populations of perennial plains grasses tend to be large and widely cross-pollinated by wind. In perennial herbaceous dicotyledons, outcrossing breeding systems are common, and the basic chromosome numbers are often higher than in related annuals.

Stebbins (1958) compared 201 species of the tribe Cichorieae (= tribe Lactuceae, family Compositae) with respect to habitat, breeding system, and basic chromosome number. The habitats range from forests and moist meadows through forest clearings and open mountain slopes to deserts and cleared fields. Stebbins describes these habitat types as stable, intermediate, and unstable, respectively; they correspond to the categories of closed, semi-open, and open habitats in Salisbury's system.

The results of Stebbins's survey are summarized in Table 43. The table shows that outcrossing in the Cichorieae prevails in stable and intermediate habitats but is not frequent in unstable habitats. The combination of outcrossing and high basic chromosome numbers, which together produce the most open recombination systems in the tribe, occurs mainly in closed and semi-open habitats (Stebbins, 1958).

We have, then, plants with open recombination systems producing excess numbers of seeds in closed communities. Their populations contain stored potential variability, and have genetic systems to release this variability, yet the populations remain relatively constant. It is probable that the constancy is maintained to a large extent by stabilizing selection.

This hypothesis can be tested by observing the geographical and ecological distribution of natural hybrids between species with open recombination systems. The species-distinguishing characters provide visible markers of release or nonrelease of variability. I will mention one example which is typical.

Two common species of sage, *Salvia apiana* and *S. mellifera,* occur sympatrically over thousands of square miles in coastal California. Their habitat is a closed or semi-open sage scrub community. The two species are isolated by various external and internal mechanisms that are insufficient to block hybridization entirely, as shown by the occasional occurrence of hybrids and introgressive hybrid derivatives. The products of natural hybridization occur only where the originally stable sage scrub community has been disrupted by human activities such as clearing or fire. In the undisturbed natural scrub community the two species remain uncontaminated by hybridization despite sympatric contacts (Epling, 1947; Anderson, 1953; Anderson and Anderson, 1954; Grant and Grant, 1964).

TABLE 43
CORRELATION BETWEEN BREEDING SYSTEMS, BASIC CHROMOSOME NUMBER, AND HABITAT IN SPECIES OF THE TRIBE CICHORIEAE (COMPOSITAE) (STEBBINS, 1958)

Breeding system and basic chromosome number	No. of species	Type of habitat		
		Stable	Intermediate	Unstable
Cross-fertilizing, $x = 9$ or 8	98	47	44	7
Cross-fertilizing, $x = 7$ or 6	28	2	26	0
Cross-fertilizing, $x = 5, 4,$ or 3	18	4	13	1
Self-fertilizing	57	6	28	23

We have good indirect evidence in this and other similar cases that much potential variability is actually released during seed formation but is suppressed by conditions in a closed community. The suppressor is probably stabilizing selection. The important task of holding a given adaptive gene combination together from generation to generation is carried out not primarily by the recombination system, which is open, but by stabilizing selection.

These generalizations may not be widely applicable in closed plant communities in the tropics. Baker (1959) maintains that the situations are quite different in tropical forests and in temperate forests. In the latter the replacement of old trees by new seedlings involves intraspecific competition and selection. In a tropical forest composed of numerous codominant tree species, by contrast, the competition preceding seedling establishment necessarily has a strong interspecific component. This and other factors might be expected to favor restrictions on recombination within each competing tree species in the tropical forest (Baker, 1959). Such species, unlike their north-temperate counterparts, may not be able to afford open recombination systems. This interesting and plausible idea warrants further study.

COLONIZING SPECIES IN HIGHER PLANTS A species which plays a pioneering role in an open community has demands on its reproductive potential which are quite different from those in climax or subclimax species.

On reaching an unoccupied territory, a colonizing species must build up a population as quickly as possible. Space is available for most, and in extreme cases all, of the zygotes the foundation population can produce. It is desirable that the maximum possible number of these zygotes carry a well-adapted gene combination. Therefore it is also desirable that the breakdown of any existing adaptive gene combination by recombination be minimized during the period of population expansion. The adaptive gene combinations are held together not so much by stabilizing selection as by the recombination system in colonizing species.

Recombination should be restricted in colonizing plants, but it cannot be dispensed with entirely because it is necessary for assembling the favorable gene combination in the first place. The opposing requirements for recombination and for the preservation of gene combinations intact during periods of population growth can be reconciled if the events of recombination are distributed sporadically in time. Then the formation of recombination types can be followed by a few generations of selection, and selection can be followed in turn by multiplication of the best-fitted types. Restricted and closed recombination systems are indeed characterized by a sporadic distribution as well as a reduced flow of variability.

We are not contending that colonizing plants with restricted or closed recombination systems always have genotypically uniform populations. In extreme cases they do; in other cases the populations are variable, but the amount of variability is less than that typically found in plants with open recombination systems. S. Jain (personal communication, 1974) has recently and rightly questioned the assumption, sometimes made, that genotypic uniformity is necessarily advantageous for colonization. The problem of the optimum level or levels of variability in a plant population colonizing a new habitat requires much further study.

We find closed recombination systems in agamospermous plants which have colonized areas exposed by the retreat of Pleistocene glaciers (e.g., Rubus section Eubatus), lava flows (Crepis section Psilochaenia), and other open habitats. Uniform populations are built up and perpetuated by asexual seed reproduction. But the agamospermous reproduction is interrupted periodically by events of sexual reproduction which produce new variant types (see Grant, 1971, chs. 18 and 21 for review).

Restricted recombination systems are characteristic of annual herbs of open natural habitats in mediterranean and desert regions and weedy annuals of cleared fields. Such colonizing annuals often have colonial populations, autogamy or a mixture of autogamy and cross-pollination, low basic chromosome numbers, strong internal barriers to interspecific hybridization, and other restrictive regulatory factors (for reviews see Stebbins, 1957; Grant, 1958, 1971).

We have examined the ecological distribution of recombination-promoting breeding systems and chromosome numbers in the Cichorieae (Compositae). Let us now look at the ecological distribution of the restrictive forms of these same regulatory factors in this tribe. Table 43 shows that self-fertilization prevails in unstable and intermediate (= open and semi-open) habitats. Basic chromosome numbers tend to be lower in species of open or semi-open habitats than in those of closed communities. Most of the species of Cichorieae in open habitats and a high proportion of those in semi-open places are annuals (Stebbins, 1958).

It is a remarkable fact that very similar combinations of restrictive regulatory factors have developed in the colonizing annual members of different angiosperm families. The Gramineae, Cruciferae, Leguminosae, Onagraceae, Polemoniaceae, Boraginaceae, Labiatae, and Compositae furnish examples of such convergence in the recombination system.

Take the annual genus Clarkia (Onagraceae) and the annual sections of Gilia (Polemoniaceae), for example. There are numerous similarities between the two phylads in life form, range of phenotypic plasticity, floral mechanism, breeding system, mode of pollination, population structure, variation pattern, chromosome number, and strength and distribution of sterility barriers. The most characteristic differences in the genetic systems of the two genera are the greater frequencies of translocations, supernumerary chromosomes, and aneuploidy in Clarkia. The likenesses clearly outnumber the differences, but the similar features are by no means family characteristics. Clarkia and Gilia both differ from the perennial climax and subclimax members of their respective families in these features. The common factor underlying the similarities is the mode of life of colonizing annuals.

It is perhaps worth noting that the science of genetics owes a large debt to colonizing annual plants with restricted recombination systems. Mendel's peas, Johannsen's beans, De Vries' Oenotheras, and East's Nicotianas are among the experimental plants which exemplify this genetic system.

COLONIZING SPECIES IN OTHER KINGDOMS The capacity for rapid expansion of uniform populations is associated with restricted recombination systems in various colonizing organisms outside the plant kingdom. A few examples will be mentioned here to show that restricted recombination in colonizing species is a widespread if not general phenomenon.

Paramecium is hermaphroditic and either cross-fertilizing or self-fertilizing, depending on the species. *Paramecium bursaria* is outcrossing, while *P. aurelia* and *P. caudatum* are predominantly autogamous. In the latter groups, periods of inbreeding alternate with events of cross-fertilization. The inbreeding paramecia tend to have a shorter generation time than the outbreeding forms, as in higher plants. Asexual fission also occurs in Paramecium and is a means of clonal multiplication (Sonneborn, 1957).

The ability of bacteria to build large, uniform populations rapidly by asexual fission is well known. The sequence of asexual cell divisions is interrupted at rare intervals by parasexual conjugation and exchange of genetic material. The new recombination types can then multiply asexually again.

Among arthropods two examples are the planktonic crustacean, Daphnia (Brooks, 1957), and the leaf-inhabiting hemipteran, Aphis (Imms, 1957). Both organisms have a sexual phase in their life history, but population growth is achieved by parthenogenesis. In Aphis the parthenogenetic reproduction begins with the advent of spring. Successive asexual generations consisting of parthenogenetic females enable the aphids to increase in numbers at a pace matching the growth of foliage of their plant hosts. Sexual females and males appear toward the end of the growing season and produce fertilized eggs which overwinter and start a new cycle of parthenogenetic females the following spring (see Imms, 1957).

A land snail, *Rumina decollata,* native in the Mediterranean region, was introduced into warm parts of North America in the nineteenth century and successfully colonized various areas of the southern United States and Mexico. It turns out to be self-fertilizing. Its North American populations are also very uniform genetically, and in fact essentially monomorphic, as shown by electrophoretic analysis of proteins representing 25 loci. Populations of the same species in its native range exhibit some polymorphism. This suggests that the North American populations are descendants of a single homozygous founder stock which increased greatly in numbers by self-fertilization (Selander and Kaufman, 1973).

In all these cases the organism is small and the generation time short. The population is capable of rapid expansion to exploit its ecological opportunities. Genetic recombination is brought about by sexual or parasexual processes at some point in the life history of the population. The products of the sexual or parasexual stage then multiply by some alternative means which avoids further recombination: asexual fission (bacteria, Paramecium), parthenogenesis (Daphnia, Aphis), or self-fertilization (Paramecium, Rumina).

RELATED CONCEPTS IN POPULATION ECOLOGY The concepts presented in the preceding sections of this chapter were worked out between 1939 and 1958 by a series of workers who were genetically oriented. In the last few years very similar concepts have been arrived at by a younger school of population ecologists. I am referring to concepts currently being discussed under the headings of r and K selection and r and K reproductive strategies. These terms and concepts were introduced by MacArthur and Wilson (1967) and have been elaborated in later publications by members of the same general school (e.g., Pianka, 1970, 1972; Gadgil and Solbrig, 1972; MacArthur, 1972; Emlen, 1973).

The symbols r and K are borrowed from equations describing the regulation of population size; in these equations r refers to the intrinsic rate of increase and K to the carrying capacity of the environment.

The r selection occurs during mass deaths of individuals owing to adverse environmental conditions, as for example the mass mortality of temperate and boreal insect populations with the onset of winter. It favors high fecundity and rapid development. Organisms having *r*-type reproductive strategies exhibit these characteristics (e.g., many high-latitude insects).

K selection takes place when the mortality has a lower rate and a higher selective component. *K* selection favors lower fecundity, better protection of the young, slower development, greater competitive ability, and more efficient utilization of resources. Organisms with *K*-type reproductive strategies are exemplified by terrestrial vertebrates of tropical and temperate regions (MacArthur and Wilson, 1967; Pianka, 1970).

A more complete list of the organismic and environmental features associated with *r* and *K* selection is given by Pianka (1970). Pianka also points out that the *r* and *K* reproductive strategies outlined above are the extremes of a spectrum and that most organisms will exhibit some intermediate condition.

The *r*-type and *K*-type reproductive strategies are adapted to different ecological conditions. The *r* strategy is the strategy of a colonizing species. The *K* strategy is that of a species belonging to a stable and saturated community (MacArthur and Wilson, 1967; Pianka, 1970). Perhaps the *r* and *K* strategies will also be found to correlate with the position of the organism in the food chain, the *r*- types being low in the chain.

MacArthur and Wilson (1967) and their followers do not refer to the basic earlier studies of the problem. Salisbury's work (1942) is especially parallel and relevant. Nor do they refer to Schmalhausen (1949), whose treatment is also relevant to their thesis. This group thus appears to have worked in some isolation from related schools of thought.

CENTRAL VS. PERIPHERAL POPULATIONS OF A SPECIES The peripheral populations of a species have a different recombination system from the central populations in a number of known instances. This was first shown in Drosphila and later confirmed in various plant groups. The plant evidence, however, goes beyond merely confirming the Drosphila evidence; it points to a quite different pattern.

Carson compared peripheral and central populations of *Drosophila robusta* in eastern North America with respect to the frequency of inversion heterozygotes. He found a relatively high frequency of inversion heterozygotes in the central populations. Populations on the periphery of the species area had a lesser amount of inversion polymorphism in some cases and were chromosomally monomorphic in others. The inversion heterozygosity restricts gene recombination. Therefore the central populations are expected to have a more restricted recombination system and the peripheral ones a more open recombination system (Carson, 1955, 1956, 1959, 1961).

This expectation is confirmed by the results of selection experiments for polygenic characters carried out with central and peripheral populations (Carson, 1958). Other species of Drosophila show a similar pattern (Carson, 1959; Carson and Heed, 1964).

The geographical distribution of recombination systems in Drosophila is considered to be adaptive. It provides for maximum release of genetic variation by recombination on the colonizing periphery of the species area. In the central part of the range where the species dominates its habitat the generation of variation is restricted to a smaller array of tried and tested genotypes (Carson, 1955, 1956, 1959).

The geographical distribution of recombination systems in *Paeonia californica* is the reverse of that in *Drosophila robusta*. In Paeonia the complete translocation heterozygotes occur near the southern and interior margin of the species area. Plants forming bivalents or small rings, on the other hand, are most abundant in the central part of the species area (see Fig. 55 and discussion in Chapter 22). In this species, in contrast to *Drosophila robusta,* the peripheral populations exhibit the greatest restrictions on gene recombination.

Another plant species with a similar distribution of recombination systems is *Gilia ochroleuca*. This diploid annual in southern California occurs in a fairly wide range of habitats, from coastal mountains to interior deserts (Fig. 57). The central area of relatively stable woodland communities is occupied by populations which are bee-pollinated and partially outcrossing. The race in the Mojave desert on the eastern part of the species area is autogamous. A local autogamous race also occurs on the extreme northwestern margin of the species area (Grant and Grant, 1956; Grant, 1958).

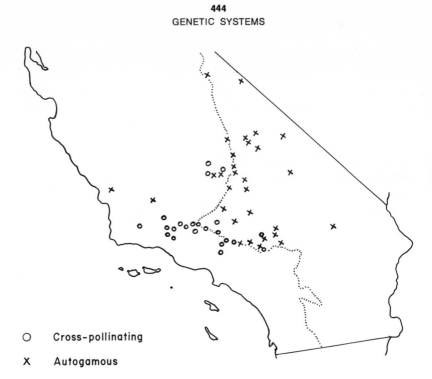

O   Cross-pollinating

X   Autogamous

Figure 57. The geographical distribution of cross-pollinating and autogamous races of *Gilia ochroleuca* (2*n* = 18) in southern California. The dotted line marks the western boundary of the desert. (Redrawn from Grant, 1958.)

Likewise in *Clarkia purpurea,* the autogamous types have the widest distribution and represent the species on the periphery of its area in Pacific North America (Lewis and Lewis, 1955).

There are reasons for believing that the distribution pattern seen in *Paeonia californica, Gilia ochroleuca,* and *Clarkia purpurea* may be common in plants. The peripheral populations of a plant species are apt to be composed of its best colonizers, and the best colonizers, in turn, are apt to have a restricted recombination system.

The problem of geographical variation in amount of recombination has been discussed by several other authors, who have reached varying and sometimes opposing opinions (Wallace, 1959; Darlington, 1963; Ford, 1963, 1964).

Darlington (1963, p. 207) concludes that an increased production of variability will generally be favored "on the periphery or colonizing margin of a species." This is in effect making a general principle of the Drosophila pattern. Elsewhere in the same work (ch. 3) Darlington recognizes the Paeonia pattern. Ford (1963) makes the useful distinction between peripheral localities of a species area in which the species populations can make colonizing advances and those in which they are at their ecological tolerance limits and can only hang on to a precarious foothold. Peripheral areas may also differ from species to species in another way of affecting the requirements for recombination. Peripheral habitats of a climax or subclimax species do not necessarily call for the same local recombination system as the peripheral habitats of a pioneering species. Distinctions such as these may help to explain the differences between the Drosophila and the Paeonia patterns.

EVOLUTIONARY CHANGES IN BASIC CHROMOSOME NUMBER

Phylogenetic trends in some of the regulatory factors of recombination can be discerned from comparative evidence in various plant groups. Such trends can be seen most clearly in the regulatory factors which are associated with morphological or endo-phenotypic features. It is safe to assume that similar trends have occurred in other regulatory factors such as population size and dispersal range.

A systematic treatment of phylogenetic changes in the recombination system in higher plants has not yet been carried out, and this is not the place to attempt so large a task. In the next two sections we merely highlight certain trends with the aid of a few examples. The examples serve to show that evolutionary changes do occur in the recombination system in higher plants, and suggest that such changes have been common in plant evolution.

Basic chromosome numbers in the tribe Cichorieae (or Lactuceae, see Vuilleumier, 1973) range from $x = 10$ to $x = 3$. The most primitive living members of the tribe are cross-fertilizing perennial herbs of forests and meadows in Eurasia (i.e., Dubyaea, Soroseris, Youngia). The wide distribution of $x = 9$ among the primitive genera of Cichorieae suggests that this is the original basic number of the tribe (Stebbins, Jenkins and Walters, 1953).

Phylogenetic decrease in basic chromosome number has taken

place repeatedly in different lines within the tribe, from $x = 9$ to $x = 8$ (Youngia), $x = 7$ (Ixeris, in part), $x = 6$ (Tragopogon), $x = 5$ (Krigia), and lower numbers. The trend reached $x = 4$ in Leontodon and Hedypnois, and $x = 3$ in Crepis and Hypochaeris. Within the large genus Crepis there is a descending aneuploid series from a primitive $n = 6$ to the derived $n = 5$, 4, and 3 (Babcock, Stebbins and Jenkins, 1937; Babcock, 1947; Stebbins, Jenkins and Walters, 1953).

The primitive karyotypes in the Cichorieae tend to be symmetrical. The chromosomes are nearly equal in size and have median or submedian centromeres. The advanced karyotypes with reduced numbers are asymmetrical. The karyotype asymmetry is in part a by-product of the unequal reciprocal translocations involved in aneuploid reduction.

Another phylogenetic trend in the Cichorieae is from perennial herbs of closed communities to annuals of open habitats. The aneuploid reduction series in the tribe are often, though not invariably, associated with these changes in life form. A number of derived annual Cichorieae have retained the original basic number, e.g., the weedy *Sonchus asper* ($2n = 18$). Many of the derived annuals, however, have reduced chromosome numbers (Babcock, 1947; Stebbins, Jenkins and Walters, 1953; Stebbins, 1958). The degree of correlation between aneuploid reduction and annual habit is shown by the data for 85 species of Crepis in Table 44.

Examples of derived annuals with reduced chromosome numbers are found in Stephanomeria ($n = 8$), Sonchus ($n = 8, 7$), Hypochaeris ($n = 6, 5, 4, 3$), and Crepis ($n = 5, 4, 3$). The reduction trend continues within the annual group. Thus the series $n = 5$, 4, 3 occurs in the annual species of Crepis. *Crepis fuliginosa* ($n = 3$) is derived from another annual close to *C. neglecta* ($n = 4$) (Tobgy, 1943).

TABLE 44
FREQUENCY DISTRIBUTION OF CHROMOSOME NUMBERS BY LIFE-FORM CLASS IN DIPLOID SPECIES OF CREPIS (COMPILED FROM DATA OF BABCOCK, 1947, pp. 50–51)

Life form	No. of species with 2n =				
	14	12	10	8	6
Perennial	3	14	12	34	
Biennial				2	
Annual			6	11	3

The Cichorieae provide the best-analyzed case of a trend, actually many individual trends, from perennial herbs of closed communities with open recombination systems to colonizing annuals with restricted recombination systems, where the variable regulatory factor involved is the basic chromosome number, which strongly affects the amount of gene linkage.

Parallel trends are known in other plant groups. The following groups exhibit aneuploid reduction series from high-number perennials to low-number annuals:

Phacelia (Hydrophyllaceae), $x = 11$ to $x = 7$ (Cave and Constance, 1947)

Eriophyllum and Pseudobahia (Compositae), $x = 8$ to $x = 3$ (Carlquist, 1956)

Polemonieae (Polemoniaceae), $x = 9$ to $x = 6$ (Grant, 1959)

Other cases are listed by Stebbins (1950, pp. 455–56).

Ascending aneuploid series are less well known in higher plants. Some students believe they are less common. A much broader base of examples will be needed before we can estimate the relative frequency of ascending and descending aneuploidy in higher plants.

One of the best-documented cases of ascending aneuploidy is Clarkia (Onagraceae), running from the ancestral $n = 7$ to the derived $n = 9$ (Lewis and Lewis, 1955; Lewis and Roberts, 1956). Another ascending series is in the Annonaceae, from $x = 7$ to $x = 9$ (Walker, 1972). The tribe Boronieae of the Rutaceae in Australia has aneuploid series starting at $x = 9$ and running up to $x = 12$ or perhaps higher, as well as down to $x = 7$ (Smith-White, 1959). A possible ascending series occurs in Machaeranthera section Psilactis (Compositae), from $n = 5$ or 4 to $n = 9$ (Turner and Horne, 1964), although other students read the series in the opposite direction.

The basic numbers in various groups of temperate woody dicotyledons are fairly high. Examples are Fagaceae ($x = 12$), Betulaceae ($x = 14$), Acer ($x = 13$), Garrya ($x = 11$), and Ulmus ($x = 14$). These numbers are higher than the modal basic numbers in perennial herbaceous dicotyledons and higher than the putative ancestral chromosome number of the dicotyledons, somewhere in the range $x = 7$–9. They are almost certainly derived numbers.

The next question is, How are these numbers derived? A long-standing hypothesis is that woody dicotyledons with $x = 11$ or higher are ancient polyploids derived from diploid ancestors which are now extinct (Stebbins, 1938, 1947, 1950, ch. 9). This hypothesis rests on considerations of the frequency and importance of polyploidy in plant evolution.

An alternative possibility should be considered, namely that the woody dicotyledons with high basic numbers are products of ascending aneuploidy (Grant, 1971, pp. 317–18). When we look at the problem from the standpoint of recombination systems rather than polyploidy, this hypothesis seems more attractive. The plants in quesion are inhabitants of temperate forests and woods. Their habitat places a selective premium on open recombination systems. They are acknowledged to possess some of the regulatory factors for open recombination, particularly large populations, outcrossing, and wide dispersal. Their high basic chromosome numbers fit in with these other factors to maximize the generation of variability by recombination.

EVOLUTIONARY CHANGES IN THE BREEDING SYSTEM The evolutionary change from partial or complete outcrossing to autogamy has occurred repeatedly in many different phyletic lines of hermaphroditic angiosperms. Derived autogamous species are well known in such families as the Leguminosae, Onagraceae, Polemoniaceae, Boraginaceae, Labiatae, Scrophulariaceae, and Compositae, to name only a few.

In the family Polemoniaceae, a changeover from cross-pollination to autogamy has taken place independently in 11 different genera. In the genus Gilia four diploid species pairs are known which consist of an outcrossing species and a closely related autogamous species. Seven other species of Gilia contain the two breeding systems on the racial level of divergence. *Gilia ochroleuca,* previously mentioned, is one example of a species containing interfertile outcrossing and autogamous races (Fig. 57) (Grant and Grant, 1965).

The autogamous species and races in the Polemoniaceae are colonizing annuals in every case. Most are reduced morphologically as compared with their cross-pollinating relatives. They usually inhabit extreme and recent environments in their territories, such as deserts or high mountains. The direction of evolution has clearly been toward autogamy as a derived condition. Furthermore, the phylogenetic trend can be viewed mainly as a changeover from an open to a restricted recombination system that takes place along with the occupation of new pioneering habitats (Grant and Grant, 1965).

Closed recombination systems as exemplified by agamospermy are also derived from open recombination systems and associated with pioneering habitats. We previously discussed the phylogenetic trend in Crepis from perennial herbs of closed communities to annuals with restricted recombination systems in open habitats. Another trend in the *Crepis occidentalis* group ends in a complex of agamospermous microspecies which occupy pioneering habitats in western North America (Babcock and Stebbins, 1938). For a general review of agamospermy see Grant (1971, chs. 18, 21).

Comparative evidence indicates that a changeover from outcrossing to inbreeding or subsexual reproduction has occurred frequently in many plant groups. This is generally agreed to be a common evolutionary trend in higher plants. Mather (1943) and Darlington (1956a, 1963, ch. 2) go a step further in contending that this trend is the only common one. The ancestors of new groups in higher plants are postulated to be species with open recombination systems; and it is only among their evolutionary derivatives that substantial restrictions on recombination are to be found (Darlington, 1963, ch. 2). Changes in the reverse direction, from an inbreeding system to a system promoting much outcrossing, are rare and exceptional (Mather, 1943; Darlington, 1963, ch. 2; Crowe, 1964).

In my opinion it is well to emphasize the widespread occurrence of evolutionary changes from open to restricted or closed recombination, with respect to the breeding system and other related regulatory factors. Such changes evidently do take place widely in higher plants. But we should not emphasize this general trend to such an extent that we obscure the existence of a complementary set of trends in the opposite direction.

Autogamy, agamospery, and obligate vegetative propagation are derived conditions in the flowering plants. But dioecism, gynodioecism, monoecism, and heterostyly are also derived. They too are widespread in their systematic distribution, indicating that the evolutionary changeover from hermaphroditism to these derived out-crossing systems has occurred many times. Furthermore, some evidence suggests that self-incompatibility in hermaphroditic angiosperms is a phylogenetically derived condition (see the discussion in Chapter 14).

INFERENCES CONCERNING THE ANCESTRAL RECOMBINATION SYSTEM The hypothesis presented here, following an earlier briefer statement (Grant, 1958), is that the ancestral angiosperms possessed a restricted recombination system. This hypothesis is preferred over the alternative hypothesis of an original open recombination system because it is consistent with theoretical expectations and with several independent types of comparative evidence.

The early angiosperms were subordinate elements in the Mesozoic vegetation, and therefore they must have existed as relatively small, semi-isolated populations. Being woody plants, their generation time was relatively long. They were diploids with basic chromosomes in the probable range of $x = 7$–9. Their reproductive organs were hermaphroditic flowers. There are reasons for believing that they were self-compatible. These features are regulatory factors making for restricted recombination.

The flowers of the early angiosperms were probably pollinated by beetles and perhaps by flies. The primitive flower-visiting insects would have carried the pollen to neighboring plants as well as to other flowers on the same plant, thus bringing about a mixture of cross- and self-pollinations. Wind pollination is inefficient in plants which exist as scattered populations of a subordinate vegetational element, whereas primitive insect pollination, with its relatively short dispersal range but channeled flow of pollen, is adapted to the same conditions. The original flower was probably an organ for securing channeled, short-range pollination. In other words, the original pollination system in the angiosperms had a restrictive effect on recombination.

When climatic and biotic changes opened up new habitats in the late Mesozoic, the angiosperms quickly rose to dominance. A rapid evolutionary response to new environmental challenges is a characteristic of colonizing species.

The inferred ancestral recombination system in the angiosperms was restricted but also flexible. Hermaphroditic, self-compatible, insect-pollinated flowers form a system capable of producing various mixtures of outcross and inbred progeny, depending on the environmental conditions. The basic chromosome number can be altered upward or downward from $x = 7$–9. The original restricted recombination system could change in the direction of either an open or a more restricted system.

Pioneering habitats have undoubtedly existed throughout the history of the angiosperms. Such environments must have selected for restricted and closed recombination systems at every stage of angiosperm history. More restricted systems have probably arisen from less restricted systems repeatedly.

A corollary of the hypothesis presented here is that all open recombination systems in the angiosperms are derived phylogenetically. Open systems would have acquired a positive adaptive value only after the angiosperms succeeded to dominance in the plant world and formed climax and subclimax communities of their own. The open recombination systems of many forest trees, woodland shrubs, and steppe grasses are just as derived and specialized as the restricted recombination systems of pioneering annual herbs.

CHAPTER TWENTY-FIVE

NEOCLASSICAL GENETICS

I

Let us turn our attention, finally, from the principal results of plant genetics to some of the distinctive characteristics of this field of science. A knowledge of what plant genetics has done, as reviewed in this book, puts us in a position to see what plant genetics is. A related larger question is the place of plant genetics in biology. I will describe, as I see it, the historical place of plant genetics in classical biology, and its actual and potential role in modern biology.

The special characteristics of plant genetics can be seen most clearly when this field is compared with other branches of genetics. Certain fields of genetics will be brought into the discussion, not to be dissected and analyzed in themselves, but to provide reference points for describing plant genetics. In each case, we are considering the fundamental research fields and not their related applied subjects, i.e., basic plant genetics but not necessarily plant breeding.

The dominant fields of genetics at present are molecular and microbial genetics. These fields share materials, methodologies, and philosophies to such an extent that they can be considered together for the purpose of our broad comparisons. My choice for the next most active and influential field of genetics today would be population genetics. Mammalian genetics and human genetics are also enjoying a boom. In any case, no matter how we rank these and other fields, plant genetics is not in a dominant position in the social hierarchy of the biological sciences today. One can look back to a period in the early decades of this century when plant genetics was in the forefront. Classical Drosophila genetics was also flourishing in that period.

We can take three of these fields—molecular genetics, population genetics, and classical Drosophila genetics—as our reference points, keeping their known characteristics in mind as we look at plant genetics directly.

II

Plant genetics as a field has produced its full share of elegant experiments, experiments which can be ranked among the best in genetics and indeed in biology. Mendel's experiment with peas and Johannsen's with beans come to mind immediately. Other experiments in the same class are those of Nilsson-Ehle with wheat and oats, Baur with Antirrhinum, East with Nicotiana, Renner with Oenothera, Michaelis with Epilobium, Harland with Gossypium, Stephens with Gossypium, and Brink with Zea. This list is by no means complete.

Many of the outstanding experiments in plant genetics are not as widely known as they should be. They have remained buried in the journal literature where they have been known only to specialists. This circumstance has hampered the task of evaluating the true worth of plant genetics. One of the purposes of the present work, as noted in the preface, is to describe some of the little-known plant genetic experiments for the benefit of a wider circle of readers. Some of the experiments described here, including ones by Michaelis, Harland, and Stephens, received their first discussion in the book literature in *The Architecture of the Germplasm* (Grant 1964).

A genetic experiment can be regarded as first-rate if it meets all of the following criteria: ingenious design, thorough execution, and good analysis of the data, pointing to a significant conclusion. Although these criteria may seem obvious enough, they have special implications when considered in the context of plant genetics.

An experiment in bacterial genetics, molecular genetics, Drosophila population genetics, or classical Drosophila genetics can be carried out in a few weeks or months in a single laboratory room. A plant genetic experiment requires a great deal more space, time, and effort. The space requirement is an experimental plot of considerable size, with accompanying greenhouse and potting shed. An experiment utilizing plants which have one or two generations per year will require several or many years of continuous effort to carry it through to completion. During this time the experimentalist must handle hundreds of plants in each growing season, usually with individual plant numbers, and must keep records about them which can be deciphered effectively in the later stage of analysis.

The large scale of plant genetic experiments has significant consequences. An experiment with bacteria or Drosophila can readily be repeated if that is necessary because of technical problems or human error; but it is always difficult and often impossible to repeat a plant genetic experiment involving several years of work. The relatively large investment of time and effort in a plant genetic experiment could be expected to affect the working habits of the plant geneticist in various ways. At the very least, it places a premium on a high degree of organizational ability in the experimentalist.

The necessary conditions for experimental work in plant genetics also have geographical side-effects. The facilities for genetic research on bacteria, Drosophila, and mice can be provided fairly easily in a laboratory building of an urban university. The same urban university can usually provide a greenhouse, a growth chamber, and perhaps even a botanic garden to accommodate a few experimental plants for cytogenetic or biosystematic studies. But the experimental plot needed for a full-scale hybridization experiment has long since transmuted into a parking lot or office building. Plant genetic research is mainly being carried out in colleges of agriculture and agricultural experiment stations away from large metropolitan centers. It is worth recalling in this connection that plant genetics, and hence classical genetics in general, had its birth and early development largely outside the universities—in gardens and agricultural stations across Europe from Czechoslovakia to England.

The long time and high risk inherent in an original experiment in plant genetics have restrictive effects of still another sort. Pioneering work in plant genetics is not geared to the tempo of modern science. From graduate-student thesis to foundation-sponsored research grant, the prescription calls for unit projects which can be completed in two or three or perhaps four years. Experiments in molecular genetics, microbial genetics, or Drosophila genetics can be fitted into the existing administrative time scale; and so can some work on plants; but plant genetic experiments of the caliber under discussion here cannot be and historically have not been.

III

Many of the fundamental concepts of general genetics have come out of experimental plant genetics; for example the following:

particulate genes	polyploidy
segregation	tetrasomic ratios
recombination of	compensating tetrasome-
unlinked genes	nullisome combinations
linkage	linked blocks of modifiers
genotype and phenotype	multifactorial linkage
cytoplasmic genes	M-V linkage
multiple factors	viability genes on rearranged
oppositional gene systems	chromosome segments
dominance	paramutation
heterosis	restricted recombination systems
permanent hybridity	in colonizing species

Other basic concepts emerged from the common ground of classical plant genetics and classical animal genetics. Some examples are:

gene mutation	inhibitor genes
complementary factors	modifier genes

The foregoing fundamental genic entities or relationships between such entities or genetic processes have been revealed either by experimental work in plant genetics primarily or by joint efforts in classical plant and animal genetics. Some of these entities or systems have turned out to be general in sexual organisms. Some are widespread but not universal. Still others are known in only a few plant groups and have not yet been sought in other kingdoms.

These genetic units and systems have been revealed not by breeding experiments alone but by the combination of well designed breeding experiments and probing analysis of the experimental results. Many plant geneticists have possessed one talent but not the other, have been thorough experimentalists but foggy analyzers, or good thinkers but poor experimentalists; and their contributions to genetics have fallen short accordingly. The great ones from Mendel onward have excelled in both areas.

This discussion leads us into the art of scientific discovery as manifested in plant genetics and classical genetics. Discovery has a variety of meanings in science. In biology one can discover a new species by exploring a little-known fauna or flora, or a new sibling species in a well-known fauna or flora, or a new cell organelle by using a more advanced type of microscope, and so on. Or one can discover the biochemical nature and action of the genetic material by the various methods of chemical assay.

The discoveries of the fundamental genic units and systems by plant geneticists and classical geneticists are in a different category. Here the operation of discovery consists of inferring the existence of a hitherto unknown and unseen entity or system of entities from the behavior of phenotypic traits in a breeding experiment. The discovery of the fundamental genic units is comparable to the discovery of oxygen in the atmosphere and of the planets in the outer solar system. This type of discovery is rightly ranked among the highest achievements in science.

IV

One may well ask, Why have so many basic concepts come out of work in plant genetics? Why, in other words, has the method of the breeding experiment in the hands of a good analytical worker been so successful when applied with plants as the experimental subjects?

Plants have several obvious virtues as experimental organisms. They can be self-pollinated or cloned, in many cases, making possible some interesting experimental manipulations. Their usually high fecundity permits large progenies to be grown from a single parental individual. But I would like to emphasize another feature. Plants have a relatively simple course of development for a higher multicellular organism.

Let us recall that genetic discovery in higher organisms consists of inferences drawn from the segregation of phenotypic characters. The organismic geneticist has to be able to read back correctly from phenotypic appearances to genotypic causes. The ability to make correct inferences depends on the relative straightforwardness of the chain of events from genetic causes to phenotypic effects. Onto-genetic development is therefore a factor affecting the chances of success in genetic reasoning.

In animals the course of development is enormously complex, with many integrative forces at work, as is well known. Here the complexity of ontogenetic development is likely to interfere with genetic analysis. The much simpler course of development in plants reduces this problem. In genetic work with plants there is a lower level of interference of ontogenetic development with factorial analysis.

V

When plant genetics and classical Drosophila genetics were at their peak, the salient problems were the mode of hereditary trans-mission, the verification and extension of the Mendelian mechanism, and the charting of genetic maps. As these problems became solved in their main essentials, interest turned to the molecular structure and biochemical action of the genetic material. This shift in interest ushered in the modern era of molecular genetics and bacterial genetics.

There are basic differences between molecular and bacterial genetics, on the one hand, and classical genetics, on the other. The methods of research in the former are those of biochemistry and microbiology. The workers are essentially biochemists and bacteri-ologists; they are not experimental breeders. The experimental or-ganisms of greatest usefulness are Neurospora, bacteria, viruses, and other microorganisms. The work is laboratory-centered exclusively. The technical jargon is quite different from that of classical genetics, and so is the theoretical superstructure.

These are not differences between sister subspecialties of the same field. They are differences between major fields of biology whose distinctness has been obscured by the common use of the blanket term "genetics." They are comparable to the differences between physics and chemistry.

The succession to dominance of molecular and microbial genetics in the modern era is not, as I see it, a result of progressive changes within a single field, but is instead a replacement of one field by another. Scientific problems came up which called for a different breed of workers. One body of knowledge became half-forgotten, or never learned in the case of the younger workers, while another distinct body of knowledge with the same surname "genetics" grew up in its place. Historical continuity was partly lost in the process.

An unfortunate by-product of this mode of historical development has been a tendency to leave classical genetics in limbo. Many modern workers with a molecular orientation consider that classical genetics has completed its mission and belongs in the history books of biology. Even its name has an archaic ring. Accordingly it is grossly underrepresented in many modern genetics textbooks, curricula, and departments, all of which only reinforces the break in continuity.

The assumption that classical genetics has completed its mission in biology is erroneous. Classical genetics successfully solved certain problems and in the process of doing so exposed others. A major problem which is only partly explored is the mode of organization of the genotype in higher organisms; that is, the spatial and functional relationships between the alleles, genes, and gene systems in a diploid or polyploid genotype, and the role of these relationships in the development and functioning of the organism. This general problem has continued to engage the attention of a relatively small number of direct descendants of the classical geneticists; let us call them neoclassical geneticists and the field neoclassical genetics.

Neoclassical genetics inherits the task of working out a system of genetics applicable to higher organisms. This task cannot be accomplished by extrapolating from the genetics of bacteria. I hasten to add that microbial genetics can contribute significantly to the picture, and has done so; but microbial genetics alone is not enough. Likewise, the biochemical and developmental genetics of individual loci in higher organisms grapple with only a small part of the whole problem.

The reductionist approach is suitable for microbial genetics, but is inadequate by itself in neoclassical genetics. In the latter field reductionism must be supplemented by a compositionist approach. This is a consequence of the great complexity and preponderant importance of gene systems and linkage systems in the genetics of higher organisms.

BIBLIOGRAPHY

Note: When several editions of a book have been cited, the later editions (and their years of publication) are listed along with the earliest edition cited.

Adams, M. W., and D. B. Shank. 1959. The relationship of heterozygosity to homeostasis in maize hybrids. *Genetics* 44: 777–86.

Allard, R. W. 1960. *Principles of Plant Breeding.* New York: Wiley.

Allard, R. W., G. R. Babbel, M. T. Clegg, and A. L. Kahler. 1972. Evidence for coadaptation in *Avena barbata. Proc. Nat. Acad. Sci. U.S.* 69: 3043–48.

Allard, R. W., S. K. Jain, and P. L. Workman. 1968. The genetics of inbreeding populations. *Advances Genet.* 14: 55–131.

Allard, R. W., and P. L. Workman. 1963. Population studies in predominantly self-pollinated species. IV. Seasonal fluctuations in estimated values of genetic parameters in lima bean populations. *Evolution* 17: 470–80.

Allen, C. E. 1940. The genotypic basis of sex-expression in angiosperms. *Bot. Rev.* 6: 227–300.

——. 1945. The genetics of bryophytes. II. *Bot. Rev.* 11: 260–87.

Allison, A. C. 1956. Sickle cells and evolution. *Sci. Amer.* Aug., p. 87.

Anderson, E. 1939a. The hindrance to gene recombination imposed by linkage: An estimate of its total magnitude. *Amer. Natur.* 73: 185–88.

——. 1939b. Recombination in species crosses. *Genetics* 24: 668–98.

——. 1949. *Introgressive Hybridization.* New York: Wiley.

——. 1953. Introgressive hybridization. *Biol. Rev.* 28: 280–307.

Anderson, E., and L. B. Abbe. 1933. A comparative anatomical study of a mutant Aquilegia. *Amer. Natur.* 67: 380–84.

Anderson, E., and B. R. Anderson. 1954. Introgression of *Salvia apiana* and *Salvia mellifera. Ann. Missouri Bot. Gard.* 41: 329–38.

Anderson, E., and W. L. Brown. 1952. Origin of corn belt maize and its genetic significance. In *Heterosis,* ed. J. W. Gowen. Ames: Iowa State Univ. Press.

Anderson, E., and K. Sax. 1936. A cytological monograph of the American species of Tradescantia. *Bot. Gaz.* 97: 433–76.

Anderson, E. G. 1935. Chromosomal interchanges in maize. *Genetics* 20: 70–83.

Apirion, D., and D. Zohary. 1961. Chlorophyll lethal in natural populations of the orchard grass (*Dactylis glomerata* L.): A case of balanced polymorphism in plants. *Genetics* 46: 393–99.

Archimowitsch, A. 1949. Control of pollination in sugar-beet. *Bot. Rev.* 15: 613–28.

Arnold, C. 1958. Selektive Befruchtung. *Ergeb. Biol.* 20: 67–96.

Ar-Rushdi, A. H. 1957. The cytogenetics of variegation in a species hybrid in Nicotiana. *Genetics* 42: 312–25.

Avdulov, N. P. 1931. Karyo-systematische Untersuchungen der Familie Gramineen. *Bull. Appl. Bot. Genet. Plant Breed.,* vol. 44, suppl.

Avery, A. G., S. Satina, and J. Rietsema. 1959. *Blakeslee: The Genus Datura.* New York: Ronald.

Avery, O. T., C. M. Macleod, and M. McCarty. 1944. Studies on the chemical nature of the substance inducing transformation of pneumococcal types. *Jour. Exp. Med.* 79: 137–58.

Babcock, E. B. 1947. The genus Crepis. *Univ. Calif. Publ. Bot.* 21: 1–198.

Babcock, E. B., and Huges, M. B. 1950. A new type of selfsterility in plants. *Proc Nat. Acad. Sci. U.S.* 36: 363–64.

Babcock, E. B., and G. L. Stebbins. 1938. The American species of Crepis: Their interrelationships and distribution as affected by polyploidy and apomixis. *Carnegie Inst. Wash. Publ.* 504.

Babcock, E. B., G. L. Stebbins, and J. A. Jenkins. 1937. Chromosomes and phylogeny in some genera of the Crepidinae. *Cytologia,* Fujii Jubilee vol., 188–210.

Baetcke, K. P., A. H. Sparrow, C. H. Nauman, and S. S. Schwemmer. 1967. The relationship of DNA content to nuclear and chromosome volumes and to radiosensitivity (LD_{50}). *Proc. Nat. Acad. Sci. U.S.* 58: 533–40.

Baker, H. G. 1948. Dimorphism and monomorphism in the Plumbaginaceae. I. A survey of the family. *Ann. Bot.* 12: 207–19.

——. 1953. Dimorphism and monomorphism in the Plumbaginaceae. II. Pollen and stigmata in the genus Limonium. *Ann. Bot.* 17: 433–45.

——. 1958. Studies in the reproductive biology of West African Rubiaceae. *Jour. West Afr. Sci. Ass.* 4: 9–24.

——. 1959. Reproductive methods as factors in speciation in flowering plants. *Cold Spring Harbor Symp. Quant. Biol.* 24: 177–91.

——. 1962. Heterostyly in the Connaraceae with special reference to *Byrsocarpus coccineus. Bot. Gaz.* 123: 206–11.

——. 1966. The evolution, functioning and breakdown of heteromorphic incompatibility systems. I. The Plumbaginaceae. *Evolution* 20: 349–68.

Baker, W. K. 1968. Position-effect variegation. *Advances Genet.* 14: 133–69.

Barton, D. W. 1951. Localized chiasmata in the differentiated chromosomes of the tomato. *Genetics* 36: 374–81.

Barton, D. W., L. Butler, J. A. Jenkins, C. M. Rick, and P. A. Young. 1955. Rules for nomenclature in tomato genetics. *Jour. Hered.* 46: 22–26.

Bateman, A. J. 1946. Genetical aspects of seed-growing. *Nature* 157: 752–55.

——. 1947. Contamination of seed crops. II. Wind pollination. *Heredity* 1: 235–46.

——. 1949. Pollinating agents and population genetics. *Hereditas,* suppl. vol. 1949, 532–33.

——. 1950. Is gene dispersion normal? *Heredity* 4: 353–63.

——. 1956. Cryptic self-incompatibility in the wallflower: *Cheiranthus cheiri* L. *Heredity* 10: 257–61.

Bateson, W., E. R. Saunders, and R. C. Punnett. 1906. *Experimental studies on the physiology of heredity.* Reports to the Evolution Committee, Royal Society, Report no. 3.

Baur, E. 1909. Das Wesen und die Erblichkeitsverhältnisse der "Varietates albomarginatae hort." von *Pelargonium zonale. Z. Indukt. Abstamm. Vererbungsl.* 1: 330–51.

——. 1910a. Propfbastarde. *Biol. Zentralbl.* 30: 497–514.

——. 1910b. Untersuchungen über die Vererbung von Chromatophorenmerkmalen bei Melandrium, Antirrhinum und Aquilegia. *Z. Indukt. Abstamm. Vererbungsl.* 4: 81–102.

——. 1911, 1914, 1919, 1930a. *Einführung in die experimentelle Vererbungslehre.* Eds. 1 (1911), 2 (1914), 3 and 4 (1919), 7–11 (1930). Berlin: Bornträger.

——. 1924. Untersuchungen über das Wesen, die Entstehung und die Vererbung von Rassenunterschieden bei *Antirrhinum majus. Bibl. Genet.* 4: 1–170.

——. 1930b. Mutationsauslösung bei *Antirrhinum majus. Z. Bot.* 23: 676–702.

Beadle, G. W. 1937. Chromosomal aberration and gene mutation in sticky chromosome plants of *Zea mays. Cytologia,* Fujii Jubilee vol., 43–56.

——. 1953. Heterosis. *Jour. Hered.* 44: 88.

——. 1955. The gene: Carrier of heredity, controller of function and agent of evolution. *Nieuwland Lect.* (Notre Dame) 7: 1–24.

——. 1957a. The role of the nucleus in heredity. In *The Chemical Basis of Heredity,* ed. W. D. McElroy and B. Glass. Baltimore: Johns Hopkins Press.

——. 1957b. *The Physical and Chemical Basis of Inheritance.* Eugene: Univ. of Ore.

——. 1962. Structure of the genetic material and the concept of the gene. In *This is Life,* ed. W. H. Johnson and W. C. Steere. New York: Holt.

Beadle, G. W., and E. L. Tatum. 1941. Genetic control of biochemical reactions in Neurospora. *Proc. Nat. Acad. Sci. U.S.* 27: 499–506.

Beermann, W. 1964. Structure and function of interphase chromosomes. *Genetics Today, Proc. 11th Internat. Congr. Genet.* (The Hague), pp. 375–83.

Belling, J. 1928. The ultimate chromomeres of Lilium and Aloe with regard to the number of genes. *Univ. Calif. Publ. Bot.* 14: 307–18.

Bemis, W. P. 1959. Selective fertilization in lima beans. *Genetics* 44: 555–62.

Benzer, S. 1955. Fine structure of a genetic region in bacteriophage. *Proc. Nat. Acad. Sci. U.S.* 41: 344–54.

——. 1957. The elementary units of heredity. In *The Chemical Basis of Heredity,* ed. W. D. McElroy and B. Glass. Baltimore: Johns Hopkins Press.

Berg, R. L. 1959. A general evolutionary principle underlying the origin of developmental homeostasis. *Amer. Natur.* 93: 103–5.

Berger, C. A., and E. R. Witkus. 1954. The cytology of *Xanthisma texanum* DC. I. Differences in the chromosome number of root and shoot. *Bull. Torrey Bot. Club* 81: 489–91.

Bergner, A. D., S. Satina, and A. F. Blakeslee. 1936. Prime types in Datura. *Proc. Nat. Acad. Sci. U.S.* 19: 103–15.

Bernström, P. 1950. Cleisto- and chasmogamic seed setting in di- and tetraploid *Lamium amplexicaule*. *Hereditas* 36: 492–506.

Beuzenberg, E. J., and J. B. Hair. 1963. Contributions to a chromosome atlas of the New Zealand flora. V. Miscellaneous families. *New Zeal. Jour. Bot.* 1: 53–67.

Blakeslee, A. F., A. D. Bergner, and A. G. Avery. 1937. Geographical distribution of prime types in Datura. *Cytologia*, Fujii Jubilee vol., 1070–93.

Böcher, T. W. 1945. Meiosis in *Anemone apennina* with special reference to chiasma localisation. *Hereditas* 31: 221–37.

Böcher, T. W., K. Larsen, and K. Rahn. 1955. Experimental and cytological studies on plant species. III. *Plantago coronopus* and allied species. *Hereditas* 41: 423–53.

Bold, H. C. 1967. *Morphology of Plants*. Ed. 2. New York: Harper.

Bonnett, O. T. 1954. The inflorescences of maize. *Science* 120: 77–87.

Boveri, Th. 1904. *Ergebnisse über Konstitution der chromatischen Substanz des Zellkerns*. Jena: Gustav Fischer.

Brewbaker, J. L. 1957. Pollen cytology and self-incompatibility systems in plants. *Jour. Hered.* 48: 271–77.

——. 1964. *Agricultural Genetics*. Englewood Cliffs, N. J.: Prentice-Hall.

Brewer, G. J., C. F. Sing, and E. R. Sears. 1969. Studies of isozyme patterns in nullisomic-tetrasomic combinations of hexaploid wheat. *Proc. Nat. Acad. Sci. U.S.* 64: 1224–29.

Brian, P. W. 1959. Effects of gibberellins on plant growth and development. *Biol. Rev.* 34: 37–84.

Bridges, C. B. 1919. Specific modifiers of eosin eye color in *Drosophila melanogaster*. *Jour. Exp. Zool.* 28: 337–84.

——. 1935. Salivary gland chromosome maps. *Jour. Hered.* 26: 60–64.

Brieger, F. G. 1950. The genetic basis of heterosis in maize. *Genetics* 35: 420–45.

Brink, R. A. 1952. Inbreeding and crossbreeding in seed development. In *Heterosis*, ed. J. W. Gowen. Ames: Iowa State Univ. Press.

——. 1956. A genetic change associated with the *R* locus in maize which is directed and potentially reversible. *Genetics* 41: 872–89.

——. 1958. Paramutation at the *R* locus in maize. *Cold Spring Harbor Symp. Quant. Biol.* 23: 379–91.

——. 1962. Phase change in higher plants and somatic cell heredity. *Quart. Rev. Biol.* 37: 1–22.

——. 1964. Genetic repression in multicellular organisms. *Amer. Natur.* 98: 193–211.

Brink, R. A., E. D. Styles, and J. D. Axtell. 1968. Paramutation: Directed genetic change. *Science* 159: 161–70.

Britten, R. J., and D. E. Kohne. 1968. Repeated sequences in DNA. *Science* 161: 529–40.

Brock, R. D. 1956. Cytological barriers in plant breeding. *Jour. Austral. Inst. Agr. Sci.* 22: 180–88.

Brooks, J. L. 1957. The species problem in freshwater animals. In *The Species Problem,* ed. E. Mayr. Washington, D.C.: American Association for the Advancement of Science.

Brown, H. S. 1961. Differential chiasma frequencies in self-pollinating and cross-pollinating species of the genus Gilia. *Aliso* 5: 67–81.

Brown, S. W. 1966. Heterochromatin. *Science* 151: 417–25.

Brown, S. W., and U. Nur. 1964. Heterochromatic chromosomes in the coccids. *Science* 145: 130–36.

Brown, W. V. 1972. *Textbook of Cytogenetics.* St. Louis, Mo.: Mosby.

Brown, W. V., and S. M. Stack. 1968. Somatic pairing as a regular preliminary to meiosis. *Bull. Torry Bot. Club* 95: 369–78.

Brücher, H. 1943. Experimentelle Untersuchungen über den Selektionswert künstlich erzeugter Mutanten von *Antirrhinum majus.* *Z. Bot.* 39: 1–47.

Buchholz, J. T. 1947. Chromosome structure under the electron microscope. *Science* 105: 607–10.

Burdick, A. B. 1954. Epistasis as a cause of heterosis in the tomato (*Lycopersicon esculentum*). *Caryologia,* suppl. vol. 1954, 1235–37.

Burnham, C. R. 1956. Chromosomal interchanges in plants. *Bot. Rev.* 22: 419–552.

Burns, J. A., and D. U. Gerstel. 1967. Flower color variation and instability of a block of heterochromatin in Nicotiana. *Genetics* 57: 155–67.

—— and ——. 1971. Inhibition of chromosome breakage and of megachromosomes by intact genomes in Nicotiana. *Genetics* 69: 211–20.

Byers, R. E., L. R. Baker, H. M. Sell, R. D. Herner, and D. R. Dilley. 1972. Ethylene: A natural regulator of sex expression of *Cucumis melo* L. *Proc. Nat. Acad. Sci. U.S.* 69: 717–20.

Cameron, D. R., and R. Moav. 1957. Inheritance in *Nicotiana tabacum.* XXVII. Pollen killer, an alien genetic locus inducing abortion of microspores not carrying it. *Genetics* 42: 326–35.

Cannon, W. B. 1932. *The Wisdom of the Body.* New York: Norton.

Carlquist, S. 1956. On the generic limits of Eriophyllum (Compositae) and related genera. *Madroño* 13: 226–39.

——. 1966. The biota of long-distance dispersal. IV. Genetic systems in the floras of oceanic islands. *Evolution* 20: 433–55.

Carlson, S. A. 1959. Comparative genetics of complex loci. *Quart. Rev. Biol.* 34: 33–67.

Carson, H. L. 1946. The selective elimination of inversion dicentric chromatids during meiosis in the eggs of *Sciara impatiens.* *Genetics* 31: 95–113.

——. 1955. The genetic characteristics of marginal populations of Drosophila. *Cold Spring Harbor Symp. Quant. Biol.* 20: 276–87.

——. 1956. Marginal homozygosity for gene arrangement in *Drosophila robusta.* *Science* 123: 630–31.

——. 1957. The species as a field for gene recombination. In *The Species Problem,* ed. E. Mayr. Washington, D.C.: American Association for the Advancement of Science.

——. 1958. Response to selection under different conditions of recombination in Drosophila. *Cold Spring Harbor Symp. Quant. Biol.* 23: 291–306.

——. 1959. Genetic conditions which promote or retard the formation of species. *Cold Spring Harbor Symp. Quant. Biol.* 24: 87–105.

——. 1961. Relative fitness of genetically open and closed experimental populations of *Drosophila robusta. Genetics* 46: 553–67.

Carson, H. L., and W. B. Heed. 1964. Structural homozygosity in marginal populations of nearctic and neotropical species of Drosophila in Florida. *Proc. Nat. Acad. Sci. U.S.* 52: 427–30.

Caspari, E. 1948. Cytoplasmic inheritance. *Advances Genet.* 2: 1–66.

Castle, W. E. 1954. Coat color inheritance in horses and other mammals. *Genetics* 39: 35–44.

Catcheside, D. G. 1939. A position effect in Oenothera. *Jour. Genet.* 38: 345–52.

——. 1947. The *P*-locus position effect in Oenothera. *Jour. Genet.* 48: 31–42.

——. 1954. The genetics of brevistylis in Oenothera. *Heredity* 8: 125–37.

Cave, M. S., and L. Constance. 1947. Chromosome numbers in the Hydrophyllaceae. III. *Univ. Calif. Publ. Bot.* 18: 449–65.

Chargaff, E. 1950. Chemical specificity of nucleic acids and mechanism of their enzymatic degradation. *Experientia* 6: 201–9.

Cherry, J. P., F. R. H. Katterman, and J. E. Endrizzi. 1971. A comparative study of seed proteins of allopolyploids of Gossypium by gel electrophoresis. *Can. Jour. Genet. Cytol.* 13: 155–58.

Chooi, W. Y. 1971a. Variation in nuclear DNA content in the genus Vicia. *Genetics* 68: 195–211.

——. 1971b. Comparison of the DNA of six Vicia species by the method of DNA-DNA hybridization. *Genetics* 68: 213–30.

Chovnick, A., A. Schalet, R. P. Kernaghan, and J. Talsma. 1962. The resolving power of genetic fine structure analysis in higher organisms as exemplified by Drosophila. *Amer. Natur.* 96: 281–96.

Clarke, C. A., and P. M. Sheppard. 1960. Super-genes and mimicry. *Heredity* 14: 175–85.

Clausen, J. 1926. Genetical and cytological investigations on *Viola tricolor* L. and *V. arvensis* Murr. *Hereditas* 8: 1–156.

——. 1951. *Stages in the Evolution of Plant Species.* Ithaca, N.Y.: Cornell Univ. Press.

Clausen, J., and W. M. Hiesey. 1958. Experimental studies on the nature of species. IV. Genetic structure of ecological races. *Carnegie Inst. Wash. Publ.* 615.

—— and ——. 1960. The balance between coherence and variation in evolution. *Proc. Nat. Acad. Sci. U.S.* 46: 494–506.

Clausen, J., D. D. Keck, and W. M. Hiesey. 1940. Experimental studies on the nature of species. I. Effect of varied environments on western North American plants. *Carnegie Inst. Wash. Publ.* 520.

——, ——, and ——. 1948. Experimental studies on the nature of species. III. Environmental responces of climatic races of Achillea. *Carnegie Inst. Wash. Publ.* 581.

Clausen, R. E. 1930. Inheritance in *Nicotiana tabacum.* X. Carmine-coral variegation. *Cytologia* 1: 358–68.

Clausen, R. E., and D. R. Cameron. 1944. Inheritance in *Nicotiana tabacum*. XVIII. Monosomic analysis. *Genetics* 29: 447–77.

—— and ——. 1950. Inheritance in *Nicotiana tabacum*. XXII. Duplicate factors for chlorophyll production. *Genetics* 35: 4–10.

Clayton, E. E. 1950. Male-sterile tobacco. *Jour. Hered.* 41: 171–75.

Clegg, M. T., and R. W. Allard. 1972. Patterns of genetic differentiation in the slender wild oat species *Avena barbata*. *Proc. Nat. Acad. Sci. U.S.* 69: 1820–24.

Cleland, R. E. 1923. Chromosome arrangements during meiosis in certain Oenotheras. *Amer. Natur.* 57: 562–66.

——. 1925. Chromosome behavior during meiosis in the pollen mother cells of certain Oenotheras. *Amer. Natur.* 59: 475–79.

——. 1936. Some aspects of the cyto-genetics of Oenothera. *Bot. Rev.* 2: 316–48.

——. 1940. Analysis of wild American races of Oenothera (Onagra). *Genetics* 25: 636–44.

——. 1949. Phylogenetic relationships in Oenothera. *Hereditas*, suppl. vol. 1949, 173–88.

——. ed. 1950. Studies in Oenothera cytogenetics and phylogeny. *Indiana Univ. Publ.*, Sci. Ser., no. 16.

——. 1962. The cytogenetics of Oenothera. *Advances Genet.* 11: 147–237.

——. 1972. *Oenothera: Cytogenetics and Evolution*. New York: Academic Press.

Cleland, R. E., and B. L. Hammond. 1950. Analysis of segmental arrangements in certain races of Oenothera. *Indiana Univ. Publ.*, Sci. Ser., no. 16: 10–72.

Coleman, L. C. 1940. The cytology of *Veltheimia viridifolia*, Jacq. *Amer. Jour. Bot.* 27: 887–95.

Colwell, R. R., and A. B. Burdick. 1959. Uptake and effect on crossing-over of ethylenediamine-tetraacetic acid (EDTA) in *Drosophila melanogaster*. *Nucleus* 2: 125–30.

Comstock, R. E. 1955. Theory of quantitative genetics: Synthesis. *Cold Spring Harbor Symp. Quant. Biol.* 20: 93–102.

Correns, K. 1900. G. Mendels Regel über das Verhalten der Nachkommenschaft der Rassenbastarde. *Ber. Deut. Bot. Ges.* 18: 158–68. (Reprinted in *Genetics* 35: 33–41, 1950.)

——. 1905. *Über Vererbungsgesetze*. Berlin: Bornträger.

——. 1909. Vererbungsversuche mit blass (gelb) grünen und buntblättrigen Sippen bei *Mirabilis jalapa, Urtica pilulifera* und *Lunaria annua*. *Z. Indukt. Abstamm. Vererbungsl.* 1: 291–329.

——. 1928. Bestimmung, Vererbung und Verteilung des Geschlechtes bei den höheren Pflanzen. *Handbuch der Vererbungswissenschaft* 2 (C): 1–138.

Cramer, P. J. S. 1954. Chimeras. *Bibliogr. Genet.* 16: 193–381.

Crane, M. B., and W. J. C. Lawrence. 1931. Inheritance of sex, colour and hairiness in the raspberry, *Rubus idaeus* L. *Jour. Genet.* 24: 243–55.

—— and ——. 1934, 1938, 1952. *The Genetics of Garden Plants*. Eds. 1, 2, and 4. London: Macmillan.

Crick, F. H. C. 1954. The structure of the hereditary material. *Sci. Amer.*, Oct., p. 54.

——. 1962. The genetic code. *Sci. Amer.*, Oct., p. 66.

Cronquist, A. 1968. *The Evolution and Classification of Flowering Plants.* Boston: Houghton.

Crow, J. F. 1948. Alternative hypotheses of hybrid vigor. *Genetics* 33: 447–87.

———. 1972. Darwinian and non-Darwinian evolution. *Proc. Sixth Berkeley Symp. Math. Statist. Prob.,* vol. 5. Berkeley: Univ. of Calif. Press.

Crowe, L. K. 1964. The evolution of outbreeding in plants. I. The angiosperms. *Heredity* 19: 435–57.

Crumpacker, D. W. 1966. Allelism and overdominance of chlorophyll genes in maize. *Genetics* 54: 1307–14.

Da Cunha, A. B. 1953. A further analysis of the polymorphism of *Drosophila polymorpha. Nature* 171: 887.

———. 1955. Chromosomal polymorphism in the Diptera. *Advances Genet.* 7: 93–138.

———. 1960. Chromosomal variation and adaptation in insects. *Annu. Rev. Entomol.* 5: 85–110.

Da Cunha, A. B., C. Pavan, J. S. Morgante, and M. C. Garrido. 1969. Studies on cytology and differentiation in Sciaridae. II. DNA redundancy in salivary gland cells of *Hybosciara fragilis* (Diptera, Sciaridae). *Genetics* 61, suppl., 335–49.

D'Amato, F., and O. Hoffmann-Ostenhof. 1956. Metabolism and spontaneous mutations in plants. *Advances Genet.* 8: 1–28.

Darlington, C. D. 1932, 1937. *Recent Advances in Cytology.* Eds. 1 and 2. London: Churchill; Philadelphia: Blakiston's.

———. 1936. The limitation of crossing over in Oenothera. *Jour. Genet.* 32: 343–51.

———. 1939, 1958. *The Evolution of Genetic Systems.* Ed. 1, London: Cambridge Univ. Press; ed. 2, New York: Basic Bks.

———. 1956a. *Chromosome Botany.* Ed. 1. London: G. Allen.

———. 1956b. Natural populations and the breakdown of classical genetics. *Proc. Roy. Soc.,* ser. B, 145: 350–64.

———. 1963. *Chromosome Botany and the Origins of Cultivated Plants.* Ed. 2. London: G. Allen; New York: Hafner.

———. 1965. *Cytology.* London: Churchill.

———. 1971a. The evolution of polymorphic systems. In *Ecological Genetics and Evolution,* ed. R. Creed. Oxford: Blackwell.

———. 1971b. *Guide to the Oxford Botanic Gardens.* Oxford: Clarendon Press.

Darlington, C. D., and L. F. La Cour. 1950. Hybridity selection in Campanula. *Heredity* 4: 217–48.

Darlington, C. D., and K. Mather. 1949. *The Elements of Genetics.* London: G. Allen.

Darlington, C. D., and G. W. Shaw. 1959. Parallel polymorphism in the heterochromatin of Trillium species. *Heredity* 13: 89–121.

Darwin, C. 1868, 1875. *The Variation of Animals and Plants under Domestication,* 2 vols. Eds. 1 and 2. London: J. Murray.

Davenport, C. B. 1908. Degeneration, albinism and inbreeding. *Science* 28: 454–55.

Day, A. 1965. The evolution of a pair of sibling allotetraploid species of Cobwebby Gilias (Polemoniaceae). *Aliso* 6: 25–75.

Dayhoff, M. O., and R. V. Eck. 1968. *Atlas of Protein Sequence and Structure*. Silver Spring, Md.: National Biomedical Research Foundation.

Demerec, M. 1931. Behaviour of two mutable genes of *Delphinium ajacis*. *Jour. Genet*. 24: 179–93.

——. 1956. A comparative study of certain gene loci in Salmonella. *Cold Spring Harbor Symp. Quant. Biol*. 21: 113–21.

Demerec, M., I. Blomstrand, and Z. E. Demerec. 1955. Evidence of complex loci in Salmonella. *Proc. Nat. Acad. Sci. U.S*. 41: 359–64.

Dempster, E. R. 1949. Effects of linkage on parental-combination and recombination frequencies in F_2. *Genetics* 34: 272–84.

De Vries, H. 1889. Intracellulare Pangenesis. Jena: Gustav Fischer.

——. 1917. *Oenothera lamarckiana* mut. *velutina*. *Bot. Gaz*. 63: 1–24.

Dobzhansky, Th. 1931. The decrease in crossing-over observed in translocations, and its probable explanation. *Amer. Natur*. 65: 214–32.

——. 1952. Nature and origin of heterosis. In *Heterosis*, ed. J. W. Gowen. Ames: Iowa State Univ. Press.

——. 1959. Evolution of genes and genes in evolution. *Cold Spring Harbor Symp. Quant. Biol*. 24: 15–30.

——. 1970. *Genetics of the Evolutionary Process*. New York: Columbia Univ. Press.

Dobzhansky, Th., and H. Levene. 1955. Genetics of natural populations. XXIV. Developmental homeostasis in natural populations of *Drosophila pseudoobscura*. *Genetics* 40: 797–808.

Dobzhansky, Th., and B. Wallace. 1953. The genetics of homeostasis in Drosophila. *Proc. Nat. Acad. Sci. U.S*. 39: 162–71.

Dorst, J. C. 1952. Two remarkable bud-sports in the potato variety Rode Star. *Euphytica* 1: 184–86.

Dounce, A. L. 1971. Nuclear gels and chromosomal structure. *Amer. Sci*. 59: 74–83.

Dubinin, N. P. 1948. Experimental investigation of the integration of hereditary systems in the processes of evolution of populations [in Russian]. *Zh. Obshch. Biol*. 9: 203–44. (English translation, University Library, Univ. of California, Los Angeles.)

Dunn, L. C. 1965. *A Short History of Genetics*. New York: McGraw-Hill.

Dunn, L. C., and E. Caspari. 1945. A case of neighboring loci with similar effects. *Genetics* 30: 543–68.

Dunn, L. C., and J. Suckling. 1955. A preliminary comparison of the fertilities of wild house mice with and without a mutant at locus *T*. *Amer. Natur*. 89: 231–33.

Durrant, A. 1962a. The environmental induction of heritable change in Linum. *Heredity* 17: 27–61.

——. 1962b. Induction, reversion and epitrophism of flax genotrophs. *Nature* 196: 1302–4.

Duvnick, D. N. 1965. Cytoplasmic pollen sterility in corn. *Advances Genet*. 13: 1–56.

Dyer, A. F. 1963. Allocyclic segments of chromosomes and the structural heterozygosity that they reveal. *Chromosoma* 13: 545–76.

East, E. M. 1910. A Mendelian interpretation of variation that is apparently continuous. *Amer. Natur.* 44: 65–82.

———. 1916. Studies on size inheritance in Nicotiana. *Genetics* 1: 164–76.

———. 1936. Heterosis. *Genetics* 21: 375–97.

———. 1940. The distribution of self-sterility in the flowering plants. *Proc. Amer. Phil. Soc.* 82: 449–518.

East, E. M., and H. K. Hayes. 1911. Inheritance in maize. *Conn. Agr. Exp. Sta. Bull.* 167.

Editors of Genetics. 1950. The Birth of Genetics. *Genetics* 35, suppl., 1–47.

Edwardson, J. R. 1956. Cytoplasmic male-sterility. *Bot. Rev.* 22: 696–738.

———. 1970. Cytoplasmic male-sterility. II. *Bot. Rev.* 36: 341–420.

Ehrman, L. 1963. Apparent cytoplasmic sterility in *Drosophila paulistorum*. *Proc. Nat. Acad. Sci. U.S.* 49: 155–57.

Ehrman, L., and D. L. Williamson. 1965. Transmission by injection of hybrid sterility to nonhybrid males in *Drosophila paulistorum:* Preliminary report. *Proc. Nat. Acad. Sci. U.S.* 54: 481–83.

Emerson, R. A. 1912. The unexpected occurrence of aleurone colors in F_2 of a cross between non-colored varieties of maize. *Amer. Natur.* 46: 612–15.

———. 1932. The present status of maize genetics. *Proc. 6th Internat. Congr. Genet.* 1: 141–52.

Emerson, R. A., G. W. Beadle, and A. C. Fraser. 1935. A summary of linkage studies in maize. *Cornell Univ. Agr. Exp. Sta. Mem.* 80.

Emerson, R. A., and E. M. East. 1913. The inheritance of quantitative characters in maize. *Nebr. Agr. Exp. Sta. Res. Bull.* 2.

Emerson, S. H. 1935. The genetic nature of DeVries's mutations in *Oenothera lamarckiana*. *Amer. Natur.* 69: 545–59.

———. 1952. Biochemical models of heterosis in Neurospora. In *Heterosis*, ed. J. W. Gowen. Ames: Iowa State Univ. Press.

Emerson, S. H., and A. H. Sturtevant. 1932. The linkage of certain genes in Oenothera. *Genetics* 17: 393–412.

Emlen, J. M. 1973. *Ecology: An Evolutionary Approach*. Reading, Mass.: Addison-Wesley.

Epling, C. 1947. Natural hybridization of *Salvia apiana* and *S. mellifera*. *Evolution* 1: 69–78.

Ernst, A. 1933. Weitere Untersuchungen zur Phänanalyse zum Fertilitätsproblem und zur Genetik heterostyler Primeln. I. *Primula viscosa*. *Arch. Julius Klaus-Stift Vererbungsforsch.* 8: 1–215.

Evans, G. M., A. Durrant, and H. Rees. 1966. Associated nuclear changes in the induction of flax genotrophs. *Nature* 212: 697–99.

Eversole, R. A., and E. L. Tatum. 1956. Chemical alteration of crossing-over frequency in Chlamydomonas. *Proc. Nat. Acad. Sci. U.S.* 42: 68–73.

Faegri, K., and L. van der Pijl. 1971. *The Principles of Pollination Ecology*. Ed. 2. Oxford: Pergamon.

Fahselt, D., and M. Ownbey. 1968. Chromatographic comparison of Dicentra species and hybrids. *Amer. Jour. Bot.* 55: 334–45.

Falconer, D. S. 1960. *Introduction to Quantitative Genetics*. New York: Ronald.

Finkner, M.D. 1954. The effect of dual pollinations in upland cotton stocks differing in genotype. *Agron. Jour.* 46: 124–28.

BIBLIOGRAPHY

Flake, R. H., E. von Rudloff, and B. L. Turner. 1969. Quantitative study of clinal variation in *Juniperus virginiana* using terpenoid data. *Proc. Nat. Acad. Sci. U.S.* 64: 487–94.

Ford, E. B. 1963. Evolutionary processes in animals. In *Chromosome Botany and the Origins of Cultivated Plants*, by C. D. Darlington, ed. 2, pp. 209–13. London: G. Allen; New York: Hafner.

———. 1964. *Ecological Genetics*. London: Methuen.

———. 1966. *Genetic Polymorphism*. London: Faber.

Fox, A. S., and S. B. Yoon. 1970. DNA-induced transformation in Drosophila: Locus specificity and the establishment of transformed stocks. *Proc. Nat. Acad. Sci. U.S.* 67: 1608–15.

Fraenkel-Conrat, H. 1956. Rebuilding a virus. *Sci. Amer.,* June, p. 42.

———. 1962. *Design and Function at the Threshold of Life: The Viruses.* New York: Academic Press.

Fraenkel-Conrat, H., B. A. Singer, and R. C. Williams. 1957. The nature of the progeny of virus reconstituted from protein and nucleic acid of different strains of tobacco mosaic virus. In *The Chemical Basis of Heredity,* ed. W. D. McElroy and B. Glass. Baltimore: Johns Hopkins Press.

Frankel, O., and A. Munday. 1962. The evolution of wheat. In *The Evolution of Living Organisms.* Symposium, Royal Society, Victoria, Australia.

Free, J. B. 1970. *Insect Pollination of Crops.* London: Academic Press.

Freese, E. 1958. The arrangement of DNA in the chromosome. *Cold Spring Harbor Symp. Quant. Biol.* 23: 13–18.

Frota-Pessoa, O. 1961. On the number of gene loci and the total mutation rate in man. *Amer. Natur.* 95: 217–22.

Fryxell, P. A. 1956. Effect of varietal mass on percentage of outcrossing in *Gossypium hirsutum* in New Mexico. *Jour. Hered.* 47: 299–301.

———. 1957. Mode of reproduction of higher plants. *Bot. Rev.* 23: 135–233.

Gadgil, M., and O. T. Solbrig. 1972. The concept of *r*- and *K*-selection: Evidence from wild flowers and some theoretical considerations. *Amer. Natur.* 106: 14–31.

Gairdner, A. E. 1929. Male sterility in flax. II. A case of reciprocal crosses differing in F_1. *Jour. Genet.* 21: 117–24.

Gajewski, W. 1953. Some observations on disturbances of floral development in Geum species and hybrids. *Acta Soc. Bot. Pol.* 22: 587–604.

Galston, A. W., and P. J. Davies. 1970. *Control Mechanisms in Plant Development.* Englewood Cliffs, N.J.: Prentice-Hall.

Garrod, A. E. 1923. *Inborn Errors of Metabolism.* Oxford: Oxford Univ. Press.

Geitler, L. 1937. Cytogenetische Untersuchungen an natürlichen Populationen von *Paris quadrifolia. Z. Indukt. Abstamm. Vererbungsl.* 73: 182–97.

———. 1938. Weitere cytogenetische Untersuchungen an natürlichen Populationen von *Paris quadrifolia. Z. Indukt. Abstamm. Vererbungsl.* 75: 161–90.

Gerstel, D. U. 1950. Self-incompatibility studies in guayule. II. Inheritance. *Genetics* 35: 482–506.

———. 1953. Chromosomal translocations in interspecific hybrids of the genus Gossypium. *Evolution* 7: 234–44.

———. 1956. Segregation in new allopolyploids of Gossypium. I. The R_1 locus in certain New World–wild American hexaploids. *Genetics* 41: 31–44.

Gerstel, D. U. 1960. Segregation in new allopolyploids of Nicotiana. I. Comparison of 6x *(N. tabacum × tomentosiformis)* and 6x *(N. tabacum × otophora). Genetics* 45: 1723–34.

——. 1963. Segregation in new allopolyploids of Nicotiana. II. Discordant ratios from individual loci in 6x *(N. tabacum × N. sylvestris). Genetics* 48: 677–89.

——. 1966. Evolutionary problems in some polyploid crop plants. *Hereditas,* suppl. vol. 1966, pt. 2, pp. 481–504.

Gerstel, D. U., and J. A. Burns. 1966. Flower variegation in hybrids between *Nicotiana tabacum* and *N. otophora. Genetics* 53: 551–67.

—— and ——. 1970. The effect of the *Nicotiana otophora* genome on chromosome breakage and megachromosomes in *N. tabacum × N. otophora* derivatives. *Genetics* 66: 331–38.

Gerstel, D. U., and L. L. Phillips. 1957. Segregation in new allopolyploids of Gossypium. II. Tetraploid combinations. *Genetics* 42: 783–97.

—— and ——. 1958. Segregation of synthetic amphiploids in Gossypium and Nicotiana. *Cold Spring Harbor Symp. Quant. Biol.* 23: 225–37.

Gerstel, D. U., and P. A. Sarvella. 1956. Additional observations on chromosomal translocations in cotton hybrids. *Evolution* 10: 408–14.

Gibor, A., and S. Granick. 1964. Plastids and mitochondria: Inheritable systems. *Science* 145: 890–97.

Gibson, J. B., and J. M. Thoday. 1962. Effects of disruptive selection. VI. A second chromosome polymorphism. *Heredity* 17: 1–26.

Glass, B. 1955. Pseudoalleles. *Science* 122: 233.

Godley, E. J. 1955a. Monoecy and incompatibility. *Nature* 176: 1176–77.

——. 1955b. Breeding systems in New Zealand plants. I. Fuchsia. *Ann. Bot.* 19: 549–59.

——. 1964. Breeding systems in New Zealand plants. III. Sex ratios in some natural populations. *New Zeal. Jour. Bot.* 2: 205–12.

Goldschmidt, R. B. 1938. *Physiological Genetics.* New York: McGraw-Hill.

——. 1940. *The Material Basis of Evolution.* New Haven, Conn.: Yale Univ. Press.

——. 1949. Heterochromatic heredity. *Hereditas,* suppl. vol. 1949, 244–55.

——. 1955. *Theoretical Genetics.* Berkeley: Univ. of Calif. Press.

Goodspeed, T. H., and P. Avery. 1939. Trisomic and other types in *Nicotiana sylvestris. Jour. Genet.* 38: 381–458.

—— and ——. 1941. The twelfth primary trisomic type in *Nicotiana sylvestris. Proc. Nat. Acad. Sci. U.S.* 27: 13–14.

Gottlieb, L. D. 1971. Gel electrophoresis: New approach to the study of evolution. *Bioscience* 21: 939–43.

Gowen, J. W., ed. 1952. *Heterosis.* Ames: Iowa State Univ. Press. Reprinted by Hafner, New York, 1964.

Grant, A., and V. Grant. 1956. Genetic and taxonomic studies in Gilia. VIII. The Cobwebby Gilias. *Aliso* 3: 203–87.

Grant, K. A., and V. Grant. 1964. Mechanical isolation of *Salvia apiana* and *Salvia mellifera* (Labiatae). *Evolution* 18: 196–212.

Grant, V. 1950a. The flower constancy of bees. *Bot. Rev.* 16: 379–98.

——. 1950b. Genetic and taxonomic studies in Gilia. I. *Gilia capitata. Aliso* 2: 239–316.

——. 1954. Genetic and taxonomic studies in Gilia. IV. *Gilia achilleaefolia*. *Aliso* 3: 1–18.

——. 1956a. The genetic structure of races and species in Gilia. *Advances Genet.* 8: 55–87.

——. 1956b. Chromosome repatterning and adaptation. *Advances Genet.* 8: 89–107.

——. 1958. The regulation of recombination in plants. *Cold Spring Harbor Symp. Quant. Biol.* 23: 337–63.

——. 1959. *Natural History of the Phlox Family*. The Hague: Nijhoff.

——. 1963. *The Origin of Adaptations*. New York: Columbia Univ. Press.

——. 1964. *The Architecture of the Germplasm*. New York: John Wiley.

——. 1966a. Linkage between viability and fertility in a species cross in Gilia. *Genetics* 54: 867–80.

——. 1966b. Block inheritance of viability genes in plant species. *Amer. Natur.* 100: 591–601.

——. 1967. Linkage between morphology and viability in plant species. *Amer. Natur.* 101: 125–39.

——. 1971. *Plant Speciation*. New York: Columbia Univ. Press.

Grant, V., and K. A. Grant. 1965. *Flower Pollination in the Phlox Family*. New York: Columbia Univ. Press.

Green, M. M. 1953. The beadex locus in *Drosophila melanogaster:* Genetic analysis of the mutant Bx^{r49k}. *Z. Indukt. Abstamm. Vererbungsl.* 85: 435–49.

——. 1955. Phenotypic variation and pseudo-allelism at the forked locus in *Drosophila melanogaster*. *Proc. Nat. Acad. Sci. U.S.* 41: 375–79.

Green, M. M., and K. C. Green. 1949. Crossing-over between alleles at the lozenge locus in *Drosophila melanogaster*. *Proc. Nat. Acad. Sci. U.S.* 35: 586–91.

—— and ——. 1956. A cytogenetic analysis of the lozenge pseudoalleles in Drosophila. *Z. Indukt. Abstamm. Vererbungsl.* 87: 708–21.

Greenblatt, I. M. 1966. Transposition and replication of modulator in maize. *Genetics* 53: 361–69.

Gregory, D. P., and W. M. Klein. 1960. Investigations of meiotic chromosomes of six genera in the Onagraceae. *Aliso* 4: 505–21.

Grun, P. 1970. Cytoplasmic sterilities that separate the cultivated potato from its putative diploid ancestors. *Evolution* 24: 750–58.

Grun, P., M. Aubertin, and A. Radlow. 1962. Multiple differentiation of plasmons of diploid species of Solanum. *Genetics* 47: 1321–33.

Gustafsson, Å. 1943. The genesis of the European blackberry flora. *Lunds Univ. Årsskr.* 39: 1–200.

——. 1951. Mutations, environment and evolution. *Cold Spring Harbor Symp. Quant. Biol.* 16: 263–81.

Haga, T. 1953. Meiosis in Paris. II. Spontaneous breakage and fusion of chromosomes. *Cytologia* 18: 50–66.

Haga, T., and M. Kurabayashi. 1954. Genom and polyploidy in the genus Trillium. V. Chromosomal variation in natural populations of *Trillium kamtschaticum* Pall. *Mem. Fac. Sci. Kyushu Univ.*, ser. E, 1: 159–85.

Hagberg, A. 1953. Heterozygosity in erectoides mutations in barley. *Hereditas* 39: 161–78.

Hagberg, A., N. Nybom, and Å. Gustafsson. 1952. Allelism of erectoides mutations in barley. *Hereditas* 38: 510–12.

Hagen, C. W. 1950. A contribution to the cytogenetics of the genus Oenothera with special reference to certain forms from South America. *Indiana Univ. Publ.,* sci. ser., 16: 305–48.

Hair, J. B., E. J. Beuzenberg, and B. Pearson. 1967. Contributions to a chromosome atlas of the New Zealand flora. IX. *New Zeal. Jour. Bot.* 5: 185–94.

Håkansson, A. 1947. Contributions to a cytological analysis of the species differences of *Godetia amoena* and *G. whitneyi. Hereditas* 33: 235–60.

Haldane, J. B. S. 1954a. *The Biochemistry of Genetics.* London: G. Allen.

——. 1954b. The statics of evolution. In *Evolution as a Process,* ed. J. Huxley. London: G. Allen.

Hamrick, J. L., and R. W. Allard. 1972. Microgeographical variation in allozyme frequencies in *Avena barbata. Proc. Nat. Acad. Sci. U.S.* 69: 2100–04.

Haney, W. J., J. B. Gartner, and G. B. Wilson. 1953. The effect of light on the expression of heterosis in *Antirrhinum majus* L. *Jour. Hered.* 44: 10–12.

Hanson, E. D. 1954. Studies on kappa-like particles in sensitives of *Paramecium aurelia,* variety 4. *Genetics* 39: 229–39.

Harborne, J. B., ed. 1972. *Phytochemical Ecology.* New York: Academic Press.

Harland, S. C. 1936. The genetical conception of the species. *Biol. Rev.* 11: 83–112.

Harrison, B. J., and K. Mather. 1950. Polygenic variability in chromosomes of *Drosophila melanogaster* obtained from the wild. *Heredity* 4: 295–312.

Hartmann, H. T., and D. E. Kester. 1968. *Plant Propagation.* Ed. 2. Englewood Cliffs, N.J.: Prentice-Hall.

Haskell, G. 1961. Seedling morphology in applied genetics and plant breeding. *Bot. Rev.* 27: 382–421.

Hayes, W. 1968. *The Genetics of Bacteria and Their Viruses.* Ed. 2. London: Blackwell.

Hecht, A. 1950. Cytogenetic studies of Oenothera, subgenus Raimannia. *Indiana Univ. Publ.,* sci. ser., 16: 255–304.

Heiken, A. 1958. Aberrant types in the potato. *Acta Agr. Scand.* 8: 319–58.

Hershey, A. D. 1953. Inheritance in bacteriophage. *Advances Genet.* 5: 89–106.

——. 1956. The organization of genetic material in bacteriophage T2. *Brookhaven Symp. Biol.* 8: 6–14.

Hershey, A. D., and M. Chase. 1952. Independent functions of viral protein and nucleic acid in growth of bacteriophage. *Jour. Gen. Physiol.* 36: 39–56.

Heslop-Harrison, J. 1957. The experimental modification of sex expression in flowering plants. *Biol. Rev.* 32: 38–90.

——. 1968. Ribosome sites and *S* gene action. *Nature* 218: 90–91.

Hiesey, W. M. 1953a. Comparative growth between and within climatic races of Achillea under controlled conditions. *Evolution* 7: 297–316.

——. 1953b. Growth and development of species and hybrids of Poa under controlled temperatures. *Amer. Jour. Bot.* 40: 205–21.

Hiesey, W. M., M. A. Nobs, and O. Björkman. 1971. Experimental studies on the nature of species. V. Biosystematics, genetics, and physiological ecology of the Erythranthe section of Mimulus. *Carnegie Inst. Wash. Publ.* 628.

Hill, J. 1967. The environmental induction of heritable changes in *Nicotiana rustica* parental and selection lines. *Genetics* 55: 735–54.

Hiorth, G. 1940. Eine Serie multipler Allele für Blütenzeichnungen bei *Godetia amoena*. *Hereditas* 26: 441–53.

——. 1942. Zur Genetik und Systematik der *amoena*-Gruppe der Gattung Godetia. *Z. Indukt. Abstamm. Vererbungsl.* 80: 289–349.

——. 1963. *Quantitative Genetik*. Berlin: Springer Verlag.

Hoagland, M. B. 1959. Nucleic acids and proteins. *Sci. Amer.*, Dec., p. 55.

Hoar, C. S. 1931. Meiosis in *Hypericum punctatum* Lam. *Bot. Gaz.* 92: 396–406.

Hoar, C. S., and E. J. Haertl. 1932. Meiosis in the genus Hypericum. *Bot. Gaz.* 93: 197–204.

Hoff, V. J. 1962. An analysis of outcrossing in certain complex-heterozygous Euoenotheras. I. Frequency of outcrossing. *Amer. Jour. Bot.* 49: 715–21.

Hooker, A. L., and K. M. S. Saxena. 1971. Genetics of disease resistance in plants. *Ann. Rev. Genet.* 5: 407–24.

Horowitz, N. H. 1950. Biochemical genetics of Neurospora. *Advances Genet.* 3: 33–71.

Howard, H. W. 1970. *Genetics of the Potato*. New York: Springer-Verlag.

Huether, C. A. 1968. Exposure of natural genetic variability underlying the pentamerous corolla constancy in *Linanthus androsaceus* ssp. *androsaceus*. *Genetics* 60:123–46.

——. 1969. Constancy of the pentamerous corolla phenotype in natural populations of Linanthus. *Evolution* 23: 572–88.

Hutchinson, J. B., and D. A. Lawes. 1953. A note on the estimation of natural crossing in cotton. *Empire Cotton Grower's Rev.* 30: 192–93.

Hyde, B. B. 1953. Differentiated chromosomes in *Plantago ovata*. *Amer. Jour. Bot.* 40: 809–15.

Iltis, H. 1932. *Life of Mendel*. Transl. New York: Norton.

Imam, A. G., and R. W. Allard. 1965. Population studies in predominantly self-pollinated species. VI. Genetic variability between and within natural populations of wild oats from differing habitats in California. *Genetics* 51: 49–62.

Imms, A. D. 1957. *A General Textbook of Entomology*. Ed. 9. London: Methuen.

Ingram, V. M. 1956. A specific chemical difference between the globins of normal human and sickle-cell anaemia haemoglobin. *Nature* 178: 792–94.

——. 1963. *The Hemoglobins in Genetics and Evolution*. New York: Columbia Univ. Press.

Jackson, R. C. 1957. New low chromosome number for plants. *Science* 128: 1115–16.

——. 1971. The karyotype in systematics. *Ann. Rev. Ecol. Systemat.* 2: 327–68.

Jacob, F., and J. Monod. 1959. Gènes de structure et gènes de régulation dans la biosynthèse des protéines. *Compt. Rend. Acad. Sci.* (Paris) 249: 1282–84.

—— and ——. 1961. On the regulation of gene activity. *Cold Spring Harbor Symp. Quant. Biol.* 26: 193–211.

Jain, S. K. 1959. Male sterility in flowering plants. *Bibliogr. Genet.* 18: 101–66.

——. 1960. Cytogenetics of rye (Secale spp.). *Bibliogr. Genet.* 19: 1–86.

——. 1967. Population dynamics of a gametophyte factor controlling selective fertilization. *Genetica* 38: 485–503.

Jain, S. K. 1968. Gynodioecy in *Origanum vulgare:* Computer simulation of a model. *Nature* 217: 764–65.

Jain, S. K., and R. W. Allard. 1960. Population studies in predominantly self-pollinated species. I. Evidence for heterozygote advantage in a closed population of barley. *Proc. Nat. Acad. Sci. U.S.* 46: 1371–77.

James, S. H. 1965. Complex hybridity of *Isotoma petraea.* I. The occurrence of interchange heterozygosity, autogamy and a balanced lethal system. *Heredity* 20: 341–53.

——. 1970. Complex hybridity in *Isotoma petraea.* II. Components and operation of a possible evolutionary mechanism. *Heredity* 25: 53–78.

Japha, B. 1939. Die Meiosis von Oenothera. II. *Z. Bot.* 34: 321–69.

Jensen, W. A. 1973. Fertilization in flowering plants. *Bioscience* 23: 21–27.

Jessop, A. P., and D. G. Catcheside. 1965. Interallelic recombination at the *his-1* locus in *Neurospora crassa* and its genetic control. *Heredity* 20: 237–56.

Jinks, J. L. 1955. A survey of the genetical basis of heterosis in a variety of diallel crosses. *Heredity* 9: 223–38.

Jinks, J. L., and K. Mather. 1955. Stability in development of heterozygotes and homozygotes. *Proc. Roy. Soc.,* ser. B., 143: 561–78.

Johannsen, W. 1903. *Ueber Erblichkeit in Populationen und in reinen Linien.* Jena: Gustav Fischer.

——. 1909, 1926. *Elemente der exakten Erblichkeitslehre.* Eds. 1 and 3. Jena: Gustav Fischer.

——. 1911. The genotype conception of heredity. *Amer. Natur.* 45: 129–59.

John, B., and K. Lewis. 1965. *The meiotic system.* Vol. 6 of *Protoplasmologia.* Vienna and New York: Springer-Verlag.

Johnson, B. L. 1953. Evidence for irregularity in crossing over of the *S* locus in the eversporting type of *Matthiola incana* (L.) R. Br. *Genetics* 38: 229–43.

——. 1972. Seed protein profiles and the origin of the hexaploid wheats. *Amer. Jour. Bot.* 59: 952–60.

Jones, D. A. 1966. On the polymorphism of cyanogenesis in *Lotus corniculatus* L. Selection by animals. *Can. Jour. Genet. Cytol.* 8: 556–67.

Jones, D. F. 1917. Dominance of linked factors as a means of accounting for heterosis. *Proc. Nat. Acad. Sci. U.S.* 3: 310–12.

——. 1932. The interaction of specific genes determining sex in dioecious maize. *Proc. 6th Internat. Congr. Genet.* 2: 104–7.

——. 1956. Genic and cytoplasmic control of pollen abortion in maize. *Brookhaven Symp. Biol.* 9: 101–12.

Jones, D. F., and P. C. Mangelsdorf. 1925. The improvement of naturally cross-pollinated plants by selecting in self-fertilized lines. I. The production of inbred strains of corn. *Conn. Agr. Exp. Sta. Bull.* 266: 347–418.

Jones, R. N., and H. Rees. 1968. Nuclear DNA variation in Allium. *Heredity* 23: 591–605.

Jones, W. N. 1934. *Plant Chimeras and Graft Hybrids.* London: Methuen.

Jørgensen, C. A. and M. B. Crane. 1927. Formation and morphology of Solanum chimeras. *Jour. Genet.* 18: 247–72.

Jukes, T. H. 1966. *Molecules and Evolution.* New York: Columbia Univ. Press.

Kappert, H., and W. Rudorf, eds. 1958–1962. *Handbuch der Pflanzenzüchtung,* 6 vols. Ed. 2. Berlin: Paul Parey.

Kaufman, B. P., and M. R. McDonald. 1956. Organization of the chromosome. *Cold Spring Harbor Symp. Quant. Biol.* 21: 233–46.

Kernaghan, R. P., and L. Ehrman. 1970. An electron microscopic study of the etiology of hybrid sterility in *Drosophila paulistorum.* I. Mycoplasma-like inclusions in the testes of sterile males. *Chromosoma* 29: 291–304.

Kerner, A. 1894–1895. *The Natural History of Plants,* 2 vols. Transl. London: Blackie and Son.

Kiang, Y. T., and W. J. Libby. 1972. Maintenance of a lethal in a natural population of *Mimulus guttatus. Amer. Natur.* 106: 351–67.

Kihara, H. 1958. Fertility and morphological variation in the substitution and restoration backcrosses of the hybrids, *Triticum vulgare × Aegilops caudata. Proc. 10th Internat. Congr. Genet. (Montreal)* 1: 142–71.

Kihlman, B. A. 1961. Biochemical aspects of chromosome breakage. *Advances Genet.* 10: 1–59.

———. 1970. Sub-chromatid exchanges and the strandedness of chromosomes. *Hereditas* 65: 171–86.

Kohel, R. J. 1969. Phenotypic stability of homozygous parents and their F_1 hybrids in Upland cotton, *Gossypium hirsutum* L. *Crop Sci.* 9: 85–88.

Komai, T. 1950. Semi-allelic genes. *Amer. Natur.* 84: 381–92.

Kornberg, A. 1961. *Enzymatic Synthesis of DNA.* New York: Wiley.

Krizenecky, J., and B. Nemec, eds. 1965. *Fundamenta Genetica.* Prague: Czechoslovak Academy of Sciences.

Kuckuck, H., and A. Mudra. 1950. *Lehrbuch der allgemeinen Pflanzenzüchtung.* Zurich: Hirzel.

Kurabayashi, M. 1958. Evolution and variation in Japanese species of Trillium. *Evolution* 12: 286–310.

Laird, C. D., and B. J. McCarthy. 1968. Nucleotide sequence homology within the genome of *Drosophila melanogaster. Genetics* 60: 323–34.

Lamm, R. 1936. Cytological studies in inbred rye. *Hereditas* 22: 217–40.

———. 1941. Varying cytological behavior in reciprocal Solanum crosses. *Hereditas* 27: 202–08.

Lammerts, W. E. 1932. Inheritance of monosomics in *Nicotiana rustica. Genetics* 17: 689–96.

Lamprecht, H. 1941. Die Artgrenze zwischen *Phaseolus vulgaris* L. und *multiflorus* Lam. *Hereditas* 27: 51–175.

———. 1944. Die genisch-plasmatische Grundlage der Artbarriere. *Agr. Hort. Genet.* 2: 75–142.

———. 1954. Selektive Befruchtung im Lichte des Verhaltens interspezifischer Gene in Linien und Kreuzungen. *Agr. Hort. Genet.* 12: 1–37.

———. 1956. Die Art-Kreuzung *Chrysanthemum carinatum* Schousb. × *Chr. coronarium* L. Zur genisch-plasmatischen Grundlage der Artbarriere. *Agr. Hort. Genet.* 14: 203–54.

———. 1961a. Die Genenkarte von Pisum bei normaler Struktur der Chromosomen. *Agr. Hort. Genet.* 19: 360–401.

———. 1961b. The gene-map of Pisum. *Rep. Plant Breed. Inst. Weibullsholm* (Sweden), 1961, 1–7.

Lamprecht, H. 1962. Studien zur Vererbung des Höhenwachstums bei Pisum sowie Koppelungsstudien. *Agr. Hort. Genet.* 20: 23–62.

Landolt, E. 1957. Physiologische und ökologische Untersuchungen an Lemnaceen. *Ber. Schweiz. Bot. Ges.* 67: 271–410.

Langlet, O. 1936. Studien über die physiologische Variabilität der Kiefer und deren Zusammenhang mit dem Klima. Beiträge zur Kenntnis der Ökotypen von *Pinus sylvestris* L. *Medd. Statens Skogsförsöksanstalt* 29: 219–470.

Langridge, J. 1962. A genetic and molecular basis for heterosis in Arabidopsis and Drosophila. *Amer. Natur.* 96: 5–27.

Laughnan, J. R. 1948. The action of allelic forms of the gene *A* in maize. I. Studies of variability, dosage and dominance relations. The divergent character of the series. *Genetics* 33: 488–517.

——. 1952. The action of allelic forms of the gene *A* in maize. IV. On the compound nature of A^b and the occurence and action of its A^d derivatives. *Genetics* 37: 375–95.

——. 1955. Structural and functional bases for the action of the *A* alleles in maize. *Amer. Natur.* 89: 91–103.

Lawrence, W. J. C., and J. R. Price. 1940. The genetics and chemistry of flower colour variation. *Biol. Rev.* 15: 35–58.

Lawrence, W. J. C., and R. Scott-Moncrieff. 1935. The genetics and chemistry of flower colour in Dahlia: A new theory of specific pigmentation. *Jour. Genet.* 30: 155–226.

Lea, D. E. 1955. *Actions of Radiations on Living Cells.* Ed. 2. Cambridge: Cambridge Univ. Press.

Lefevre, G. 1971. Salivary chromosome bands and the frequency of crossing over in *Drosophila melanogaster. Genetics* 67: 497–513.

Lehmann, E. 1932. Der Anteil von Kern und Plasma an den reziproken Verschiedenheiten von Epilobium-Bastarden. *Z. Pflanzenzücht.* 17: 157–72.

Leiter, E. H., D. A. LaBrie, A. Bergquist, and R. P. Wagner. 1971. In vitro mitochondrial complementation in *Neurospora crassa. Biochem. Genet.* 5: 549–61.

Leng, E. R. 1960. Long-term selection of corn for oil and protein content. *Ill. Agr. Exp. Sta. Rep.,* mimeographed.

Lerner, I. M. 1954. *Genetic Homeostasis.* London: Oliver.

——. 1958. *The Genetic Basis of Selection.* New York: Wiley.

Levin, D. A. 1972. Plant density, cleistogamy, and self-fertilization in natural populations of *Lithospermum caroliniense. Amer. Jour. Bot.* 59: 71–77.

Levin, D. A., and D. E. Berube. 1972. Phlox and Colias: The efficiency of a pollination system. *Evolution* 26: 242–50.

Levin, D. A., G. Howland, and E. Steiner. 1972. Protein polymorphism and genic heterozygosity in a population of the permanent translocation heterozygote, *Oenothera biennis. Proc. Nat. Acad. Sci. U.S.* 69: 1475–77.

Levin, D. A., and H. W. Kerster. 1973. Assortative pollination for stature in *Lythrum salicaria. Evolution* 27: 144–52.

Levin, D. A., and B. A. Schaal. 1970. Reticulate evolution in Phlox as seen through protein electrophoresis. *Amer. Jour. Bot.* 57: 977–87.

Levine, R. P. 1952. Adaptive responses of some third chromosome types of *Drosophila pseudoobscura. Evolution* 6: 216–33.

——. 1955. Chromosome structure and the mechanism of crossing over. *Proc. Nat. Acad. Sci. U.S.* 41: 727–30.

Levy, M., and D. A. Levin. 1971. The origin of novel flavonoids in Phlox allotetraploids. *Proc. Nat. Acad. Sci. U.S.* 68: 1627–30.

Lewis, D. 1941. Male sterility in natural populations of hermaphrodite plants. *New Phytol.* 40: 56–63.

——. 1942. The evolution of sex in flowering plants. *Biol. Rev.* 17: 46–67.

——. 1949. Incompatibility in flowering plants. *Biol. Rev.* 24: 472–96.

——. 1954a. Comparative incompatibility in angiosperms and fungi. *Advances Genet.* 6: 235–85.

——. 1954b. Report of the department of genetics. *Annu. Rep. John Innes Hort. Inst.* 45: 12–17.

Lewis, D., and L. K. Crowe. 1956. The genetics and evolution of gynodioecy. *Evolution* 10: 115–25.

Lewis, E. B. 1945. The relation of repeats to position effect in *Drosophila melanogaster*. *Genetics* 30: 137–66.

——. 1950. The phenomenon of position effect. *Advances Genet.* 3: 73–115.

——. 1951. Pseudoallelism and gene evolution. *Cold Spring Harbor Symp. Quant. Biol.* 16: 159–74.

——. 1952. The pseudoallelism of white and apricot in *Drosophila melanogaster*. *Proc. Nat. Acad. Sci. U.S.* 38: 953–61.

——. 1955. Some aspects of position pseudoallelism. *Amer. Natur.* 89: 73–89.

Lewis, H. 1953. The mechanism of evolution in the genus Clarkia. *Evolution* 7: 1–20.

——. 1954. Quantitative variation in wild genotypes of Clarkia. *Internat. Union Biol. Sci.,* ser. B, 15: 114–22.

Lewis, H., and M. E. Lewis. 1955. The genus Clarkia. *Univ. Calif. Publ. Bot.* 20: 241–392.

Lewis, H., P. H. Raven, C. S. Venkatesh, and H. L. Wedberg. 1958. Observations of meiotic chromosomes in the Onagraceae. *Aliso* 4: 73–86.

Lewis, H., and M. R. Roberts. 1956. The origin of *Clarkia lingulata*. *Evolution* 10: 126–138.

Lewis, H., and J. Szweykowski. 1964. The genus Gayophytum (Onagraceae). *Brittonia* 16: 343–91.

Lewis, K. R., and B. John. 1963. *Chromosome Marker.* Boston: Little.

—— and ——. 1966. The meiotic consequences of spontaneous chromosome breakage. *Chromosoma* 18: 287–304.

Lewis, W. H. 1970. Chromosomal drift, a new phenomenon in plants. *Science* 168: 1115–16.

Lewis, W. H., R. L. Oliver, and T. J. Luikart. 1971. Multiple genotypes in individuals of *Claytonia virginica*. *Science* 172: 564–65.

L'Heritier, P. 1948. Sensitivity to CO_2 in Drosophila—a review. *Heredity* 2: 325–48.

Lima-de-Faria, A. 1952. Chromomere analysis of the chromosome complement of rye. *Chromosoma* 5: 1–68.

Lindstrom, E. W. 1936. Genetics of polyploidy. *Bot. Rev.* 2: 197–215.

Little, T. M. 1945. Gene segregation in autotetraploids. *Bot. Rev.* 11: 60–85.

——. 1958. Gene segregation in autotetraploids. II. *Bot. Rev.* 24: 318–39.

Loegering, W. Q., and E. R. Sears. 1963. Distorted inheritance of stem-rust resistance in Timstein wheat caused by a pollen-killing gene. *Can. Jour. Genet. Cytol.* 5: 65–72.

Longley, A. E. 1945. Abnormal segregation during megasporogenesis in maize. *Genetics* 30: 100–13.

Longo, G. P., and J. G. Scandalios. 1969. Nuclear gene control of mitochondrial malic dehydrogenase in maize. *Proc. Nat. Acad. Sci. U.S.* 62: 104–11.

Lotsy, J. P. 1916. *Antirrhinum rhinanthoides* mihi, une nouvelle espèce Linnéenne, obtenue expérimentalement. *Arch. Néerl. Sci. Nat.,* ser. 3B, 3: 195–204.

Löve, A. 1943. Cytogenetic studies on Rumex subgenus Acetosella. *Hereditas* 30: 1–136.

——. 1949. Mutations at the crater of Hekla in eruption. *Hereditas,* suppl. vol. 1949, 621–22.

Löve, A., and N. Sarkar. 1956. Cytotaxonomy and sex determination of *Rumex paucifolius. Can. Jour. Bot.* 34: 261–68.

MacArthur, R. H. 1972. *Geographical Ecology.* New York: Harper.

MacArthur, R. H., and E. O. Wilson. 1967. *The Theory of Island Biogeography.* Princeton, N. J.: Princeton Univ. Press.

McClintock, B. 1932. A correlation of ring-shaped chromosomes with variegation in *Zea mays. Proc. Nat. Acad. Sci. U.S.* 18: 677–81.

——. 1938. The production of homozygous deficient tissues with mutant characteristics by means of the aberrant behavior of ring-shaped chromosomes. *Genetics* 23: 315–76.

——. 1941. The stability of broken ends of chromosomes in *Zea mays. Genetics* 26: 234–82.

——. 1953. Induction of instability at selected loci in maize. *Genetics* 38: 579–99.

——. 1956. Controlling elements and the gene. *Cold Spring Harbor Symp. Quant. Biol.* 21: 197–216.

——. 1961. Some parallels between gene control systems in maize and bacteria. *Amer. Natur.* 95: 265–77.

Mackey, J. 1954. Neutron and X-ray experiments in wheat and a revision of the speltoid problem. *Hereditas* 40: 65–180.

McMinn, H. E. 1951. Studies in the genus Diplacus (Scrophulariaceae). *Madroño* 11: 33–128.

Makino, S. 1951. *An Atlas of the Chromosome Numbers in Animals.* Ames: Iowa State Univ. Press.

Malinowski, E. 1955. Hybrid vigour in Phaseolus and Petunia. *Bull. Acad. Pol. Sci.,* II, 3: 181–88.

Mangelsdorf, P. C. 1958. The mutagenic effect of hybridizing maize and teosinte. *Cold Spring Harbor Symp. Quant. Biol.* 23: 409–21.

Mangelsdorf, P. C., and G. S. Fraps. 1931. A direct quantitative relationship between vitamin A in corn and the number of genes for yellow pigmentation. *Science* 73: 271–72.

Mangelsdorf, P. C., and W. C. Galinat. 1964. The tunicate locus in maize dissected and reconstituted. *Proc. Nat. Acad. Sci. U.S.* 51: 147–50.

Mangelsdorf, P. C., and D. F. Jones. 1926. The expression of Mendelian factors in the gametophyte of maize. *Genetics* 11: 423–55.

Margulis, L. 1970. *Origin of Eukaryotic Cells.* New Haven, Conn.: Yale Univ. Press.

Markert, C. L., and H. Ursprung. 1971. *Developmental Genetics.* Englewood Cliffs, N. J.: Prentice-Hall.

Marshall, D. R., and R. W. Allard. 1969. The genetics of electrophoretic variants in Avena. I. The esterase E_4, E_9, E_{10}, phosphatase P_5 and anodal peroxidase APX_5 loci in *A. barbata. Jour. Hered.* 60: 17–19.

Mather, K. 1941. Variation and selection of polygenic characters. *Jour. Genet.* 41: 159–93.

——. 1943. Polygenic inheritance and natural selection. *Biol. Rev.* 18: 32–64.

——. 1944. The genetical activity of heterochromatin. *Proc. Roy. Soc.,* ser. B. 132: 308–32.

——. 1949. *Biometrical Genetics: The Study of Continuous Variation.* London: Methuen.

——. 1950. The genetical architecture of heterostyly in *Primula sinensis. Evolution* 4: 340–52.

——. 1953. Genetical control of stability in development. *Heredity* 7: 297–336.

——. 1955. Response to selection. *Cold Spring Harbor Symp. Quant. Biol.* 20: 158–65.

Mather, K., and B. J. Harrison. 1949. The manifold effect of selection. *Heredity* 3: 1–52, 131–62.

Mayo, O. 1970. The role of duplications in evolution. *Heredity* 25: 543–53.

Mayr, E. 1963. *Animal Species and Evolution.* Cambridge: Harvard Univ. Press.

Mazia, D. 1954. The particulate organization of the chromosome. *Proc. Nat. Acad. Sci. U.S.* 40: 521–27.

Mendel, G. 1866. Versuche über Pflanzen-Hybriden. *Verhandl. Naturforsch. Brünn* 4: 1–47. Reprinted in *Historiae Naturalis Classica,* J. Cramer, ed. Codicote, England, Cramer, 1960.

——. 1965. *Experiments in Plant Hybridization.* Translation and reprint, commentary by R. A. Fisher, ed. J. H. Bennett. Edinburgh and London: Oliver.

Mertens, T. R., A. B. Burdick, and F. R. Gomes. 1956. Phenotypic stability in rate of maturation of heterozygotes for induced chlorophyll mutations in tomato. *Genetics* 41: 791–803.

Mettler, L. E., and T. G. Gregg. 1969. *Population Genetics and Evolution.* Englewood Cliffs, N. J.: Prentice-Hall.

Meurman, O., and E. Therman. 1939. Studies on the chromosome morphology and structural hybridity in the genus Clematis. *Cytologia* 10: 1–14.

Meyer, V. G. 1970. A facultative gymnosperm from an interspecific cotton hybrid. *Science* 169: 886–88.

Michaelis, P. 1929. Über den Einfluss von Kern und Plasma auf die Vererbung. *Biol. Zentralbl.* 49: 302–16.

——. 1933. Entwicklungsgeschichtlich-genetische Untersuchungen an Epilobium. II. Die Bedeutung des Plasmas für die Pollenfertilität des *Epilobium luteum-hirsutum* Bastardes. *Z. Indukt. Abstamm. Vererbungsl.* 65: 1–71, 353–411.

Michaelis, P. 1950. Manifestationswechsel beim Zusammenwirken chromosomaler und plasmatischer Erbfaktoren. *Naturwissenschaften* 21: 494.

——. 1951. Plasmavererbung und Heterosis. *Z. Pflanzenzücht.* 30: 250–75.

——. 1953. Der Nachweis einer Plasmavererbung beim Weidenröschen. *Umschau,* 9.

——. 1954. Cytoplasmic inheritance in Epilobium and its theoretical significance. *Advances Genet.* 6: 287–401.

Miller, R. J., and D. E. Koeppe. 1971. Southern corn leaf blight: Susceptible and resistant mitochondria. *Science* 173: 67–69.

Moav, R. 1961. Genetic instability in Nicotiana hybrids. II. Studies of the *Ws* (*pbg*) locus of *N. plumbaginifolia* in *N. tabacum* nuclei. *Genetics* 46: 1069–87.

Moav, R., and D. R. Cameron. 1960. Genetic instability in Nicotiana hybrids. I. The expression of instability in *N. tabacum* × *N. plumbaginifolia.* *Amer. Jour. Bot.* 47: 87–93.

Moens, P. B. 1964. Chiasma distribution and the segregation of markers in chromosome 2 of the tetraploid tomato. *Genetics* 49: 123–33.

Moll, R. H., M. F. Lindsey, and H. F. Robinson. 1964. Estimates of genetic variances and level of dominance in maize. *Genetics* 49: 411–23.

Moll, R. H., J. H. Lonnquist, J. V. Fortuno, and E. C. Johnson. 1965. The relationship of heterosis and genetic divergence in maize. *Genetics* 52: 139–44.

Mooring, J. 1958. A cytogenetic study of *Clarkia unguiculata*. I. Translocations. *Amer. Jour. Bot.* 45: 233–42.

——. 1961. The evolutionary role of translocations in *Clarkia unguiculata* (Onagraceae). *Recent Advances Bot.,* 853–58.

Moos, J. R. 1955. Comparative physiology of some chromosomal types in *Drosophila pseudoobscura*. *Evolution* 9: 141–51.

Morgan, T. H. 1926. *The Theory of the Gene.* Ed. 1. New Haven, Conn: Yale Univ. Press. Reprinted, New York, Hafner, 1964.

Morgan, T. H., A. H. Sturtevant, H. J. Muller, and C. B. Bridges. 1915. *The Mechanism of Mendelian Heredity.* New York: Holt.

Mukai, T., and A. B. Burdick. 1959. Single gene heterosis associated with a second chromosome recessive lethal in *Drosophila melanogaster*. *Genetics* 44: 211–32.

Muller, H. J. 1935. On the dimensions of chromosomes and genes in dipteran salivary glands. *Amer. Natur.* 69: 405–11.

——. 1947. The gene. *Proc. Roy. Soc.,* ser B, 134: 1–37.

——. 1959. The mutation theory re-examined. *Proc. 10th Internat. Congr. Genet. (Montreal),* 1: 306–17.

Muller, H. J., and A. A. Prokofyeva. 1935. The individual gene in relation to the chromomere and chromosome. *Proc. Nat. Acad. Sci. U.S.* 21: 16–26.

Müntzing, A. 1930. Outlines to a genetic monograph of the genus Galeopsis. *Hereditas* 13: 185–341.

——. 1959. A new category of chromosomes. *Proc. 10th Internat. Congr. Genet.* (Montreal), 1: 453–67.

——. 1961. *Genetics: Basic and Applied.* Ed. 2. Stockholm: LTs Förlag.

——. 1963. A case of preserved heterozygosity in rye in spite of long-continued inbreeding. *Hereditas* 50: 377–413.

Müntzing, A., and S. Akdik. 1948. Cytological disturbances in the first inbred generations of rye. *Hereditas* 34: 485–509.

Munz, P. A. 1965. Onagraceae. *North American Flora,* ser. II, pt. 5.

Murr, S. M., and G. L. Stebbins. 1971. An albino mutant in *Plantago insularis* requiring thiamine phosphate. I. Genetics. *Genetics* 68: 231–58.

Neel, J. V. 1949. The inheritance of sickle cell anemia. *Science* 110: 64–66.

Negi, S. S., and H. P. Olmo. 1972. Certain embryological and biochemical aspects of cytokinin SD 8339 in converting sex of a male *Vitis vinifera* (*sylvestris*). *Amer. Jour. Bot.* 59: 851–57.

Neuffer, M. G. 1955. Dosage effect of multiple *Dt* loci on mutation of *a* in the maize endosperm. *Science* 121: 399–400.

Neuffer, M. G., L. Jones, and M. S. Zuber. 1968. *The mutants of maize.* Madison, Wis.: Crop Science Society.

Newman, L. J. 1966. Bridge and fragment aberrations in *Podophyllum peltatum*. *Genetics* 53: 55–63.

——. 1967. Meiotic chromosomal aberrations in wild populations of *Podophyllum peltatum*. *Chromosoma* 22: 258–73.

Nilsson-Ehle, H. 1909. Kreuzungsuntersuchungen an Hafer und Weizen. *Lunds Univ. Årsskr.,* ser. 2, 5 (2): 1–122.

——. 1911. Kreuzungsuntersuchungen an Hafer und Weizen. *Lunds Univ. Årsskr.,* Afd. 2, 7 (6): 1–84.

Nishiyama, I. 1928. On hybrids between *Triticum spelta* and two dwarf wheat plants with 40 somatic chromosomes. *Bot. Mag. Tokyo* 42: 154–77.

Nobs, M. A., and W. M. Hiesey. 1957. Studies on differential selection in Mimulus. *Carnegie Inst. Wash. Yearb.* 56: 291–92.

—— and ——. 1958. Performance of Mimulus races and their hybrids in contrasting environments. *Carnegie Inst. Wash. Yearb.* 57: 270–72.

Novitski, E., and G. Braver. 1954. An analysis of crossing over within a heterozygous inversion in *Drosophila melanogaster*. *Genetics* 39: 197–209.

Oehlkers, F. 1940. Bastardierungsversuche in der Gattung Streptocarpus Lindley. III. Neue Ergebnisse über die Genetik von Wuchsgestalt und Geschlechtsbestimmung. *Ber. Deut. Bot. Ges.* 58: 76–91.

Oehlkers, F., and C. Harte. 1943. Über die Aufhebung des Gonen- und Zygotenausfalls bei Oenothera. *Flora* 37: 106–24.

Ohno, S. 1970. *Evolution by Gene Duplication.* New York: Springer-Verlag.

Olby, R. C. 1966. *Origins of Mendelism.* London: Constable.

Olmsted, C. E. 1944. Growth and development in range grasses. IV. Photoperiodic responses in twelve geographic strains of side-oats grama. *Bot. Gaz.* 106: 46–74.

Ornduff, R. 1964. The breeding system of *Oxalis suksdorfii*. *Amer. Jour. Bot.* 51: 307–14.

——. 1966a. The origin of dioecism from heterostyly in Nymphoides (Menyanthaceae). *Evolution* 20: 309–14.

——. 1966b. The breeding system of *Pontederia cordata* L. *Bull. Torrey Bot. Club* 93: 407–16.

——. 1970a. Incompatibility and the pollen economy of *Jepsonia parryi*. *Amer. Jour. Bot.* 57: 1036–41.

——. 1970b. Heteromorphic incompatibility in *Jepsonia malvifolia*. *Bull. Torrey Bot. Club* 97: 258–61.

Ornduff, R. 1971. The reproductive system of *Jepsonia heterandra*. *Evolution* 25: 300–311.

——. 1972. The breakdown of trimorphic incompatibility in Oxalis section Corniculatae. *Evolution* 26: 52–65.

Ownbey, M., and G. D. McCollum. 1953. Cytoplasmic inheritance and reciprocal amphiploidy in Tragopogon. *Amer. Jour. Bot.* 40: 788–96.

Pandey, K. K. 1960. Evolution of gametophytic and sporophytic systems of self-incompatibility in angiosperms. *Evolution* 14: 98–115.

——. 1968. Compatibility relationships in flowering plants: Role of the S-gene complex. *Amer. Natur.* 102: 475–89.

——. 1969. Elements of the S-gene complex. IV. S-allele polymorphism in Nicotiana species. *Heredity* 24: 601–19.

——. 1970. Elements of the S-gene complex. VI. Mutations of the self-incompatibility gene, pseudo-compatibility and origin of new self-incompatibility alleles. *Genetica* 41: 477–516.

Paris, C. D., W. J. Haney, and G. B. Wilson. 1960. A survey of the interactions of genes for flower color. *Mich. State Univ. Agr. Exp. Sta. Bull.* 281.

Paroda, R. S., and H. Rees. 1971. Nuclear DNA variation in Eu-Sorghums. *Chromosoma* 32: 353–63.

Patterson, J. T., W. Stone, S. Bedichek, and M. Suche. 1934. The production of translocations in Drosophila. *Amer. Natur.* 68: 359–69.

Pavan, C. 1964. Modern concept of chromosome structure and function. *Triangle* 6: 287–93.

Pelling, C. 1966. A replicative and synthetic chromosomal unit—the modern concept of the chromomere. *Proc. Roy. Soc.,* ser. B, 164: 279–89.

Pelton, J. S. 1964. Genetic and morphogenetic studies of angiosperm single-gene dwarfs. *Bot. Rev.* 30: 479–512.

Peterson, P. A. 1958. Cytoplasmically inherited male sterility in Capsicum. *Amer. Natur.* 92: 111–19.

Phillips, L. L. 1962. Segregation in new allopolyploids of Gossypium. IV. Segregation in New World × Asiatic and New World × wild American hexaploids. *Amer. Jour. Bot.* 49: 51–57.

Phillips, L. L., and D. U. Gerstel. 1959. Segregation in new allopolyploids of Gossypium. III. Leaf shape segregation in hexaploid hybrids of New World cottons. *Jour. Hered.* 50: 103–8.

Phinney, B. O. 1956. Growth response of single-gene dwarf mutants in maize to gibberellic acid. *Proc. Nat. Acad. Sci. U.S.* 42: 185–89.

Pianka, E. R. 1970. On r- and K-selection. *Amer. Natur.* 104: 592–97.

——. 1972. r and K selection or b and d selection? *Amer. Natur.* 106: 581–88.

Pijl, L. van der. 1972. *Principles of Dispersal in Higher Plants.* Ed. 2. Berlin: Springer-Verlag.

Pontecorvo. G. 1953. The genetics of *Aspergillus nidulans.* *Advances Genet.* 5: 141–238.

——. 1958. *Trends in Genetic Analysis.* New York: Columbia Univ. Press.

Powers, L. 1944. An expansion of Jones's theory for the explanation of heterosis. *Amer. Natur.* 78: 275–80.

Prazmo, W. 1965. Cytogenetic studies on the genus Aquilegia. III. Inheritance of the traits distinguishing different complexes in the genus Aquilegia. *Acta Soc. Bot. Pol.* 34: 403–37.

Preer, J. R. 1950. Microsopically visible bodies in the cytoplasm of the "killer" strains of *Paramecium aurelia*. *Genetics* 35: 344–62.

Putt, E. D. 1954. Cytogenetic studies of sterility in rye. *Can. Jour. Agr. Sci.* 34: 81–119.

Quinby, J. R., and R. E. Karper. 1946. Heterosis in Sorghum resulting from the heterozygous condition of a single gene that affects duration of growth. *Amer. Jour. Bot.* 33: 716–21.

Quisenberry, J. E., and R. J. Kohel. 1971. Phenotypic stability of cotton. *Crop Sci.* 11: 827–29.

Rabideau, G. S., W. G. Whaley, and C. Heimsch. 1950. The absorption and distribution of radioactive phosphorus in two maize inbreds and their hybrid. *Amer. Jour. Bot.* 37: 93–99.

Raff, R. A., and H. R. Mahler. 1972. The non-symbiotic origin of mitochondria. *Science* 177: 575–82.

Raffel, D., and H. J. Muller. 1940. Position effect and gene divisibility considered in connection with three strikingly similar scute mutations. *Genetics* 25: 541–83.

Rana, R. S., and H. K. Jain. 1965. Adaptive role of interchange heterozygosity in the annual Chrysanthemum. *Heredity* 20: 21–29.

Raven, P. H., and D. P. Gregory. 1972. Observations of meiotic chromosomes in Gaura (Onagraceae). *Brittonia* 24: 71–86.

Raven, P. H., and H. Lewis. 1960. Observations on the chromosomes and relationships of Hauya and Xylonagra. *Aliso* 4: 483–84.

Ray, P. M., and H. F. Chisaki. 1957a. Studies on Amsinckia. II. Relationships among the primitive species. *Amer. Jour. Bot.* 44: 537–44.

—— and ——. 1957b. Studies on Amsinckia. III. Aneuploid diversification in the Muricatae. *Amer. Jour. Bot.* 44: 545–54.

Redei, G. P. 1962. Single locus heterosis. *Z. Vererbungsl.* 93: 164–70.

Rees, H. 1961. Genotypic control of chromosome form and behaviour. *Bot. Rev.* 27: 288–318.

——. 1972. DNA in higher plants. *Brookhaven Symp. Biol.* 23: 394–418.

Rees, H., F. M. Cameron, M. H. Hazarika, and G. H. Jones. 1966. Nuclear variation between diploid angiosperms. *Nature* 211: 828–30.

Rees, H., and G. M. Evans. 1966. A correlation between the localisation of chiasmata and the replication pattern of chromosomal DNA. *Exp. Cell Res.* 44: 161–64.

Rees, H., and S. Hassouna. 1964. Mineral metabolism and hybrid vigour in rye. *Ann. Bot.* 28: 101–11.

Rees, H., and J. B. Thompson. 1955. Localisation of chromosome breakage at meiosis. *Heredity* 9: 399–407.

—— and ——. 1956. Genotypic control of chromosome behaviour in rye. III. Chiasma frequency in homozygotes and heterozygotes. *Heredity* 10: 409–24.

Rendel, J. M. 1953. Heterosis. *Amer. Natur.* 87: 129–38.

Renner, O. 1917. Artbastarde und Bastardarten in der Gattung Oenothera. *Ber. Deut. Bot. Ges.* 35: 21–26.

——. 1924. Die Scheckung der Oenotherenbastarde. *Biol. Zentralbl.* 44: 309–36.

Renner, O. 1925. Untersuchungen über die faktorielle Konstitution einiger komplexheterozygotischer Oenotheren. *Bibl. Genet.* 9: 1–168.

——. 1928. Über Koppelungswechsel bei Oenothera. *Z. Indukt. Abstamm. Vererbungsl.*, suppl. 2, vol. 1928: 1216–20.

——. 1936. Zur Kenntnis der nichtmendelnden Buntheit der Laublätter. *Flora* 30: 218–90.

——. 1937. Zur Kenntnis der Plastiden- und Plasmavererbung. *Cytologia*, Fujii Jubilee vol., 644–55.

——. 1941. Über die Entstehung homozygotischer Formen aus komplexheterozygotischen Oenotheren. *Flora* 35: 201–38.

Reuther, W., L. D. Batchelor, and H. J. Webber, eds. 1968. *The Citrus Industry.* Vol. 2, *Anatomy, Physiology, Genetics, and Reproduction.* Ed. 2. Berkeley: Univ. of Calif. Press.

Rhoades, M. M. 1933. The cytoplasmic inheritance of male sterility in *Zea mays. Jour. Genet.* 27: 71–93.

——. 1941. The genetic control of mutability in maize. *Cold Spring Harbor Symp. Quant. Biol.* 9: 138–44.

——. 1942. Preferential segregation in maize. *Genetics* 27: 395–407.

——. 1943. Genic induction of an inherited cytoplasmic difference. *Proc. Nat. Acad. Sci. U.S.* 29: 327–29.

——. 1945. On the genetic control of mutability in maize. *Proc. Nat. Acad. Sci. U.S.* 31: 91–95.

——. 1946. Plastid mutations. *Cold Spring Harbor Symp. Quant. Biol.* 11: 202–7.

——. 1950. Meiosis in maize. *Jour. Hered.* 41: 59–67.

——. 1952. Preferential segregation in maize. In *Heterosis,* ed. J. W. Gowen. Ames: Iowa State Univ. Press.

——. 1954. Chromosomes, mutations, and cytoplasm in maize. *Science* 120: 115–20.

——. 1955. The cytogenetics of maize. *Agronomy* 5: 123–219.

Rhoades, M. M., and E. Dempsey. 1953. Cytogenetic studies of deficient-duplicate chromosomes derived from inversion heterozygotes in maize. *Amer. Jour. Bot.* 40: 405–24.

—— and ——. 1966. The effect of abnormal chromosome 10 on preferential segregation and crossing over in maize. *Genetics* 53: 989–1020.

Rhoades, M. M., and B. McClintock. 1935. The cytogenetics of maize. *Bot. Rev.* 1: 292–325.

Richmond, R. C., and J. R. Powell. 1970. Evidence of heterosis associated with an enzyme locus in a natural population of Drosophila. *Proc. Nat. Acad. Sci. U.S.* 67: 1264–67.

Rick, C. M. 1947. Partial suppression of hair development indirectly affecting fruitfulness and the proportion of cross-pollination in a tomato mutant. *Amer. Natur.* 81: 185–202.

——. 1959. Non-random gene distribution among tomato chromosomes. *Proc. Nat. Acad. Sci. U.S.* 45: 1515–19.

——. 1963. Differential zygotic lethality in a tomato species hybrid. *Genetics* 48: 1497–1507.

——. 1966. Abortion of male and female gametes in the tomato determined by allelic interaction. *Genetics* 53: 85–96.

——. 1971a. The tomato *Ge* locus: Linkage relations and geographic distribution of alleles. *Genetics* 67: 75–85.

——. 1971b. Some cytogenetic features of the genome in diploid plant species. *Stadler Genet. Symp.* 2: 153–74.

Rick, C. M., and P. G. Smith. 1953. Novel variation in tomato species hybrids. *Amer. Natur.* 87: 359–73.

Rieger, R. 1965. Bau und Funktion der Chromosomen. *Biol. Rundsch.* 2: 265–79.

Rieger, R., and A. Michaelis. 1967. *Chromosomenmutationen.* Jena: Gustav Fischer.

Rieger, R., A. Michaelis, and M. M. Green. 1968. *A Glossary of Genetics and Cytogenetics.* Ed. 3. Berlin and New York: Springer-Verlag.

Ris, H. 1957. Chromosome structure. In *The Chemical Basis of Heredity,* ed. W. D. McElroy and B. Glass. Baltimore: Johns Hopkins Press.

——. 1962. Interpretation of ultrastructure in the cell nucleus. *Symp. Internat. Soc. Cell Biol.* 1: 69–88.

Ris, H., and B. L. Chandler. 1963. The ultrastructure of genetic systems in prokaryotes and eukaryotes. *Cold Spring Harbor Symp. Quant. Biol.* 28: 1–8.

Ris, H., and D. F. Kubai. 1970. Chromosome structure. *Annu. Rev. Genet.* 4: 263–94.

Robbins, W. J. 1941. Growth of excised roots and heterosis in tomato. *Amer. Jour. Bot.* 28: 216–25.

Roberts, H. F. 1929. *Plant Hybridization Before Mendel.* Princeton, N.J.: Princeton Univ. Press. Reprinted by Hafner, New York, 1965.

Roemer, Th., and W. Rudorf, eds. 1941–1958. *Handbuch der Pflanzenzüchtung.* Ed. 1. 6 vols. Berlin: Paul Parey.

Roman, H. 1948. Selective fertilization in maize. *Genetics* 33: 122.

Rothschild, M. 1972. Some observations on the relationships between plants, toxic insects and birds. In *Phytochemical Ecology,* ed. J. B. Harborne. New York and London: Academic Press.

Rüdenberg, L. 1965. Effects of phosphorus and calcium deficiencies on meiosis in tomato. *Phyton* 22: 119–25.

Rudkin, G. T. 1965. The relative mutabilities of DNA in regions of the X chromosome of *Drosophila melanogaster. Genetics* 52: 665–81.

Russell, W. A., and C. R. Burnham. 1950. Cytogenetic studies of an inversion in maize. *Sci. Agr.* 30: 93–111.

Sager, R. 1972. *Cytoplasmic Genes and Organelles.* New York: Academic Press.

Salisbury, E. J. 1942. *The Reproductive Capacity of Plants.* London: G. Bell.

Sansome, F. W. 1933. Chromatid segregation in *Solanum lycopersicum. Jour. Genet.* 27: 105–32.

Sansome, F. W., and J. Philp. 1939. *Recent Advances in Plant Genetics.* Ed. 2. London: Churchill; Philadelphia: Blakiston's.

Sanyal, P., and A. N. Dutta. 1954. Natural crossing in *Hibiscus sabdariffa. Sci. Cult.* 19: 407.

Sarkissian, I. V., and R. C. Huffaker. 1962. Depression and stimulation by chloramphenicol of development of carboxylating enzyme activity in inbred and hybrid barley. *Proc. Nat. Acad. Sci. U.S.* 48: 735–43.

Sarkissian, I. V., and H. K. Srivastava. 1967. Mitochondrial polymorphism in maize. II. Further evidence of correlation of mitochondrial complementation and heterosis. *Genetics* 57: 843–50.

Sax, K. 1931. Chromosome ring formation in *Rhoeo discolor*. *Cytologia* 3: 36–53.

——. 1940. An analysis of X-ray induced chromosomal aberrations in Tradescantia. *Genetics* 25: 41–68.

Sax, K., and K. Mather. 1939. An X-ray analysis of progressive chromosome splitting. *Jour. Genet.* 37: 483–90.

Schmalhausen, I. I. 1949. *Factors of Evolution*. Transl. Philadelphia: Blakiston's.

Schnick, S. M., T. Mukai, and A. B. Burdick. 1960. Heterozygote viability of a second chromosome recessive lethal in *Drosophila melanogaster*. *Genetics* 45: 315–29.

Schuler, J. F. 1954. Natural mutations in inbred lines of maize and their heterotic effect. I. Comparison of parent, mutant and their F_1 hybrid in a highly inbred background. *Genetics* 39: 908–22.

Schultz, J. 1929. The minute reaction in the development of *Drosophila melanogaster*. *Genetics* 14: 366–419.

Schwanitz, F. 1956. Grossmutationen. *Umschau* 56: 45–48.

——. 1957. Spornbildung bei einem Bastard zwischen drei Digitalis-Arten. *Biol. Zentralbl.* 76: 226–31.

Schwanitz, F., and H. Schwanitz. 1955. Eine Grossmutation bei *Linaria maroccana* L.: mut. *gratioloides*. *Beitr. Biol. Pflanzen* 31: 473–97.

Schwemmle, J., ed. 1938. Genetische und Zytologische Untersuchungen an Eu-Oenotheren. *Z. Indukt. Abstamm. Vererbungsl.* 75: 358–800.

——. 1968. Selective fertilization in Oenothera. *Advances Genet.* 14: 225–324.

Schwemmle, J., and W. Koepchen. 1953. Weitere Untersuchungen zur selektiven Befruchtung. *Z. Indukt. Abstamm. Vererbungsl.* 85: 307–46.

Scossiroli, R. E. 1954. Effectiveness of artificial selection under irradiation of plateaued populations of *Drosophila melanogaster*. *Internat. Union Biol. Sci.*, ser. B, 15: 42–66.

Sears, E. R. 1944. Cytogenetic studies with polyploid species of wheat. II. Additional chromosomal aberrations in *Triticum vulgare*. *Genetics* 29: 232–46.

——. 1959. The systematics, cytology and genetics of wheat. *Handbuch der Pflanzenzüchtung* 2: 164–87.

Sehgal, S. M. 1963. *Effects of teosinte and "Tripsacum" introgression in maize*. Cambridge: Publications of the Bussey Institute, Harvard University.

Selander, R. K., and D. W. Kaufman. 1973. Self-fertilization and genetic population structure in a colonizing land snail. *Proc. Nat. Acad. Sci. U.S.* 70: 1186–90.

Shacklette, H. T. 1964. Flower variation of *Epilobium angustifolium* L. growing over uranium deposits. *Can. Field Natur.* 78: 32–42.

Shannon, M. P., T. C. Kaufman, and B. H. Judd. 1970. Lethality patterns of mutations in the zeste-white region of *Drosophila melanogaster*. *Genetics* 64, suppl., 58.

Shepherd, K. W., and G. M. E. Mayo. 1972. Genes conferring specific plant disease resistance. *Science* 175: 375–80.

Shull, G. H. 1914. Duplicate genes for capsule form in *Bursa bursa-pastoris*. *Z. Indukt Abstamm. Vererbungsl.* 12: 97–149.

——. 1952. Beginnings of the heterosis concept. In *Heterosis*, ed. J. W. Gowen. Ames: Iowa State Univ. Press.

Silow, R. A. 1939. The genetics of leaf shape in diploid cottons and the theory of gene interaction. *Jour. Genet.* 38: 229–76.

Simmonds, N. W. 1945. Meiosis in tropical *Rhoeo discolor*. *Nature* 155: 731.

——. 1965a. Chimeral potato mutants. *Jour Hered.* 56: 139–42.

——. 1965b. Mutant expression in diploid potatoes. *Heredity* 20: 65–72.

——. 1966. *Bananas.* Ed. 2. London: Longmans.

——. 1969. Genetical bases of plant breeding. *Jour. Rubber Res. Inst. Malaya* 21: 1–10.

Simpson, D. M. 1954. Natural cross-pollination in cotton. *U.S. Dept. Agr. Tech. Bull.* 1094.

Simpson, D. M., and E. N. Duncan. 1956. Cotton pollen dispersal by insects. *Agron. Jour.* 48: 305–8.

Simpson, G. G., and W. S. Beck. 1965. *Life: An Introduction to Biology.* New York: Harcourt.

Sinnott, E. W., L. C. Dunn, and Th. Dobzhansky. 1958. *Principles of Genetics.* Ed. 5. New York: McGraw-Hill.

Sinnott, E. W., and D. Hammond. 1930. Factorial balance in the determination of fruit shape in Cucurbita. *Amer. Natur.* 64: 509–24.

Sirks, M. J. 1956. *General Genetics.* Ed. 5, transl. The Hague: Nijhoff.

Skalinska, M. 1929. Das Problem des Nichterscheinens des väterlichen Typus in der Spaltung der partiell sterilen Aquilegia-Species-Bastarde. *Acta Soc. Bot. Pol.* 6: 138–64.

Slizynska, H. 1938. Salivary chromosome analysis of the white-facet region of *Drosophila melanogaster. Genetics* 23: 291–99.

Smith, E. B. 1970. Pollen competition and relatedness in Haplopappus section Isopappus (Compositae). II. *Amer. Jour. Bot.* 57: 874–80.

Smith, F. H., and Q. D. Clarkson. 1956. Cytological studies of interspecific hybridization in Iris, subsection Californicae. *Amer. Jour. Bot.* 43: 582–88.

Smith, H. H. 1952a. Relation between sterility and morphological characters in an interspecific Nicotiana cross. *Genetics* 37: 26–38.

——. 1952b. Fixing transgressive vigor in *Nicotiana rustica*. In *Heterosis*, ed. J. W. Gowen. Ames: Iowa State Univ. Press.

——. 1962. Studies on the origin, inheritance and mutation of genic-cytoplasmic male sterility in Nicotiana. *Genetics* 47: 985–86.

——. 1968. Recent cytogenetic studies in the genus Nicotiana. *Advances Genet.* 14: 1–54.

——. 1971. Broadening the base of genetic variability in plants. *Jour. Hered.* 62: 265–76.

Smith, H. H., and K. Daly. 1959. Discrete populations derived by interspecific hybridization in Nicotiana. *Evolution* 13: 476–87.

Smith, H. H., and S. A. Sand. 1957. Genetic studies on somatic instability in cultures derived from hybrids between *Nicotiana langsdorffii* and *N. sanderae. Genetics* 42: 560–82.

Smith-White, S. 1959. Cytological evolution in the Australian flora. *Cold Spring Harbor Symp. Quant. Biol.* 24: 273–89.

Snow, R. 1960. Chromosomal differentiation in *Clarkia dudleyana*. *Amer. Jour. Bot.* 47: 302–9.

Snow, R., and M. P. Dunford. 1961. A study of interchange heterozygosity in a population of *Datura meteloides*. *Genetics* 46: 1097–1110.

Sonneborn, T. M. 1950a. The cytoplasm in heredity. *Heredity* 4: 11–36.

——. 1950b. Partner of the genes. *Sci. Amer.*, Nov., 1950.

——. 1957. Breeding systems, reproductive methods, and species problems in Protozoa. In *The Species Problem,* ed. E. Mayr. Washington, D.C.: American Association for the Advancement of Science.

Sparrow, A. H., H. J. Price, and A. G. Underbrink. 1972. A survey of DNA content per cell and per chromosome of prokaryotic and eukaryotic organisms: Some evolutionary considerations. *Brookhaven Symp. Biol.* 23: 451–94.

Srb, A. M., and N. H. Horowitz. 1944. The ornithine cycle in Neurospora and its genetic control. *Jour. Biol. Chem.* 154: 129–39.

Stack, S. M., and W. V. Brown. 1969. Somatic and premeiotic pairing of homologues in *Plantago ovata*. *Bull. Torrey Bot. Club* 96: 143–49.

Stadler, L. J. 1942. Some observations on gene variability and spontaneous mutation. *Spragg Mem. Lect.* (Michigan State Univ.), 3: 3–15.

Stadler, L. J., and F. M. Uber. 1942. Genetic effects of ultraviolet radiation in maize. IV. Comparisons of monochromatic radiations. *Genetics* 27: 84–118.

Stebbins, G. L. 1938. Cytological characteristics associated with the different growth habits in the dicotyledons. *Amer. Jour. Bot.* 25: 189–98.

——. 1947. Types of polyploids: Their classification and significance. *Advances Genet.* 1: 403–29.

——. 1950. *Variation and Evolution in Plants.* New York: Columbia Univ. Press.

——. 1957. Self fertilization and population variability in the higher plants. *Amer. Natur.* 91: 337–54.

——. 1958. Longevity, habitat, and release of genetic variability in higher plants. *Cold Spring Harbor Symp. Quant. Biol.* 23: 365–78.

——. 1959. Genes, chromosomes, and evolution. In *Vistas in Botany,* ed. W. Turrill. London: Pergamon.

——. 1966. Chromosomal variation and evolution. *Science* 152: 1463–69.

——. 1968.Gene action, mitotic frequency, and morphogenesis in higher plants. *Develop. Biol.,* suppl. 1, pp. 113–35.

——. 1971. *Chromosomal Evolution in Higher Plants.* London: Arnold.

Stebbins, G. L., and S. Ellerton. 1939. Structural hybridity in *Paeonia californica* and *P. brownii*. *Jour. Genet.* 38: 1–36.

Stebbins, G. L., and S. K. Jain. 1960. Developmental studies of cell differentiation in the epidermis of monocotyledons. I. Allium, Rhoeo, and Commelina. *Develop. Biol.* 2: 409–26.

Stebbins, G. L., J. A. Jenkins, and M. S. Walters, 1953. Chromosomes and phylogeny in the Compositae, tribe Cichorieae. *Univ. Calif. Publ. Bot.* 26: 401–29.

Stebbins, G. L., and E. Yagil. 1966. The morphogenetic effects of the hooded gene in barley. I. The course of development in hooded and awned genotypes. *Genetics* 54: 727–41.

Steffensen, D. 1953. Induction of chromosome breakage at meiosis by a magnesium deficiency in Tradescantia. *Proc. Nat. Acad. Sci. U.S.* 39: 613–20.

——. 1955. Chromosome breakage with a calcium deficiency in Tradescantia. *Proc. Nat. Acad. Sci. U.S.* 41: 155–60.

——. 1959. A comparative view of the chromosome. *Brookhaven Symp. Biol.* 12: 103–24.

Steiner, E. 1956. New aspects of the balanced lethal mechanism in Oenothera. *Genetics* 41: 486–500.

Stephens, J. C., and R. F. Holland. 1954. Cytoplasmic male-sterility for hybrid sorghum seed production. *Agron. Jour.* 46: 20–23.

Stephens, S. G. 1945a. A genetic survey of leaf shape in New World cottons—a problem in critical identification of alleles. *Jour. Genet.* 46: 313–30.

——. 1945b. The modifier concept: A developmental analysis of leaf-shape 'modification' in the New World cottons. *Jour. Genet.* 46: 331–44.

——. 1948. A biochemical basis for the pseudo-allelic anthocyanin series in Gossypium. *Genetics* 33: 191–214.

——. 1949. The cytogenetics of speciation in Gossypium. I. Selective elimination of the donor parent genotype in interspecific backcrosses. *Genetics* 34: 627–37.

——. 1950. The internal mechanism of speciation in Gossypium. *Bot. Rev.* 16: 115–49.

——. 1951a. Possible significance of duplication in evolution. *Advances Genet.* 4: 247–65.

——. 1951b. "Homologous" genetic loci in Gossypium. *Cold Spring Harbor Symp. Quant. Biol.* 16: 131–41.

——. 1953. The breeding system in cotton and its possible significance in methods of cotton improvement (Abstract). Raleigh, N.C.: North Carolina Acad. Sci., 1953.

——. 1955. Symposium on pseudoallelism and the theory of the gene: summary, synthesis and critique. *Amer. Nat.* 89: 117–22.

Stephens, S. G., and M. D. Finkner. 1953. Natural crossing in cotton. *Econ. Bot.* 7: 257–69.

Stern, C. 1943. Genic action as studied by means of the effects of different doses and combinations of alleles. *Genetics* 28: 441–75.

——. 1949a, 1960. *Principles of Human Genetics.* Eds. 1 and 2. San Francisco: Freeman.

——. 1949b. Gene and character. In *Genetics, Paleontology, and Evolution,* ed. G. Jepsen, E. Mayr, and G. G. Simpson. Princeton, N.J.: Princeton Univ. Press.

——. 1954. Two or three bristles. *Amer. Sci.* 42: 213–47.

——. 1968. *Genetic Mosaics and Other Essays.* Cambridge: Harvard Univ. Press.

Stern, C., G. Carson, M. Kinst, E. Novitski, and D. Uphoff. 1952. The viability of heterozygotes for lethals. *Genetics* 37: 413–49.

Stern, C., and E. R. Sherwood, eds. 1966. *The Origin of Genetics.* San Francisco: Freeman.

Stubbe, H. 1952. Über einige theoretische und praktische Fragen der Mutationsforschung. *Abhandl. Sächsisch. Akad. Wiss., Math.-Natur. Kl.* (Leipzig), 47: 3–23.

——. 1953. Über mono- und di-gen bedingte Heterosis bei *Antirrhinum majus* L. *Z. Indukt. Abstamm. Vererbungs.* 85: 450–78.

——. 1959. Considerations on the genetical and evolutionary aspects of some mutants of Hordeum, Glycine, Lycopersicon and Antirrhinum. *Cold Spring Harbor Symp. Quant. Biol.* 24: 31–40.

Stubbe, H. 1965. *Kurze Geschichte der Genetik bis zur Wiederentdeckung der Vererbungsregeln Gregor Mendels.* Ed. 2. Jena: Gustav Fischer.

———. 1966. *Genetik und Zytologie von Antirrhinum L. sect. Antirrhinum.* Jena: Gustav Fischer.

Stubbe, H., and F. von Wettstein. 1941. Über die Bedeutung von Klein- und Grossmutationen in der Evolution. *Biol. Zentralbl.* 61: 265–97.

Sturtevant, A. H. 1918. An analysis of the effects of selection. *Carnegie Inst. Wash. Publ.* 264.

———. 1921. A case of rearrangement of genes in Drosophila. *Proc. Nat. Acad. Sci. U.S.* 7: 235–37.

———. 1965. *A History of Genetics.* New York: Harper.

Sturtevant, A. H., and G. W. Beadle. 1936. The relations of inversions in the X chromosomes of *Drosophila melanogaster* to crossing over and disjunction. *Genetics* 21: 554–604.

——— and ———. 1939. *An Introduction to Genetics.* Philadelphia: Saunders.

Suomalainen, E. 1947. On the cytology of the genus Polygonatum group Alternifolia. *Ann. Acad. Sci. Fenn.,* ser. 4, 13: 1–65.

Sutton, W. S. 1902. On the morphology of the chromosome group in *Brachystola magna. Biol. Bull.* 4: 24–39.

———. 1903. The chromosomes in heredity. *Biol. Bull.* 4: 231–51.

Swanson, C. P. 1957. *Cytology and Cytogenetics.* Englewood Cliffs, N.J.: Prentice-Hall.

Symposium. 1955. Pseudoallelism and the theory of the gene. *Amer. Natur.* 89: 65–122.

Taylor, J. H. 1963. The replication and organization of DNA in chromosomes. In *Molecular Genetics,* ed. J. H. Taylor. New York: Academic Press.

Ter-Avanesjan, D. V. 1949. The role of number of pollen grains per flower in fertilization in plants [in Russian]. *Bull. Appl. Bot. Genet. Plant Breed.* 28: 119–33. (English abstr. in *Plant Breed. Abstr.* 25: 491, 1955.)

Ter-Avanesjan, D. V., and F. G. Nigmatullin. 1959. The role of the amount of pollen in inheritance of characters and properties in wheat hybrids [in Russian]. *Selek. Semenovod.* (Moscow) 1: 71–73. (English abstr. in *Plant Breed. Abstr.* 30: 284, 1960.)

Thoday, J. M. 1958a. Homeostasis in a selection experiment. *Heredity* 12: 401–15.

———. 1958b. Natural selection and biological progress. In *A Century of Darwin,* ed. S. A. Barnett. Cambridge: Harvard Univ. Press.

———. 1961. Location of polygenes. *Nature* 191: 368–70.

Thomas, C. A. 1971. The genetic organization of chromosomes. *Annu. Rev. Genet.* 5: 237–56.

Thompson, R. C., T. W. Whitaker, and G. W. Bohn. 1958. Natural cross-pollination in lettuce. *Proc. Amer. Soc. Hort. Sci.* 72: 403–9.

Tilney-Basset, R. A. E. 1963. The structure of periclinal chimeras. *Heredity* 18: 265–85.

Timofeeff-Ressovsky, N. W. 1934a. The experimental production of mutations. *Biol. Rev.* 9: 411–57.

——. 1934b. Über den Einfluss des genotypischen Milieus und der Aussenbedingungen auf die Realisation des Genotyps. *Nachr. Ges. Wiss. Göttingen,* N.F., 1: 53–106.

——. 1940. Mutations and geographical variation. In *The New Systematics,* ed. J. Huxley. Oxford: Oxford Univ. Press.

Ting, Y. C. 1965. Spontaneous chromosome inversions of Guatemalan teosintes *(Zea mexicana). Genetica* 36: 229–42.

Tischler, G. 1951–1953. *Allgemeine Pflanzenkaryologie.* Vols. 2 and 3 of *Handbuch der Pflanzenanatomie* (Berlin).

Tobgy, H. A. 1943. A cytological study of *Crepis fuliginosa, C. neglecta,* and their F_1 hybrid, and its bearing on the mechanism of phylogenetic reduction in chromosome number. *Jour. Genet.* 45: 67–111.

Tschermak, E. 1900. Über künstliche Kreuzung bei *Pisum sativum. Ber. Deut. Bot. Ges.* 18: 232–39. Reprinted in *Genetics* 35: 42–47, 1950.

Turesson, G. 1922. The genotypical response of the plant species to its habitat. *Hereditas* 3: 211–350.

——. 1925. The plant species in relation to habitat and climate: Contributions to the knowledge of genecological units. *Hereditas* 6: 147–236.

Turner, B. L., and D. Horne. 1964. Taxonomy of Machaeranthera sect. Psilactis (Compositae-Astereae). *Brittonia* 16: 316–31.

Uhl, C. H. 1965. Chromosome structure and crossing over. *Genetics* 51: 191–207.

——. 1970. Chromosomes of Graptopetalum and Thompsonella (Crassulaceae), *Amer. Jour. Bot.* 57: 1115–21.

Vaarama, A. 1954. Inheritance of morphological characters and fertility in the progeny of the hybrid *Rubus idaeus* × *R. arcticus. Caryologia,* suppl. vol. 1954, 846–50.

Valentine, D. H. 1970. Cryptic polymorphism in *Primula elatior. Israel Jour. Bot.* 19: 71–76.

Vasek, F. C. 1964. Outcrossing in natural populations. I. The Breckinridge Mountain population of *Clarkia exilis. Evolution* 18: 213–18.

——. 1965. Outcrossing in natural populations. II *Clarkia unguiculata. Evolution* 19: 152–56.

——. 1967. Outcrossing in natural populations. III. The Deer Creek population of *Clarkia exilis. Evolution* 21: 241–48.

Vickery, R. K., and R. L. Olson. 1956. Flower color inheritance in the *Mimulus cardinalis* complex. *Jour. Hered.* 47: 194–99.

Vuilleumier, B. S. 1967. The origin and evolutionary development of heterostyly in the angiosperms. *Evolution* 21: 210–26.

——. 1973. Lactuceae. *Jour. Arnold Arboretum* 54: 42–93.

Waddington, C. H. 1957. *The Strategy of the Genes.* London: G. Allen.

——. 1962. *New Patterns in Genetics and Development.* New York: Columbia Univ. Press.

Wagner, R. P. 1969. Genetics and phenogenetics of mitochondria. *Science* 163: 1026–31.

——. 1972. Evolution and genetic significance of mitochondria. *Stadler Genet. Symp.* 4: 39–55.

Wagner, R. P., and H. K. Mitchell. 1955, 1964. *Genetics and Metabolism.* Eds. 1 and 2. New York: Wiley.

Walker, J. W. 1972. Chromosome numbers, phylogeny, phytogeography of the Annonaceae and their bearing on the (original) basic chromosome number of angiosperms. *Taxon* 21: 57–65.

Wall, J. R., and T. L. York. 1957. Inheritance of seedling cotyledon position in Phaseolus species. *Jour. Hered.* 48: 71–74.

Wallace, B. 1959. Influence of genetic systems on geographical distribution. *Cold Spring Harbor Symp. Quant. Biol.* 24: 193–204.

——. 1963. Modes of reproduction and their genetic consequences. In *Statistical Genetics and Plant Breeding*. National Academy of Sciences and National Research Council (Washington, D.C.), publ. no. 982.

Wallace, B., and Th. Dobzhansky. 1959. *Radiation, Genes, and Man.* New York: Holt.

Walters, J. L. 1942. Distribution of structural hybrids in *Paeonia californica. Amer. Jour. Bot.* 29: 270–75.

——. 1952. Heteromorphic chromosome pairs in *Paeonia californica. Amer. Jour. Bot.* 39: 145–51.

——. 1956. Spontaneous meiotic chromosome breakage in natural populations of *Paeonia californica. Amer. Jour. Bot.* 43: 342–54.

Walters, M. 1957. Studies of spontaneous chromosome breakage in interspecific hybrids of Bromus. *Univ. Calif. Publ. Bot.* 28: 335–447.

Watson, J. D. 1970. *Molecular Biology of the Gene.* Ed. 2. New York: Benjamin.

Watson, J. D., and F. H. C. Crick. 1953. Molecular structure of nucleic acids. *Nature* 171: 737–38.

Watts, L. E. 1958. Natural cross-pollination in lettuce, *Lactuca sativa* L. *Nature* 181: 1084.

Weismann, A. 1883. Über die Vererbung. Jena: Gustav Fischer.

——. 1893. *The Germ-plasm: A Theory of Heredity.* Transl. London: Walter Scott.

Westergaard, M. 1958. The mechanism of sex determination in dioecious flowering plants. *Advances Genet.* 9: 217–81.

——. 1964. Studies on the mechanism of crossing over. I. Theoretical considerations. *Compt. Rend. Lab. Carlsberg* 34: 359–405.

Wettstein, F. von. 1925. Genetische Untersuchungen an Moosen (Musci und Hepaticae). *Bibliogr. Genet.* 1: 1–38.

Whaley, W. G. 1944. Heterosis. *Bot. Rev.* 10: 461–98.

——. 1952. Physiology of gene action in hybrids. In *Heterosis,* ed. J. W. Gowen. Ames: Iowa State Univ.

Whaley, W. G., C. Heimsch, and G. S. Rabideau. 1950. The growth and morphology of two maize inbreds and their hybrid. *Amer. Jour. Bot.* 37: 77–84.

Whitehouse, H. L. K. 1950. Multiple-allelomorph incompatibility of pollen and style in the evolution of the angiosperms. *Ann. Bot.* 14: 199–216.

——. 1959. Cross- and self-fertilization in plants. In *Darwin's Biological Work,* ed. P. R. Bell. Cambridge: Cambridge Univ. Press.

——. 1967. A cycloid model for the chromosome. *Jour. Cell Sci.* 2: 9–22.

Whittingham, A. D., and G. L. Stebbins. 1969. Chromosomal rearrangements in *Plantago insularis* Eastw. *Chromosoma* 26: 449–68.

Wigan, L. G. 1949. The distribution of polygenic activity on the X chromosome of *Drosophila melanogaster. Heredity* 3: 53–66.

Wilkins, M. F. H., A. R. Stokes, and H. R. Wilson. 1953. Molecular structure of deoxypentose nucleic acids. *Nature* 171: 738–40.

Williamson, D. L., and L. Ehrman. 1967. Induction of hybrid sterility in non-hybrid males of *Drosophila paulistorum*. *Genetics* 55: 131–40.

Wilson, E. B. 1896. *The Cell in Development and Heredity*. Ed. 1. New York: Macmillan.

Wimber, D. E. 1968. The nuclear cytology of bivalent and ring-forming Rhoeos and their hybrids. *Amer. Jour. Bot.* 55: 572–74.

Winge, Ø. 1955. On interallelic crossing over. *Heredity* 9: 373–84.

Wolf, C. B. 1948. Taxonomic and distributional studies of the New World cypresses. *Aliso* 1: 1–250.

Wolstenholme, D. R. 1965. A DNA and RNA-containing cytoplasmic body in *Drosophila melanogaster* and its relation to flies. *Genetics* 52: 949–75.

Woodworth, C. M., E. R. Leng, and R. W. Jugenheimer. 1952. Fifty generations of selection for protein and oil in corn. *Agron. Jour.* 44: 60–65.

Wright, S. 1959. Genetics, the gene, and the hierarchy of biological sciences. *Proc. 10th Internat. Congr. Genet.* (Montreal), 1: 475–89.

Yampolsky, C., and H. Yampolsky. 1922. Distribution of sex forms in the phanerogamic flora. *Bibl. Genet.* 3: 1–62.

Yarnell, S. H. 1962. Cytogenetics of the vegetable crops. III. Legumes. (A) Garden peas, *Pisum sativum* L. *Bot. Rev.* 28: 465–537.

Yoon, C. H. 1955. Homeostasis associated with heterozygosity in the genetics of time of vaginal opening in the house mouse. *Genetics* 40: 297–309.

Yu, C. P., and T. S. Chang. 1948. Further studies on the inheritance of anthocyanin pigmentation in asiatic cotton. *Jour. Genet.* 49: 46–56.

Yunis, J. J., and W. G. Yasmineh. 1971. Heterochromatin, satellite DNA, and cell function. *Science* 174: 1200–9.

Zarzycki, K., and J. Rychlewski. 1972. Sex ratios in Polish natural populations and in seedling samples of *Rumex acetosa* L. and *R. thyrsiflorus* Fing. *Acta Biol. Cracoviensia,* ser. Bot. 15: 135–51.

AUTHOR INDEX

AUTHOR INDEX

ORGANISM INDEX

SUBJECT INDEX

Adaptive gene combinations, 147, 374 ff., 388 ff.
Adaptive value, 94 ff., 146, 374 ff.
Alcaptonuria, 36
Anisomery, 179 ff.
Arginine, 34, 156–57
Autogamy, 118, 124, 397, 407, 418–19, 422, 430, 439, 441, 443–44, 448–49
Auxin, 65, 276

B chromosomes, 230, 239, 312
Balanced lethals, 239, 388, 401, 405
Balanced polygenic systems, 205 ff.
Block inheritance, 353 ff.
Blood types, 103
Breeding systems, 251–52, 418 ff.
Bud sports, 287–88

Canalization, 44, 60, 119 ff.
Cell differentiation, 51 ff.
Certation, 232
Chiasmata, 82 ff., 417–18, 430
Chimeras, 279 ff.
Chromomere, 72, 85, 89, 134
Chromosomes: B, 230, 239, 312; bands, 72–73, 86, 89, 134, 197, 380 ff.; breaks, 83, 291 ff., 316; fine structure, 83 ff., 89, 197; morphology, 303 ff.; number, 304 ff., 340–41, 421 ff., 430, 436–37, 439, 445 ff., 450–51; sex, 256 ff.; silent regions, 365; size, 304 ff., 315; sticky, 291–92; structural rearrangements, 316 ff., 351–52

Chromosome theory of heredity, 13 ff.
Cis-trans test, 76 ff.
Cistron, 87 ff., 132
Cleistogamy, 418
Codominance, 102–3
Codon, 71, 88
Colonizing populations, 433–34, 437 ff., 450
Complementary factors, 115, 157–58, 166, 249–50
Compound loci, *see* Gene clusters, Pseudoalleles
Controlling element, 132, 149 ff., 154
Crossing-over, 14 ff., 82 ff., 318
Cytoplasmic genes, 116, 132, 210 ff., 267

Dauermodifikation, 212, 226
Deficiencies, 318 ff.
Differential segments, 403
Dioecism, 251 ff., 263
Dispersal range, 424–27
DNA: hybridization, 201; quantity, 198, 313–15; redundant, 132, 198; repetitive, 132, 201 ff., 315; structure, 26 ff., 88
Dominance, 101 ff., 114, 124, 178, 187–88
Duplicate factors, 170–71, 188, 247 ff.
Duplications, 198, 318 ff., 369–70

Ecological factors affecting recombination system, 432 ff.
Endopolyploidy, 197, 312